Studies
in the History of Mathematics and
Physical Sciences

14

Studies in the History of
Mathematics and Physical Sciences

continued after index

Gerd Graßhoff

The History of Ptolemy's Star Catalogue

With 113 Illustrations

Springer-Verlag
New York Berlin Heidelberg
London Paris Tokyo Hong Kong

Gerd Graßhoff
Universität Hamburg
Philosophisches Seminar
D-2000 Hamburg 13
FRG

Series Editor

Gerald Toomer
History of Mathematics Department
Brown University
Providence, RI 02912, USA

Library of Congress Cataloging-in-Publication Data

Grasshoff, Gerd.
  The history of Ptolemy's star catalogues / Gerd Grasshoff.
      p.   cm.—(Studies in the history of mathematics and physical
  sciences ; 14)
  Includes bibliographical references.
  ISBN 0-387-97181-5 (U.S.)
    1. Stars—Catalogs—History.   2. Ptolemy fl. 2nd cent. Almagest.
  I. Title.
  QB65.G685   1990
  523.8′0212—dc20   89-21997

Printed on acid-free paper.

Photocomposed copy prepared from the author's LaTeX file.
Printed and bound by R. R. Donnelley & Sons, Inc., Harrisonburg, Virginia.
Printed in the United States of America.

9 8 7 6 5 4 3 2 1

ISBN 0-387-97181-5 Springer-Verlag New York Berlin Heidelberg
ISBN 3-540-97181-5 Springer-Verlag Berlin Heidelberg New York

To
B.-S. and the Elephant

Table of Contents

List of Figures

List of Tables

Introduction

Ptolemy's *Almagest* shares with Euclid's *Elements* the glory of being the scientific text longest in use. From its conception in the second century up to the late Renaissance, this work determined astronomy as a science. During this time the Almagest was not only a work on astronomy; the subject was defined as what is described in the Almagest. The cautious emancipation of the late middle ages and the revolutionary creation of the new science in the 16th century are not conceivable without reference to the Almagest. This text lifted European astronomy to the high standard of knowledge on which the new science flourished. Before, the Ptolemaic models of the orbits of the sun, the moon, and the planets had been refined by Arabic astronomers. They provided the structural elements with which Copernicus and Kepler ushered in the era of modern astronomy. The Almagest survived the destruction of its epicyclic representation of the planetary orbits in the conceptual traces left behind in the theories of its successors. The clear separation of the sidereal from the tropical year, the celestial coordinate systems, the concepts of time, the forms of the constellations, and brightness classifications of celestial objects are, among many other things, still part of the astronomical canon even today.

The scientific interest of the star catalogue in the seventh and eighth books of the Almagest lasted longer than any other part. As late as the beginning of the 18th century the Royal Astronomer Edmund Halley used the catalogue and discovered through a comparison with his own observations the proper motion of the fixed stars. Three centuries before Tycho Brahe had been the first European to revise the star catalogue and replace the Ptolemaic coordinates with his own. In the longitudes of the stars of the Almagest Tycho recognized a large systematic error of one degree. Tycho was one of the first to suspect that the star catalogue of the Almagest is not the product of Ptolemy's own accomplishments as observer, as the text would have us believe, but had been obtained through a simple conversion of measurements made by the most illustrious of Ptolemy's predecessors: Hipparchus. This speculation could have been promptly confirmed or discredited if the presumed Hipparchan catalogue had existed, permitting a direct comparison with the coordinates of the Almagest. As in the case of Euclid's Elements, however, the comprehensive material of the Almagest had the effect of rendering older works obsolete for scientific use. There was no real necessity to laboriously copy out older, scientifically outdated texts, which had served as sources for the Almagest, and to preserve them for the generations to come. For this reason the only work of Hipparchus that has been handed down in its entirety is one early commentary on the astronomy of Aratus and Eudoxus. Whether a Hipparchan catalogue must have existed or not is one of the central questions of this book.

Without even the smallest documentary fragment available it can be misleading to conceive of a Hipparchan compilation of stellar coordinates in the form of a modern star catalogue. Even if it could be proven that Hipparchus had star coordinates of a reasonable accuracy documented in an unknown form and left them for Ptolemy's exploitation, this would in no way prove the existence of a catalogue as it appears in the Almagest. Thus two distinct historical questions arise: (i) did Hipparchus record the stellar positions in the form of a catalogue at all and, if so, (ii) in what type of coordinate system were the positions given? Instead of a catalogue one could imagine a celestial globe as the documentary medium, and instead of an ecliptical coordinate system Hipparchus might have used e.g. polar distances and expressions equivalent to right ascensions. The actual Hipparchan observations of the stellar positions could be carried out as declination measurements in conjunction with the times of the meridian transit of a star. In what follows we will therefore speak of a *star register* when we refer to the more general types of documentation of the stellar positions. Star catalogues contain only tables or lists of star names with their positions and brightnesses.

After Tycho's allegations, the problem arose of deciding between the historical links of two catalogues, of which only one is preserved. In the large arena for possible interpretations thereby created, a centuries-long dispute developed: in one camp Ptolemy was labelled a plagiarist and a forger, in the other he was considered to be the greatest astronomer of antiquity, with an irreproachable integrity.

"There are no secrets as such, there are only uninformed people of all degrees", writes Christian Morgenstern. But what about the historian of science who was not informed about the compilation of the Ptolemaic star catalogue by eye-witness reports? If the only one who is fully informed is he who experienced or shared the secret at first hand, or who heard about it from someone involved and became informed in this way, then it follows that historians constantly struggle without any real hope against the mechanisms of forgetting and suppression and that they have to restrict themselves to be archivists of contemporary reports. The secret of the falling apple which inspired Newton – who should know the secret better than the boy next door who was searching at the time for the first ripe fruit on the tree-limbs? Or Galileo's balls falling from the Leaning Tower of Pisa – who could more truly attest to it than the beggar at the entrance to the church who had been driven from his place by the experimental mania of the new era in physics? And who should better be able to testify to the secret of the source of the Ptolemaic star catalogue than the assistant who carefully copied the manuscripts? It appears as though the solution of riddles should be easiest for those who lived closest to the past events and who had access to the widest variety of contemporary reports.

The *history of the interpretation* of the Ptolemaic star catalogue shows that this picture is deceiving. It was not those who used the star catalogue right after Ptolemy who understood the riddle best. The efforts at understanding, as we intend to show through the thematic succession of the chapters, are the product of a historical process of growing insights, the result of an expanding series of arguments, sources and interpretive strategies which implant their solutions in the increasingly clearer and more complex picture of the historical epoch in question. And they are dependent on our conception of the type of events that can occur in history.

The first half of this work, therefore, covers the main theses of those authors who are significant for the current discussion, in their historical order within the framework of the subject. Although this way of presentation leads to repetitions of the theme discussed in particular cases (for example, the calculation of the precession constant by Ptolemy), one should always notice how the context of discussion changes the perspective of interpretation. As a rule the presentations will not be accompanied by a critical discussion depending on the argumentation in the second half of the book, in order that the sequence of argument in which the contributions were made public shall not be disturbed. The main purpose of this type of presentation is, for instance, to separate the opinion of Laplace or Tycho regarding the origin of the Ptolemaic star catalogue from interpretations of a later period with other historical and methodological backgrounds, as, for instance, the analysis of Vogt. Only in the second half of this study are the historical interpretations integrated into the current debate about the origin of the Ptolemaic star catalogue.

The first chapter describes the time in which the Almagest defined the standards of astronomy. The first section of it summarizes without commentary Ptolemy's explication of the astronomy of the fixed stars in the form in which it served as the foundation for the following generations of astronomers and as it appears to the reader who faces the text for the first time. The second section highlights the difficulties that the Arabic astronomers had with the star catalogue and especially with the determination of the precession constant. It explains why the systematic errors of the longitudes could not have attracted their attention and that as a consequence the historical problems of the star catalogue had necessarily to remain outside their ken.

The second chapter examines the period in which the accusation of forgery was raised and the historical evaluation of the Almagest was undertaken solely against the backdrop of the modern concept of science with its high estimation of empirical data. The chief figures here are Tycho and Delambre. The third chapter summarizes the reaction of the historians since the beginning of this century in which, inspired particularly by Vogt's contributions, an attempt was made to rehabilitate Ptolemy as an accomplished observer.

The fourth and fifth chapters evaluate the previous arguments. The investigation uses the new critical revisions of the catalogue edited by Toomer and Kunitzsch with newly recalculated positions of the identified stars. A catalogue with the Ptolemaic data along with the accurate positions of the stars is printed in Appendix A.

The critique of Vogt's previously uncontested thesis illustrates quite clearly that the errors of a large group of stars from the Almagest correlate significantly with his reconstructed Hipparchan coordinates. With that result, Vogt's claim to have proven the independent observation of the two star registers is discarded. Even more, strong correlations of errors in the two star registers provide evidence for originally common observations. Still, if one relies solely on an analysis of Vogt's reconstructed coordinates, it cannot be excluded that common errors in the observation and evaluation methods could generate correlating errors in two independent catalogues.

If other possible causes for the correlating coordinate errors can be excluded, then it is evident that early Hipparchan coordinates were used in Ptolemy's compilation of the star catalogue. A list of stellar longitudes first published by Gundel can go to show that as far as the coordinates are concerned, ecliptical longitudes of a Hipparchan origin are at work. It exhibits the existence of truly Hipparchan ecliptical longitudes. However, the question remains open whether Hipparchus recorded his star register in an ecliptical coordinate system.

Chapter five extends the critical evaluation of the previous arguments. The detailed analysis of the coordinate errors reveals interesting structures in the star catalogue. The stellar positions were not determined independently of one another, but were rather observed in groups relative to a number of reference stars so that the positional errors of the reference stars were carried over to the positions of the other related stars. Furthermore, one can find a periodic error in longitude that was caused by the inaccuracies of the Ptolemaic/Hipparchan solar theory. It proves that the measurement procedures involve theoretical calculations of the solar longitude, and that the number of reference stars with a longitude directly related to the position of the sun must be reasonably large. The mean error in longitude points up the fact that the epoch of observation for the coordinates given in the Almagest and measured by Hipparchus coincides with the epoch of the Aratus Commentary – provided that Ptolemy used Hipparchan coordinates with an additional 2°40′ on the longitudes.

The clearest evidence for the impact of Hipparchan observations on Ptolemy's coordinates can be garnered by a new type of error analysis. The new method avoids reconstructing the Hipparchan coordinates and then comparing the positional errors of the two registers. When instead the Hipparchan data of the Aratus Commentary are compared directly with the values of the phenomena as calculated from the coordinates of the Almagest, the common observational basis of the two sources becomes obvious.

Consequently one has to assume that a substantial proportion of the Ptolemaic star catalogue is grounded on those Hipparchan observations which Hipparchus already used for the compilation of the second part of his Commentary on Aratus. Although it cannot be ruled out that coordinates resulting from genuine Ptolemaic observations are included in the catalogue, they could not amount to more than half of the catalogue.

Finally, the last chapter argues that the assimilation of Hipparchan observations can no longer be discussed under the aspect of plagiarism. Ptolemy, whose intention was to develop a comprehensive theory of celestial phenomena, had no access to the methods of data evaluation using arithmetical means with which modern astronomers can derive from a set of varying measurement results the one representative value needed to test a hypothesis. For methodological reasons, then, Ptolemy was forced to choose from a set of measurements the one value corresponding best to what he had to consider as the most reliable data. When an intuitive selection among the data was no longer possible – which can occur quite often even with careful measurements – Ptolemy had to consider those values as "observed" which could be confirmed by theoretical predictions. Scientific theories are refuted when no measurement confirms the prediction. For this reason many observations in the Almagest appear as if they are constructed from the theory alone: in other words, they look like fabrications. This misinterpretation ignores the fact that the selection of observation values is a very legitimate and even necessary step for the construction of complex theories. The ancient understanding of "observation" does not include data evaluation of the modern type. Rather, it expresses the particular property of a certain type of theoretical statement that its truth value can be confirmed or refuted by the result of measurement procedures.

Seen in this context it can no longer be surprising that the Hipparchan stellar coordinates, interpreted by Ptolemy as theoretical statements, were accorded more credibility compared with the positions which he had himself observed, and that Ptolemy, even if he observed all the positions with the astrolabe, had to compile the star catalogue of the Almagest from those coordinates that could be derived from a Hipparchan star register. The dispute about the scientific respectability of Ptolemy is nothing more than an argumentative dead-end arising from a misinterpretation of the concept of observation in ancient astronomy.

The history of the Ptolemaic star catalogue, conceived as the history of the interpretations by its readers, changes into a history about the complex genesis of the star catalogue in the Almagest.

All quotations are translated by me if not noted otherwise. L. Schäfer, Ch. Scriba, and A. Kleinert made it possible to work on the book at Hamburg University. More than anybody else I am indebted to Otto Neugebauer: his HAMA initiated my work, and his enthusiasm for the subject was a permanent source of inspiration over years. For numerous corrections and comments I thank G. Toomer. P. Kunitzsch patiently discussed with me all my philological questions concerning the star catalogue. I had endless discussions about astronomical subjects with Ch. Münkel, L. Wisotzki, and J. Jahn. S. Pramesa helped me translate my German manuscript. For valuable comments and other assistance I thank J. Dobrzycki, W. Duerbeck, B. Goldstein, B. Idlavas, H. Schwan, W. Seitter, Th. Spitzley, the Rechenzentrum of Hamburg University and the Institute for Advanced Study in Princeton.

1. The Stars of the Almagest

1.1 The Documents

1.1.1 Persons

The most fruitful period of ancient Greek astronomy was the time of Hipparchus and Ptolemy. Up to then Babylonian astronomy succeeded in predicting solar and lunar eclipses with great precision. Its major concern, the calculation of the visibility conditions of the moon and planets, could be achieved by comprehensive algebraic schemes to a high degree of accuracy.

Hipparchus was probably the first to combine the numerical precision of Babylonian astronomy with Greek geometrical models. In his person the two different astronomical traditions merged to form the powerful astronomical theories that followed. It has been shown that many of the Hipparchan basic parameters are of Babylonian origin.[1] His solar theory was taken over by Ptolemy as well as his determination of the length of the year and the essential parameters in the lunar theory. Ptolemy reports that Hipparchus did not succeed in formulating a satisfactory planetary theory, although he did refute the planetary theories of his predecessors. Hipparchus provided the empirical basis and the methodological standards for Ptolemy's construction of the astronomical theories.

Several documents mention Nicaea in Asia Minor as the birth place of Hipparchus. The main source of our knowledge about the astronomer Hipparchus is the major work of his successor Claudius Ptolemy, the Almagest. Hipparchan observations and theoretical considerations are frequently quoted by Ptolemy and they provide a general framework for Hipparchus' period of scientific activity. The earliest mentioned Hipparchan observations are determinations of the equinoxes in book III of the Almagest. Ptolemy assigns an observation of an autumnal equinox on 26/27 September −146 to Hipparchus himself.[2] He also cites a list of observations of autumnal equinoxes, the earliest of 27 September −161, which Hipparchus "considers to have been very accurately observed".[3] Since Ptolemy does not unambiguously state that Hipparchus actually had observed these himself, one has to consider the

---

[1]Toomer, G. J. (1978), Hipparchus, in: *Dictionary of Scientific Biography*, ed. C. C. Gillispie, New York, vol. XV, pp. 211ff.

[2]Ptolemy, C. (1984), *Ptolemy's Almagest*, trans. and annot. by G. J. Toomer, London, p. 138.

[3]Ptolemy, C. (1984), pp. 133f.

year −146 as the earliest documented reference. The latest Hipparchan observation quoted in the Almagest is an observation of the moon on 7 July −126.[4]

Of a similar Hipparchan observation on 2 May −126 Ptolemy says: "Now Hipparchus records that he observed the sun and the moon with his instruments in Rhodes ...".[5] All the other Hipparchan observations in the Almagest refer to Rhodes with a geographical latitude of $\varphi = 36°$, too. It is only in Ptolemy's partly preserved treatise "On the Phases of the Fixed Stars and Related Weather Prognostications" that the Hipparchan observations are attributed to a place called "Bithynia", which is the kingdom in which Nicaea was located.[6] From this evidence it is plausible that Hipparchus lived most of his scientific career on Rhodes.[7] For the most part we have only indirect access to the scientific contributions of Hipparchus. The only preserved Hipparchan text, titled "Commentary on the Phenomena of Aratus and Eudoxus",[8] contains a detailed criticism of the older Greek texts on fixed stars by Aratus, Eudoxus and Attalus. The second book with Hipparchus' own account of the phenomena related to fixed stars reveals his extensive study of the stellar positions. Almost all other references to the scientific contributions of Hipparchus are found either in Ptolemy's quotations or must be reconstructed from calculation schemes and observations.

The biography of Ptolemy is as fragmentary as that of Hipparchus.[9] The observations in the Almagest, Ptolemy's main work, cover a time between +127 and +141. Since the Almagest is quoted in the other major Ptolemaic texts, the "Tetrabiblos", the "Handy Tables", the "Planetary Hypotheses" and the "Geography", it has to predate them. Ptolemy attributes several observations dating between +127 and +132 to the "mathematician Theon", who could be either his colleague or his teacher in Alexandria. Also, all the other Ptolemaic observations refer to Alexandria in Lower Egypt and there is no evidence that Ptolemy ever worked at other places. Ptolemy formulated his astronomical theories as they endured in their main features for the next millennium. He developed the planetary theory and refined the lunar theory. Thus, using the Hipparchan solar theory, he was able to predict eclipses accurately. Furthermore, the central importance of the Almagest as a systematic mathematical formulation of the astronomical knowledge cannot be underestimated. Easy tabulations for the major mathematical procedures made the Almagest the comprehensive and practical astronomical handbook for following generations of astronomers.

1.1.2 Methodological Background

The Almagest opens its astronomical exposition with two introductory chapters where Ptolemy discusses the rank of astronomy among the sciences and the method-

---

[4]Ptolemy, C. (1984), p. 230.

[5]Ptolemy, C. (1984), pp. 227.

[6]Ptolemy, C. (1898–52), *Claudii Ptolemaei opera quae exstant omnia*, ed. J. L. Heiberg et. al., Opera Astronomica Minora, vol. II, pp. 3–67.

[7]For a summary of Hipparchus' biography and scientific work cf. Toomer, G. J. (1978), *Hipparchus*, in: Gillispie, C. C. (ed.) (1970–80), *Dictionary of Scientific Biography*, New York, vol. XV, pp. 207–224.

[8]Hipparchus (1894), *Hipparchi in Arati et Eudoxi Phenomena Commentarium*, ed. and German trans. C. Manitius, Leipzig.

[9]Toomer, G. J. (1975), *Ptolemy*, in: Gillispie, C. C. (ed.) (1970–80), *Dictionary of Scientific Biography*, New York, vol. XI, pp. 186–208.

ological structure of his book. In the tradition of Aristotelian metaphysics, Ptolemy counts astronomy as part of mathematics whose methods provide "sure and unshakable knowledge", thereby being distinguished from physics, whose investigation of the material world, due to the "unstable and unclear nature of matter" can offer no real hope "... that the philosophers will ever be agreed about them."[10]

The wish to deduce astronomical laws with mathematical rigor imposes a methodological order on astronomy for Ptolemy that is reflected in the thematic structuring of the Almagest.[11]

> "We shall try to note down everything which we think we have discovered up to the present time; we shall do this as concisely as possible and in a manner which can be followed by those who have already made some progress in the field. For the sake of completeness in our treatment we shall set out everything useful for the theory of the heavens in the proper order, but to avoid undue length we shall merely recount what has been adequately established by the ancients. However, those topics which have not been dealt with [by our predecessors] at all, or not as usefully as they might have been, will be discussed at length, to the best of our ability."

With the certainty of the deductive form of argumentation Ptolemy first of all develops the auxiliary mathematical and astronomical hypotheses in order to formulate the astronomical theories in their logical order. The solar theory is the fundamental astronomical hypothesis for all others in the Almagest. No measurement of the position of the other celestial objects is possible without it. All positional data, whether obtained by use of the astrolabe, a meridian instrument, an eclipse or the times of rising and setting, are based upon the position of the sun. Consequently, Ptolemy first outlines in the Almagest a theory of the motion of the sun and proceeds with the closely related lunar theory before continuing with the fixed stars and the planets. The order of subjects relates to a tree of definitions with the most general and fundamental definition at the top and all other subsequently defined concepts below.[12]

> "Secondly, we have to go through the motion of the sun and of the moon, and the phenomena accompanying these [motions]; for it would be impossible to examine the theory of the stars thoroughly without first having a grasp of these matters. Our final task in this way of approach is the theory of the stars. Here too it would be appropriate to deal first with the sphere of the so-called 'fixed stars', and follow that by treating the five 'planets', as they are called."

Ptolemy's systematic astronomy hereby resembles the Aristotelian methodology as it is developed in the "Analytica Posteriora".[13]

---

[10] Ptolemy, C. (1984), p. 36.
[11] Ptolemy, C. (1984), p. 37.
[12] Ptolemy, C. (1984), p. 37.
[13] Cf. Aristotle (1984).

The chapters in the Almagest concerning the fixed stars maintained their unquestioned validity until the beginning of modern astronomy. They will be the subject of the following pages. Ptolemy's crucial statements will be mostly quoted without further interpretation, which could anticipate later discussion.

1.1.3 The Almagest on Fixed Stars

The celestial phenomena of the fixed stars are discussed in the seventh and eighth books of the Almagest. In the eleven chapters of these two books Ptolemy develops a unified theory of the fixed star phenomena which allows the calculation of all the important configurations and apparent motions of the stars for any given time. The thematic sequence and the subjects emphasized by Ptolemy provide a glimpse into the ancient astronomy of the stellar motions. For example, the detailed discussion of the question whether the celestial sphere rotates uniformly, especially in connection with the precession motion, clearly illustrates that these views were not yet a part of the canonical knowledge of astronomy at the time the Almagest was written (ca. +150).

Rigorously adhering to the principles of a deductive mode of argumentation, the chapter on the fixed stars begins with the demonstration that the motion of the stars can be treated as the motion of a sphere with constant distances between the stars.

VII.1 (Seventh book, chapter 1): That the fixed stars always maintain the same position relative to each other.

Ptolemy compares the alignments of the stars in the constellations as they are reported by Hipparchus with his own observations without finding any difference. This proves immediately that there is no relative motion of the stars; hence all motions of the stars can be represented by a superposition of rotations of a sphere. This conclusion places Ptolemy even by his own testimony in opposition to the early Hipparchus who initially promoted the hypothesis that "only the stars in the vicinity of the zodiac effect had a rearward motion, as Hipparchus proposes in the first hypothesis he puts forward".[14] Ptolemy's reveals the procedure of comparison:[15]

> "If one were to match the above alignments too against the diagrams forming the constellations on Hipparchus' celestial globe, he would find that the positions of the [relevant stars] on the globe resulting from the observations made at the time [of Hipparchus], according to what he recorded, are very nearly the same as at present."

The quoted passage does not unambiguously state whether Ptolemy actually resorted to a Hipparchan celestial globe or whether he drew Hipparchus' observational data on a globe for a direct comparison with his observations.[16] In another passage Ptolemy tells the reader that "observations recorded by Hipparchus, which are our chief source for comparison, have been handed down to us in a thoroughly satisfactory form."[17] In contrast to the excellent recordings of Hipparchus the older

---

[14]Ptolemy, C. (1984), p. 322.
[15]Ptolemy, C. (1984), p. 327.
[16]Ptolemy, C. (1984), cf. Toomer's footnote p. 327.
[17]Ptolemy, C. (1984), p. 321.

observations done by Aristyllos and Timocharis are neither well observed nor carefully "worked out".[18] All Hipparchan sources mentioned by Ptolemy are lost today. A copper globe allegedly belonging to Ptolemy, which reportedly had been found as late as 1043 in a library in Cairo, was never seen again.[19]

VII.2: That the sphere of the fixed stars, too, performs a rearward motion along the ecliptic.

In the first chapter Ptolemy presents his proof that the fixed stars move on a rigid sphere. With the available observational accuracy of about 10 minutes of arc and records of astronomical data over a period of several centuries, the motion of the sphere of fixed stars could no longer be described solely through the daily rotation. A second slower rotation around the pole of the ecliptic, later called the "precession motion", must be added.

Hipparchus was the first who realized the necessity of a second motion of the celestial sphere. He did not discover the additional motion by analysis of star observations, but through the determination of the equinoxes. The dates of the equinoxes are one of the most fundamental astronomical parameters in ancient science. The parameters of the solar theory, itself fundamental to all other theories, are derived from equinox observations, as is the definition of the coordinate systems. The equinox is defined by the moment when the sun's path on the zodiac intersects the celestial equator. Only at that moment are the lengths of day and night equal. Because of the precession motion the ecliptical longitude of the equinox was increasing by $1^{\circ}.38$ per century in the time of Hipparchus and Ptolemy. Ptolemy recounts Hipparchus' discovery of precession:[20]

> "For Hipparchus too, in his work 'On the displacement of the solstitial and equinoctial points', adducing lunar eclipses from among those accurately observed by himself, and from those observed earlier by Timocharis, computes that the distance by which Spica is in advance of the autumnal [equinoctial] point is about 6° in his own time, but was about 8° in Timocharis' time. For his final conclusion is expressed as follows: 'If, then, Spica, for example, was formerly 8°, in zodiacal longitude, in advance of the autumnal [equinoctial] point, but is now 6° in advance', and so forth. Furthermore he shows that in the case of almost all the other fixed stars for which he carried out the comparison, the rearward motion was the same amount."

Ptolemy demonstrates the precession motion by observations with a spherical astrolabe whose construction he describes in full detail in the first chapter of the fifth book. The spherical astrolabe (fig. 1.1) consists of a ring system which rotates freely on two axes pointing to the ecliptical and equatorial poles.[21] The ecliptical coordinates of a celestial object can be read off directly from the graduation on the ecliptic ring and the inner ring.

[18]Ptolemy, C. (1984), p. 321.

[19]Sezgin, F. (1978), *Geschichte des arabischen Schrifttums*, Leiden, vol. VI, p. 84.

[20]Ptolemy, C. (1984), p. 327.

[21]Adapted from Ptolemy, C. (1963), *Handbuch der Astronomie*, German trans. and annot. by K.

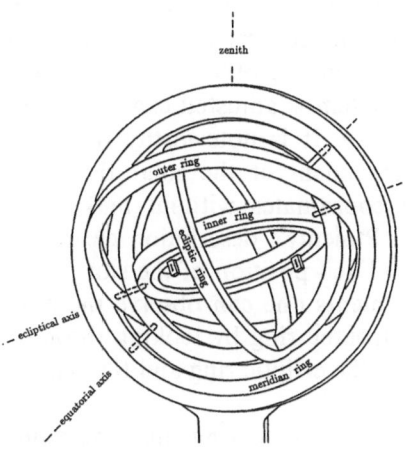

Figure 1.1: Spherical astrolabe.

In relation to other methods of position measurement the spherical astrolabe is a complicated instrument. It would be much easier to determine the position of a celestial object by measuring either the horizontal coordinates, i.e. the height above the horizon and the azimuth at a given time, or the declination and right ascension. All these coordinates could easily be measured by observations of the meridian transit, when an object culminates on the north-south meridian. Of course, an observer has to wait until a star culminates during the night for such an observation, but that would be no serious objection to an astronomical program devoted to compiling a star catalogue of the entire visible sky. That Ptolemy prefers the astrolabe for the measurement of stellar positions of his catalogue could be motivated by two reasons: The positions in the star catalogue are given in the ecliptical coordinate system. With the astrolabe one can read off the ecliptical coordinates directly from the graduation rings. This, firstly, avoids complicated and laborious transformations to the ecliptical coordinate system, as in the case of declination measurements, and secondly allows direct observational control of any catalogued pair of ecliptical coordinates. It is only with the help of an astrolabe that the coordinates of the star catalogue can be considered empirical data which could be directly "observed".

To take a measurement the astrolabe was set up in such a way that the meridian ring lay in the plane of the north-south meridian with an axial inclination corresponding to the geographical latitude of the observational site. Before one can read off the ecliptical coordinates from the astrolabe, the system of rings must be adjusted in such a way that the ecliptic ring is parallel to the plane of ecliptic at the moment of observation.

For daylight observations, one turns the whole ring system until the outer

Manitius, introduction and corr. by O. Neugebauer, 2 vols., Leipzig, vol. I, p. 255.

ring marks the solar longitude on the graduation of the ecliptic ring. As Ptolemy describes it, one has to calculate the longitude of the sun and sets the instrument accordingly. This setting can be controlled by adjusting the ring system so that the sun casts a shadow exactly on the other side of the ecliptic and the outer ring. At this moment the ring system is adjusted exactly to the position of the ecliptical coordinate system in the sky. It is noteworthy that the control measurement of the solar position is independent of the solar theory. This means that Ptolemy could measure ecliptical coordinates without making use of the theory and its possible errors. However, Ptolemy's description clearly requires the adjustment of the instrument to the calculated position of the sun.

After the initial adjustments one can observe the moon or another object with a rotation of the inner astrolabe ring by looking through the diopter. Its ecliptical longitude can be read from the ecliptic ring and the ecliptical latitude from the inner ring.

At night the cage of the ecliptic ring must be adjusted to either the known ecliptical longitude of the moon or a reference star. First, the inner ring is turned until it intersects the outer ring at the known longitude of the moon or the reference star. Then the cage of the ecliptic ring is rotated until the object of reference is visible in the plane of the inner ring. At that moment the astrolabe is adjusted, interestingly, without making use of the latitude of the reference objects. The alignment of the ecliptic ring is not without its difficulties and has to be continuously corrected following the daily motion of the sky. The speed of the latter requires an extraordinary observational talent and constant correction of the set-up, especially when the position of several stars in a row is to be measured.

For the determination of the precession motion Ptolemy evaluates an observation of Regulus, the brightest star of the constellation Leo, on 23 February +139. According to the Almagest, Ptolemy measured the position of the moon at sunset and then, half an hour later, the position of Regulus relative to the moon.[22] From this result Ptolemy was able to obtain the ecliptical longitude of Regulus and to establish an increase of the value by 2°40′ since the time of Hipparchus. Hence, Ptolemy confirms the Hipparchan value of one degree per century.

At the end of the second chapter Ptolemy mentions that he checked the motion of the fixed star sphere in the direction of the zodiac signs with observations of the star Spica.[23]

> "In the same way we took sightings of Spica and the brightest among those stars near the ecliptic, from the moon, and then [having done that], were in a better position to use those stars to take sightings of the rest. We [thus] find that their distances relative to each other are, again, very nearly the same as those observed by Hipparchus, but their individual distances from the solstitial or equinoctial points are in each case about

---

[22]Ptolemy, C. (1984), p. 328. There are many difficulties in the numerical details of Ptolemy's evaluation, as analyzed later. Manitius repeats the calculation and finds that Ptolemy did not consider the change of the parallax, in spite of his own considerations in the chapters on the theory of the moon. With a correct calculation Ptolemy would have found a precession of 2°30′ instead of the reported 2°40′. However, this small correction cannot account for the difference from the accurate precession value of 3°40′. Cf. Ptolemy, C. (1963), vol. II, pp. 397ff.

[23]Ptolemy, C. (1984), p. 328.

$2\frac{2}{3}°$ farther to the rear than those derivable from what Hipparchus recorded."

Ptolemy cites the lost Hipparchan text "On the length of the year" in which Hipparchus gives an estimation of the precession constant:[24]

"For if the solstices and equinoxes were moving, from that cause, not less than $\frac{1}{100}$ th of a degree in advance [i.e. in the reverse order] of the signs, in the 300 years they should have moved not less than 3°."

VII.3: That the rearward motion of the sphere of the fixed stars, too, takes place about the poles of the ecliptic.

Until now Ptolemy has only demonstrated that the longitude of Regulus and Spica increased by 2°40′ over the period of 265 years between his observations and those of Hipparchus. The axis of the rotation is not yet unambiguously fixed.

In the third chapter the orientation of the motion is confirmed by a comparison of the ecliptical latitudes of Spica for the time of Hipparchus and Ptolemy. In the case that the precession motion rotates around the pole of the ecliptic, the ecliptical latitudes of the stars should not show any measurable changes during the time for which historical records are available.

Ptolemy reports that Hipparchus had already recognized the orientation of the precession motion around the pole of the ecliptic in another text, entitled "On the displacement of the solstitial and equinoctial points", though Hipparchus seemed uncertain of the result since he could base his calculations only on the unreliable observations of the astronomers of the school of Timocharis. Ptolemy can be more certain of his findings, because he can rely on the accurate measurements of Hipparchus. He finds no significant change in the ecliptical latitudes at all.

However, the exact value of the precession motion is confirmed by evaluating the declinations of bright stars in comparison with older observations. The declinations of the stars are easily obtained through the observation of the meridian transit: as soon as the star passes the north-south meridian at the site of observation, the altitude h_c of the star over the horizon is measured and, through an uncomplicated arithmetical operation with the geographical latitude φ, one obtains the declination δ of the star as

$$\delta = h_c + \varphi - 90°. \tag{1.1}$$

The simple way of measuring declinations might be the reason that these measurements were not only recorded by Hipparchus, but also in the older astronomical school of Timocharis.

Ptolemy, for his part, records the declinations of two sets of nine stars in both the northern and southern part of the sky for the time of Timocharis, Hipparchus and himself, and he calculates from a subset of six stars the precession motion. From all the quoted calculations Ptolemy obtains results confirming a precession motion

---

[24]Ptolemy, C. (1984), p. 329. As Toomer remarks, the mentioned time difference of 300 years refers to the solstice observation of Meton (−431), reported in book III (solar theory) of the Almagest; Ptolemy, C. (1984), p. 138.

of one degree per century which, although identical with the minimal Hipparchan value, deviates substantially from the accurate value of $1^{\circ}38$.[25]

VII.4: On the method used to record [the positions of] the fixed stars.

The fourth chapter introduces the fixed star catalogue. An appropriate coordinate system must be chosen so that the positions of the stars can be calculated without too great mathematical complications for any given time. Interesting phenomena like rising and setting times, meridian transits, or the relative positions of two celestial objects to each other should be derivable with little effort. If only the daily revolution had to be considered for the celestial motions, the equatorial coordinate system would be most convenient for a catalogue of stars. From the declination of a star one can calculate the maximal altitude over the horizon and the circumstances of rising and setting. Together with the right ascension, all positions of a star in the sky could be derived with only limited inconvenience.

The discovery of the precession motion then made it clear that the motions of the stars are not so easy to represent. The accuracy of 10 minutes of arc in the position entries would require the correction of the star catalogue after only a short period of time. With a precession of $1^{\circ}38$ per century, an adjustment is necessary in extreme cases only after little more that 10 years. With these more complex motions one needs a coordinate system with which the precession motion could be integrated in a particularly simple way.

It has been demonstrated in the second chapter of the seventh book that the precession motion is a rotation around the pole of the ecliptic.[26] When the coordinate system is arranged in such a way that its poles coincides with the poles of the precession motion, an uncomplicated conversion of the star coordinates to the respective epoch is possible: the ecliptical latitude of a star remains unaffected by the precession motion, and the ecliptical longitude increases at a constant value with time by the precession constant. This enormous advantage requires that Ptolemy's star catalogue be compiled in ecliptical coordinates. For example, from a star catalogue for a given epoch, the ecliptical coordinates can be calculated by a simple addition of the precession to the ecliptical longitudes. The accurate value was $1^{\circ}38$ during Ptolemy's time. The Almagest, though, takes over the Hipparchan minimal value of one degree per century. This had to cause problems for the following generation of astronomers who used the star catalogue of the Almagest reduced to their epoch.[27]

In the following we quote at full length Ptolemy's important statements on the method by which the data of the catalogue were obtained:[28]

> "So we thought it appropriate, in making our observations and
> records of each of the above fixed stars, and of the others too, to
> give their positions, as observed in our time, in terms of longitude

---

[25] *Explanatory Supplement to the Astronomical Ephemeris and the American Ephemeris and Nautical Almanac*, Her Majesty's Stationary Office (1961), London, pp. 28ff.

[26] This is valid for the limited period between Hipparchus and Ptolemy with changes of the latitude of less than 1'. Cf. Explanatory Supplement, pp. 28ff.

[27] If not problems for the accuracy of the stellar position, then problems for a sound theory of precession motion.

[28] Ptolemy, C. (1984), pp. 339f.

and latitude ... Hence, again using the same instrument (because the astrolabe rings in it are constructed to rotate about the poles of the ecliptic), we observed as many stars as we could sight down to the sixth magnitude. [We proceeded as follows.] We always arranged the first of the above-mentioned astrolabe rings [to sight] one of the bright stars whose position we had previously determined by means of the moon, setting the ring to the proper graduation on the ecliptic [ring for that star], then set the other ring, which was graduated along its entire length and could also be rotated in latitude toward the poles of the ecliptic, to the required star, so that at the same time as the control star was sighted [in its proper position], this star too was sighted through the hole on its own ring. For when these conditions were met, we could readily obtain both coordinates of the required star at the same time by means of its astrolabe ring: the position in longitude was defined by the intersection of that ring and the ecliptic [ring], and the position in latitude by the arc of the astrolabe ring cut off between the same intersection and the upper sighting-hole.

In order to display the arrangement of stars on the solid globe according to the above method, we have set it out below in the form of a table in four sections. For each star (taken by constellation), we give, in the first section, its description as a part of the constellation; in the second section, its position in longitude, as derived from observation, for the beginning of the reign of Antoninus ([the position is given] within a sign of the zodiac, the beginning of each quadrant of the zodiac being, as before, established at [one of] the solstitial or equinoctial points); in the third section we give its distance from the ecliptic in latitude, to the north or south as the case may be for the particular star; and in the fourth, the class to which it belongs in magnitude."

Ptolemy claims very explicitly to have observed the stars of the catalogue with a spherical astrolabe in the year +137. It is precisely this statement whose truth or falsity has been granted or contested for more than one thousand years.

The position of a number of brighter reference stars then – fundamental stars – were determined with the help of the position of the moon and sun, and the positions of the remaining stars were measured relative to them. Through this method any error in the positions of the fundamental stars would have carried over to the catalogued positions of the relatively measured stars. Because of their importance, the positions of the fundamental stars must be measured with particular care, whereby the desired precession is dependent on the accuracy of the actual measuring with the astrolabe as well as on the accuracy of the measurements or calculations of the sun.

Ptolemy is aware of possible errors in observation. In the first chapter of the third book the concept of the length of the year is discussed. There he criticizes an incorrectly evaluated measurement of Hipparchus and deems possible an inexact observation or calculation of the lunar position.[29] In addition the influence of

---

[29] "It is more plausible to suppose, either that the distances of the moon from the nearest stars at the eclipses have been too crudely estimated, or that there has been an error or inaccuracy in the determinations of the moon's parallax with respect to its apparent position, or of the motion of the sun

optical illusions on the accuracy of the estimates of position is well known.[30]
Ptolemy considers measurements made with the astrolabe to be reliable.[31]

Ptolemy gives no indication of an earlier star catalogue comparable to that in the
Almagest, though he must have had access to extensive records of stellar data in the
constellations. As for the grouping and configuration of the constellations Ptolemy
admits openly to deviating from the traditional terminology of his predecessors:[32]

> "Furthermore, the descriptions which we have applied to the indi-
> vidual stars as parts of the constellation are not in every case the same
> as those of our predecessors (just as their descriptions differ from their
> predecessors'): in many cases our descriptions are different because they
> seemed to be more natural and to give a better proportioned outline to
> the figures described."

The historical question later emerged whether Ptolemy also catalogued the
positions and magnitudes of the stars independently of his predecessor.

---

from the equinox of the time of mid-eclipse." Ptolemy, C. (1984), p. 136.

[30] Ptolemy, C. (1984), p. 421.

[31] Ptolemy, C. (1984), pp. 453f: "We cannot derive this from the ancient observations [of Mercury],
but we can do so from our own observations made with the astrolabe. For it is in this situation that
one can best appreciate the usefulness of this way of making observations, since, even if those stars
with previously determined positions which are visible are not near the planet being observed (which is
generally the case with Mercury, since, for the majority of the fixed stars, it is rare that they are visible
when they are [only] as far from the sun as Mercury is), one can still determine positions of the planet
in question accurately in latitude and longitude, by sighting stars which are at a considerable distance."

[32] Ptolemy, C. (1984), p. 340.

1.2 The Arabic Revision of the Almagest

During the time between its composition and the beginning of the 16th century the Almagest strengthened its unique position as the standard work of astronomy, especially through scientific activity in the Orient.

Until its decline in the fifth century, Alexandria was the center of influence for the Ptolemaic texts. Particularly in the fourth century Alexandrian scholars produced a series of commentaries some of which are at least partially preserved. The so-called "small astronomy", an allusion to the "great astronomy" of Ptolemy, consists of a collection of mathematical and astronomical treatises supposed to serve as an introduction to the more complex parts of the Almagest.[33]

Towards the beginning of the fourth century Pappus wrote a commentary to the Almagest from which only the parts on the fifth and sixth book are still preserved. His commentary had more the character of an elucidation and added nothing to the astronomical knowledge contained in the Almagest.[34] It is still unknown whether the commentary of Pappus covered all the books of the Almagest or whether it restricted itself to a discussion of the motion of the sun and moon. A century later Theon of Alexandria included these expositions of Pappus in his comprehensive commentary.[35]

Despite the extensive commentaries on the Almagest, no critical inspection of its contents, above all of the star catalogue, is known to us from antiquity. The theoretical and practical advances of the Almagest in comparison to the alternatives of its predecessors must have been so remarkable that small numerical inaccuracies in the Ptolemaic theories could not force an astronomer to severe revisions. Theon tells of a number of astrologers before Ptolemy who did not assume a constantly increasing longitude of the spring equinox due to the precession motion, but rather a periodical oscillation over an arc of 8 degrees.[36]

It is possible that, shortly after the discovery of the precession motion by Hipparchus, inexact observation or sheer astrological speculations were the source of such theories. Besides the astronomical difficulties in developing a satisfactory

---

[33] In the introduction to his translation Manitius sketches the transmission of the Almagest: Ptolemy, C. (1963), vol. I, p. V. See also Dreyer, J. L. E. (1953), *A History of Astronomy from Thales to Kepler*, 2nd edition, New York; Suter, H. (1900), *Die Mathematiker und Astronomen der Araber und ihre Werke*, Leipzig; Sezgin, F. (1978), *Geschichte des arabischen Schrifttums*, vol. VI; The introduction of Kunitzsch, P. (1975), *Zur Kritik der Koordinatenüberlieferung im Sternkatalog des Almagest*, Göttingen; Kunitzsch, P. (1974), *Der Almagest. Die Syntaxis Mathematica des Claudius Ptolemäus in arabisch-lateinischer Überlieferung*, Wiesbaden.

[34] Rome, A. (1936/43/31), *Commentaires de Pappus et de Théon d'Alexandrie sur l'Almageste*, 3 vols., Biblioteca Apostolica Vaticana, Studi e Testi 72, 106, 54. Roma, vol. I.

[35] Rome, A. (1936/43/31), vol. II und vol. III. See also *Theonis Alexandrini in Claudii Ptolemaei Magnam Constructionem Commentariorum Lib. XI*, Basel, 1538.

[36] Dreyer, J. L. E. (1953), p. 204: "According to certain opinions ancient astrologers believe that from a certain epoch the solstitial signs have a motion of 8° in the order of the signs, after which they go back the same amount; but Ptolemy is not of this opinion, for without letting this motion enter into the calculations, these when made by the tables are always in accord with the observed places. Therefore we also advise not to use this correction; still we shall explain it. Assuming that 128 years before the reign of Augustus the greatest movement, which is 8°, having taken place forward, the stars began to move back; to the 128 years elapsed before Augustus we add 313 years to Diocletian and 77 years since his time, and of the sum (518) we take the eightieth part, because in 80 years the motion amounts to 1°. The quotient (6°28′30″) subtracted from 8° will give the quantity by which the solstitial points will be more advanced than by the tables". See also Neugebauer, O. (1975), pp. 631ff.

theory of the precession motion, the strong desire to formulate the motions of the celestial sphere into a theory preserving traditional astrological interpretations and the validity of the ancient observations, led to the construction of models with a non-linear precession motion, even after the composition of the Almagest.[37]

However, without an exact knowledge of the precession motion, the Ptolemaic coordinates of the stars cannot be adequately checked later, not even with the most accurate method of measuring. Though the stellar coordinates are a result of observations, they cannot be confirmed or corrected just by a repetition of the observations some centuries later. Obviously only an exact recalculation of the stellar positions for the time of the Almagest allows one to check the data contained therein. Even if the coordinates of a later epoch are accurately measured, the correct ecliptical longitudes of an earlier period can only be calculated when the actual motion of the spring equinox due to precession is subtracted. Therefore a test of Ptolemy's star catalogue requires an adequate theory of the precession motion.

The possibility of critically checking the Ptolemaic star catalogue arose for the first time after solid knowledge of the motion of the stellar sphere had been gained. After the decline of Alexandria as the scientific center of the ancient world in the fifth century, the religious centers of the Orient took over the tradition of the Greek sciences and with them the astronomical theories of the Almagest.[38]

At the end of the eighth century, with the flourishing of Islamic civilization, an astronomical science was developed which absorbed and revised first the Indian, then, somewhat hesitantly, the Greek tradition, and later was transmitted through Spain to medieval Europe. With the decline of the kingdoms in the Orient and their breakup into a plethora of small dynasties, astronomy received an impulse to improved formulations on a level of complexity exceeding those of the Almagest.[39] The Abbasid Caliph al-Ma'mūn initiated the heyday of the sciences as, first in Damascus and then in Baghdad (from 829), he built observatories for the testing and revision of the traditional astronomical knowledge on the basis of independent observations.[40] A small list of 24 stars with coordinates independent of the Almagest bears testimony to the observations of the astronomical school of al-Ma'mūn. The earliest translations of the Almagest known today date to that period.[41] The activity of the Islamic astronomers focused on improving the parameters in the astronomical theories without calling into question the theoretical edifice itself, namely the views offered in the Almagest. One accomplishment of this time was the accurate measuring of the meridian with $56\frac{2}{3}$ miles for $1°$ of the meridian or 20400 miles for the circumference;[42] another the improvement of the astronomical measuring methods themselves.[43]

The Ptolemaic catalogue was converted to the epoch of that time in that the

---

[37]Mercier, R. (1976/77), Studies in the Medieval Conception of Precession, 2 parts, *Archives Internationales d'Histoire des Sciences* **26** (I), **27** (II), part I, p. 209.

[38]Cf. Kunitzsch, P. (1974), pp. 1ff.

[39]Dreyer, J. L. E. (1953), p. 245.

[40]Ptolemy, C. (1963), vol. I, p. VI.

[41]Kunitzsch, P. (1974), *Der Almagest. Die Syntaxis Mathematica des Claudius Ptolemäus in arabischlateinischer Überlieferung*, Wiesbaden, pp. 6ff. Kennedy, E. S. (1956), *A Survey of Islamic Astronomical Tables*, Trans. Amer. Philos. Soc., N. S. 46.2, pp. 132ff.

[42]Nallino, C. A. (1944), *Raccolta di scritti*, vol. V, p. 421.

[43]Sezgin, F. (1978), *Geschichte des Arabischen Schrifttums*, Leiden, vol. VI, p. 20.

ecliptical longitudes of certain reference stars were observed and the difference from the longitudes given in the Almagest was added to the longitudes of the other stars. After this had been carried out, Islamic astronomers possessed a comprehensive star catalogue devoid of any significant errors in its coordinates. Each systematic error in longitude of the Almagest necessarily remained unnoticed in such a procedure. This explains why Tycho Brahe's later discovery – that the longitudes of the Almagest stars are systematically one degree too small – could not be detected by these astronomers. The critical transmission of the Ptolemaic star catalogue during this time is known to us through the work of al-Battānī (d. 929), aṣ-Ṣūfī (903–986), al-Bīrūnī (d. 1048), Ibn aṣ-Ṣalāḥ (d. 1154) and Uluġ Bēg (1394–1449).[44] The comprehensive astronomical treatise of al-Battānī contain,[45] besides longer expositions on the lunar and solar theory, a number of tables among which two star catalogues can be found. One of these lists 75 stars whose equatorial coordinates were measured as fundamental coordinates for the other stars. The catalogue contains all bright stars in the same sequence as they are catalogued in the Almagest.[46] The second, even more comprehensive register includes 533 Ptolemaic stars whose ecliptical longitudes were calculated by adding 11 degrees 10' for the epoch 1 March +880 using a precession constant of 1 degree for 66 years.[47] We know from aṣ-Ṣūfī that al-Battānī had considered for his register only the Ptolemaic stars whose coordinates show no variations in the different versions of the Almagest.[48]

The extensive philological activity practised by Islamic astronomers shows that already 700 years after the writing of the Almagest a large quantity of numerical values had been corrupted through copying errors. Before a critical appraisal of the genesis of the Ptolemaic catalogue and, the sole matter of importance for the Arabic astronomers, a scientific use of the coordinates could be made, these errors had to be eliminated. In the time following a number of astronomers concentrated their work on removing improbable interpretations by a critical comparison of the existing copies with their own exact observations of the stellar positions.

The value of the precession constant itself provides clues to the procedure of its derivation.

The precession constant of 1 degree every 66 years ($54.5''/y$) is larger than the accurate value of 1 degree every 72 years ($50''/y$). It was used in the ninth and tenth century and was borrowed from the star register of Zīj al-mumtahan,[49] which was composed in the school of al-Maʾmūn at the new observatory in Baghdad around the year 830.[50] In response to a decree of the Caliph the astronomers performed observations in order to check and possibly correct the values coming down through tradition. The far too small Hipparchan/Ptolemaic value of precession of 1 degree

---

[44]Kunitzsch, P. (1974), p. 47.

[45]Nallino, C. A. (1899–1907), *Al-Battani sive Albatenii Opus astronomicum*, ed. Carolo Alphonso Nallino, 3 vols., Milano.

[46]Kunitzsch, P. (1974), p. 50.

[47]Kunitzsch, P. (1974), p. 50. As G. Toomer pointed out to me, it seems likely that al-Battanı used the constant $1\frac{1}{2}$ degrees per century.

[48]Kunitzsch, P. (1974), p. 47.

[49]Kunitzsch, P. (1974), p. 51.

[50]Suter, H. (1900), *Die Mathematiker und Astronomen der Araber und ihre Werke*, Abhandlungen zur Geschichte der Mathematischen Wissenschaften mit Einschluss ihrer Anwendungen. Heft X. Reprint New York, 1972, pp. 8&10.

per century had been recognized very early as false and for that reason it had to be newly determined by the Islamic astronomers. To keep the error as small as possible, the astronomers were forced to select the longest period of time between their own position measurements and the epoch of the older coordinates whose longitudes had increased due to the precession motion. Nothing, therefore, seemed more reasonable to them than to call upon the old Ptolemaic star register and to make a comparison between the catalogued ecliptical longitudes of the epoch +137 and the longitudes of the epoch +830 they had measured themselves. The difference should amount to exactly 10 degrees when the real precession value of $50''^{/y}$ is taken as the basis. Now the longitudes of the Ptolemaic catalogue, however, are on the average 1 degree too small. Consequently the Islamic astronomers obtained a difference of 11° instead of 10°. If the Almagest is used as the source of the observations, the precession constant that is calculated is too large by about 10 per cent. Instead of $50''^{/y}$ the astronomers of the Caliph al-Ma'mūn obtained a precession constant of $55''^{/y}$ or, expressed in other terms, one degree in 66 years.

Later, the value of the precession constant was improved still further. The astronomer Naṣīr ad-Dīn aṭ-Ṭūsī computed in 1274 a precession constant of 1 degree for 70 years ($51.4''^{/y}$) which was still larger than the accurate value and possibly included the coordinates of the Almagest in the calculation as well.[51] Two pivotal conclusions can be drawn concerning the observational practice of the Islamic astronomers and the status of the Ptolemaic star catalogue at this time.

(i) The Islamic astronomers of the ninth century carried out observations of the positions of the fixed stars whose accuracy vis- a-vis that of Hipparchus was improved and which lay in the neighborhood of 10'. The fixed star register of Zīj al-mumtahan, from which one hundred years later the astronomer aṣ-Ṣūfī borrowed the precession constant, enjoyed a very good reputation.[52] However, these star catalogues included only the most important stars and could not, therefore, replace the Almagest in any way.

(ii) Even though the Islamic astronomers were able to perform their own exact observations, they nevertheless placed full trust in the essential formulations of Ptolemy's Almagest. They could improve the value of the precession motion by using the coordinates of the Almagest for its calculation, but this very process prevented them from any critical inspection beyond the purely philological testing of their authenticity. The scope of al-Battānī's star register and the remarks of aṣ-Ṣūfī and Ibn aṣ-Ṣalāḥ tell us that they were well aware of the deficiencies of the copies. At first there was no reason to doubt the correctness of the original Ptolemaic catalogue data before the coordinates of the Almagest had been reconstructed from the ever increasing number of manuscript copies and before a proper theory of the precession motion was established.

In the history of Arabic astronomy aṣ-Ṣūfī (903–983) wrote one of the most

---

[51] If one assumes accurate position measurements in the year 1274, one should derive a precession of 1° in 68 years on the basis of the longitudes in the Almagest. It is also possible that the astronomers of that period neglected the Ptolemaic longitudes and based their calculations of the precession constant entirely on observations from early Islamic astronomy.

[52] Kunitzsch, P. (1974), p. 51. Suter, H. (1900), p. 8.

important texts on the fixed stars since Ptolemy.[53] Aṣ-Ṣūfī checked how all of the Ptolemaic constellations and stars had been handed down through tradition and also, at least partially, their agreement with the positions he had determined himself. He was the first to maintain that Ptolemy had not observed the stars of the catalogue himself, but had taken them from an older manuscript and increased the ecliptical longitudes by the value of precession in accordance with the Hipparchan value of 1 degree per one hundred years.

Aṣ-Ṣūfī presumes that Ptolemy made use of the data of Menelaus which had been obtained 41 years before the epoch of the Almagest (+137) as foundation for his catalogue, and then added 25' to the Menelaic longitudes. It is not clear why aṣ-Ṣūfī makes this claim. As he tells us, although several copies of the Almagest were available to him, he had no Greek source of Menelaus' writing.[54] Disregarding for the moment aṣ-Ṣūfī's motives for making this claim, it cannot be the longitudinal errors of the Ptolemaic stars of one degree that had prompted him to his interpretation. In 41 years the hypothetical longitudes of Menelaus increase by 34'. According to aṣ-Ṣūfī, Ptolemy would have added 25', and with that obtained longitudes only 9' too small, all of which makes up a negligible error. In spite of his allegation that the positions of the stars of his catalogue were not observed by Ptolemy himself, aṣ-Ṣūfī had obviously not yet discovered the systematic errors in longitude of the Almagest.

For a long period following this remains the unique instance of a doubt about authenticity of the Ptolemaic star catalogue. It is highly speculative whether aṣ-Ṣūfī supplied an interpretive model for Tycho Brahe's later examinations of the Ptolemaic catalogue. From the late middle ages to the 16th century the astronomer aṣ-Ṣūfī was indeed known: it is evident from two wooden engravings of the northern and southern hemispheres by Dürer on which aṣ-Ṣūfī is depicted as one of the four greatest proponents of astronomy.[55] His texts were not translated into Latin, though.[56] In several medieval manuscripts with star lists attributed to aṣ-Ṣūfī there are illustrations influenced by the Arabic tradition. The coordinates and the description of the star positions are taken from the version of the Almagest as translated by Gerhard von Cremona.[57]

For its own part Islamic astronomy formulated no critique on the accuracy of the original coordinate measurements of Ptolemy, choosing rather to restrict itself to a philological purification of the copying errors and making new and independent measurements. As late as ca. +1150 Ibn aṣ-Ṣalāḥ examined with great meticulousness the transmission of the Ptolemaic star catalogue and restored a number of coordinates that are extremely helpful today for the reconstruction of the original Ptolemaic star catalogue.[58] Despite the extensive discussion of the possible star positions at the time of Ptolemy, the text contains no remarks on the systematic errors in longitude. The author's only interest was to solve the problem

---

[53] Sezgin, F. (1978), p. 212. In French translation: Schjellerup, H. C. F. C. (1874), *Description des étoiles fixes*, St. Petersburg. Cf. Kunitzsch, P. (1974), p. 51.

[54] Björnbo, A. A. (1901), p. 202. Cf. section III.2.

[55] Sezgin, F. (1978), p. 212. The engraving is printed in Strohmaier, G. (1984), *Die Sterne des Abd ar-Rahman as-Sufi*, Hanau.

[56] Kunitzsch, personal communication.

[57] Kunitzsch, P. (1965), Sufi Latinus, *Zeitschrift der Morgenländischen Gesellschaft* **115**, pp. 65-74. Strohmaier, G. (1984), p. 12.

[58] Kunitzsch, P. (1975), *Zur Kritik der Koordinatenüberlieferung im Sternkatalog des Almagest*, Göttingen.

of restoring the original numbers by checking them against his own observations. Without knowledge of the appropriate model for the precession motion it was not possible to estimate systematic errors in the Almagest.

With the increasing number of independent observations Islamic astronomy could emancipate itself from the Ptolemaic star catalogue. Huge instruments were built to refine the precision of the measurements. Al-Bīrūnī, for example, owned a quadrant with a radius of 7.5m.[59] The high point of independent Islamic observations was reached with the fixed star observations of Uluġ Bēg who revised a fraction of the traditional star catalogue and replaced the coordinates by more accurate ones in the observatory in Samarkand.[60]

With Spain as a conduit the astronomical knowledge of the Orient quickly spread to scholarly circles in medieval Europe and was sufficiently comprehensive to permit a critical evaluation of the accuracy of the Ptolemaic star catalogue.[61]

[59]Wiedemann, E. (1970), *Aufsätze zur Arabischen Wissenschaftsgeschichte*, 2 vols., Hildesheim, vol. I, p. 559.

[60]Sezgin, F. (1978), p. 30. Knobel, E. B. (1917), *Ulughbeg's Catalogue of Stars*, Washington.

[61]Sezgin describes the influence of Islamic astronomy in Sezgin, F. (1978), pp. 37–59. See also Mercier, R. (1976/77); Dobrzycki, J. (1963), Katalog gwiazd w de Revolutionibus, Studia i Materialy z Dziejow Nauki Polskiej, Seria C, Z. 7.; Swerdlow, N. M., Neugebauer, O. (1984), *Mathematical Astronomy in Copernicus's De Revolutionibus*, New York.

2. Accusations

2.1 Tycho Brahe

The Almagest became known in Europe through the Latin translation of Gerard of Cremona in 1175. Astronomy began to assimilate Ptolemaic theory and its Arabic revisions into the emerging physical sciences and thereby laid the ground for the following rapid scientific development. In the 16th century Copernicus succeeded in overcoming the geocentric construction of Ptolemy's planetary orbits, but the methodological structure of his "De revolutionibus" was still oriented on the book that was written a millennium before.

Copernicus' star catalogue is based exclusively on the data of the Almagest.[1] Copernicus complained about the inaccuracies of the catalogue as he also complained about the lack of a viable alternative to it, but it was Tycho Brahe, the last and the most meticulous observer before the introduction of optical instruments, who was the first to lay the groundwork for a systematic appraisal of the Ptolemaic coordinate errors through his own highly precise star coordinates.

Tycho was indeed the first European to replace the Ptolemaic star catalogue with his own, far more exact positional measurements. The appearance and identification of a new star in the year 1572 inspired him, as, reportedly, a similar event had inspired Hipparchus, to assemble a new star catalogue.[2] Tycho also calculated the precession motion anew without the use of the stellar coordinates of the Almagest. This was the first step to a historical interpretation of the accomplishments of Ptolemy. The early Arabic astronomers, who were still forced to base their calculations of the precession motion on the coordinates and the times recorded in the Almagest, could in principle not discover any systematic errors in Ptolemy's longitudes.

Tycho Brahe had access to their observational material, with which he was able to justify a simple linear precession motion independently of the Almagest. After that, he was in a position to compare the Ptolemaic star catalogue with the positions recalculated from his accurate measurements.

A correct theory of the precession motion is an irreplaceable precondition for the checking of the Ptolemaic coordinates. As long as medieval astronomy still formulated and computed the spring equinox with a theory of trepidation incorporating the observations of the Almagest in its basic parameters, the systematic

---

[1]Cf. Dobrzycki, J. (1963) and Swerdlow, N. M., Neugebauer, O. (1984).
[2]Dreyer, J. L. E. (1953), p. 365.

errors of the stellar longitudes of Ptolemy could not be detected.

In the introductory comments to the chapters on the sphere of fixed stars in "Astronomiae Instauratae Progymnasmata" (1602) as well as in the introduction to his star catalogue "Stellarum Inerrantium Restitutio" (1598), Tycho sketches out the historical development of the star catalogues.[3] In the "Progymnasmata" a brief remark can be found that the star catalogue of the Almagest had been compiled through the conversion of the Hipparchan stellar coordinates.[4] In "Stellarum Inerrantium Restitutio" Tycho came to the conclusion that the lower limit of the Hipparchan precession constant used by Ptolemy for the conversion of the stellar longitudes to his epoch could in fact account for the errors in longitudes of the stars in the Almagest,[5] although Ptolemy himself was prevented from discovering these by certain systematic errors of his own methods of observation. As possible causes for the error in longitude Tycho considers an inadequate solar and lunar theory[6], the reduction of the solar longitude through the effect of refraction at sunset, and the neglect of the lunar parallax.[7] Tycho studied the Arabic astronomers and suggested historical reasons for their errors. He discovered that the error in longitude of Ptolemy's star catalogue is responsible for the large precession constant of al-Battānī.[8] Although Tycho committed himself to the thesis of a Hipparchan origin of the Ptolemaic star catalogue, his astronomical research opens the way for two different possibilities of historical interpretation.

(i) The errors in longitude of the star catalogue result from a transformation of the Hipparchan coordinates with a precession constant that is too small. A series of systematic errors, like the deficiencies in the solar and lunar theory and the disregard of the effects of refraction and parallax, must have led Ptolemy to confirm the conversions he made from the Hipparchan star register.

(ii) Because the systematic errors in the solar theory confirm the longitudes in the catalogue by later observation, these errors could also be the original cause for the inaccuracies of the fixed star catalogue. If Ptolemy had used an observational method which assumes the erroneous solar theory, it follows that the systematic errors of the star catalogue would be generated thereby.

---

[3]Brahe, T. (1913–29), *Tychonis Brahe Dani Opera omnia*, ed. J. L. E. Dreyer, 15 vols., Copenhagen, vols. II and III.

[4]Brahe, T. (1913–29), vol. II, p. 151: "Post hos Claudius etiam Ptolemaeus, circa Annum a nato Christo 140, Alexandriae quoque Aeqypti nonnulla in harum progreßione animaduertere, atque literis mandare, aggreßus est; Hipparchico tamen, circa earum adinuicem, quoad longum & latum collocationem, totaliter retento Abaco."

[5]Brahe, T. (1913–29), vol. III, pp. 335f.

[6]The maximal error of the latter Tycho estimates as 1/4 degree.

[7]Brahe, T. (1913–29), vol. III, p. 336: "Incedens enim lubrica illa uiâ & ad fallendum prona, quae a Sole per Lunam Stellarum loca monstraret, facile quartae partis unius gradus, si non dimidiae, errorem incaute admittere potuit: ueluti alibi a nobis expressiûs pandetur. Imo cum refractiones Solis iuxta Horizontem (circa quem, cum hanc pragmatiam exercebat, constituebatur) positi, ut de Parallaxibus non dicam, neglexerit, praecisionem ipsissimam non attigit, uti et saepius his alijsque de causis tam in Sole quam reliquis Planetis & Stellis fixis deuiationem aliqualem commisisse uidetur. Verum hoc non ob id refero, quod tanti artificis & de tota re Astronomica adeo praeclare meriti Viri, sine cuius operibus uix pateret ad hanc Artem accessus, traditiones eleuare praesumam: sed solummodo ut negotij subtilitatem et labyrinthos, ubi summa requiritur praecisio, maximis etiam artificibus obrepentes, aliquatenus indicem."

[8]Brahe, T. (1913–29), vol. III, p. 336.

Therefore, it is not possible to make a decision between the two interpretative alternatives based solely on the systematic errors in the stellar longitudes. For the following generations of astronomers, the specific judgment about the origin of the star catalogue was more and more based on the possibility of a coherent interpretation of the totality of astronomical claims in the Almagest.

2.2 Laplace and Lalande

Laplace doubted the Hipparchan origin of the star catalogue in his "Exposition du Système du Monde."[9] In chapter two of the fifth book he states that Hipparchus' length of the year was too large and that Ptolemy's assimilation of this theory explains why the position of the mean sun was too small by one degree at the time of the Almagest. Since the star positions are determined relative to the position of the sun using the astrolabe as described in the Almagest, the solar theory alone is capable of explaining the stellar longitudes in the catalogue.[10]

> "This remark moves us to examine whether, as generally believed, Ptolemy's star catalogue is merely the one prepared by Hipparchus adjusted to the time of the former through a yearly precession of 111″. This opinion is grounded on the fact that the systematic error of the longitudes of the stars in this catalogue disappears when one reduces it to the time of Hipparchus. However, the explanation offered by us for this error vindicates Ptolemy against the accusation that he had simply assimilated the work of Hipparchus and it appears justified to believe him when he says that he himself had observed the stars of his catalogue, even the ones belonging to the sixth magnitude."

Laplace's statements go beyond those of Tycho in that it clearly offers a coherent interpretation of the Almagest according to the principle of the greatest possible credibility.

A significantly richer historical interpretation of the Almagest, which Laplace refers to in the chapter just mentioned, had already been articulated by Lalande.[11]

The texts document a newly awakened interest in the Almagest particularly among the astronomers of the 18th century. In 1712 the Royal Astronomer Edmund Halley edited the Greek text of Ptolemy's star catalogue.[12] The vast span of time from the epoch of the Almagest turned the star catalogue into an historical witness of ancient star data which promised interesting evaluations in spite of its recognized inadequacies.

Halley compared the latitudes of the bright stars with the ecliptical latitudes of his time and so he was the first who successfully demonstrated the proper motion

---

[9]Laplace, P. S. (1796), Exposition du Système du Monde, Paris. Cited after Laplace, P. S. (1797), Darstellung des Weltsystems, Frankfurt, vol. II, pp. 253 ff.

[10]Laplace, P. S. (1797), vol. II, pp. 254f.

[11]Lalande, J. D. (1757), Mémoire sur les équations séculaires, et sur les moyens mouvemens du Soleil, de la Lune, de Saturne, de Jupiter et Mars, avec les observations de Tycho-Brahé, faites sur Mars en 1593, tirées des manuscrits de cet Auteur. Mémoires de mathématique et de physique, tirées des registres de l'Académie Royal des Sciences, de l'Année 1757. pp. 411–470.

[12]Halley, E. (1712), *Geographiae Veteris Scriptores Graeci Minores*, Oxford, vol. III.

of the stars for Sirius, Arcturus and Aldebaran.[13] In 1786 a French translation of the Almagest star catalogue was published by the Abbé Montignot, followed by the German translation of the astronomer Bode in 1795.[14] The advanced mathematical treatment of celestial mechanics in the 18th century led to a highly precise theory of celestial motions. An accurate approximation to the motions of three celestial bodies attracting each other was accomplished and the dimensions of the solar system were successfully determined through the observations of the Venus transits of 1761 and 1769. At that time the data of the Almagest were considered the testing instance by which the precision of a theory for long periods of time could be controlled.

One of the most famous astronomers of the 18th century was Joseph-Jérome Lalande, whose textbook "Traité d'astronomie" of 1764 became, with new editions in 1771 and 1792, a standard work in the field.[15] Lalande, who stood firmly in the tradition of the encyclopedists, examines the major theories of the Almagest, the conclusion of which he draws in "Mémoires de l'Académie Royale des Sciences."[16] Lalande investigates here the Ptolemaic measurements of the equinoxes and finds there an awesome deviation from the accurate values, very different from the high precision earlier obtained by Hipparchus. All of the inaccuracies of the Ptolemaic observations are, as Lalande sees it, explicable in a most natural way if all of them are interpreted as mere theoretical constructions.[17] Lalande uses five arguments to support the claim that Ptolemy had not himself made the observations in the Almagest, but had only calculated the results from the theory and then claimed them as the fruit of actual observations:

(i) Ptolemy reports the lunar eclipses of 19–20 March -199 and 12 September -199 which he evaluated for the calculation of the parameters for the lunar theory. Ptolemy criticizes the Hipparchan analysis whose calculations assume a time difference between the eclipses of 176 days, one hour and 20 minutes, and he replaces it with a time difference of 176 days and 24 minutes.[18] For Lalande, this proves that Ptolemy had undertaken certain changes in the observational data as reported by Hipparchus in order to bring the values in agreement with his theory.[19]

(ii) The Ptolemaic measurements of the equinoxes are highly distorted. According to Lalande, the measurement of 26 September +139 as well as of 22 March is incorrect by 11 hours, and it is peculiar that this error tallies with the theoretical values.

---

[13] Halley, E. (1718), Considerations on the Change of the Latitudes of some of the principal fixt Stars. *Phil. Trans.* **30**, No. 355, pp. 736–738.

[14] Ptolemy, C. (1963), vol. I, p. XXII.

[15] Hankins, T. L. (1973), Lalande, in: *Dictionary of Scientific Biography*, ed. C. C. Gillispie, New York, vol. VII, p. 580.

[16] Lalande, J. J. (1757), Mémoires de l'Académie Royale des Sciences, de l' Année 1757, Paris. Cf. Wilson, C. (1984), The Sources of Ptolemy's Parameters, *Journal for the History of Astronomy* **15**, pp. 37ff.

[17] Lalande, J. J. (1757), pp. 420f.

[18] Ptolemy, C. (1963), p. 214.

[19] Lalande, J. J. (1757), p. 420; Toomer shows that the differences in time are caused by inaccuracies in the calculation of Hipparchus. Cf. Toomer, G. J. (1973), The Chord Table of Hipparchus and the Early History of Greek Trigonometry, *Centaurus* **18**, pp. 6–28.

(iii) The third indication concerns the star catalogue. From the data of Hipparchus, al-Battānī, Tycho and his own measurements, Lalande derives a constant motion of precession of 50.5″ per year. This motion would be 2″ per year larger if the coordinates of the star catalogue are included in the reckoning. Lalande cites in his evaluation Monnier from "Institutions astronomiques" (1746), for whom it is true beyond any doubt that Ptolemy was not in a position to determine even one single fixed star position.

(iv) The Ptolemaic measurement of the lunar parallax is plagued by a substantial error of 42′, drastically exceeding the Hipparchan error of 13′.

(v) According to Lalande, Ptolemy had claimed that in early antiquity the obliquity of the ecliptic should have amounted to 24°, and that only for later periods had Ptolemy used the value 23°51′20″ taken over from Eratosthenes and Hipparchus. For Lalande this absurd thesis proves the incompetence of Ptolemy as an observer, for the investigations of Kepler showed very clearly that such a large variation could never have actually happened.[20]

Lalande did not examine the passages of the Almagest under consideration himself: rather, he refers to others who checked the accuracy of the statements in the Almagest and who discovered grave errors. After his long list of grievances about Ptolemy's observational accuracy, Lalande asks the suggestive question whether or not all this provides a sufficient reason to condemn the authenticity of the rest of Ptolemy's observations as well.[21]

In the standard astronomical text of the time, Lalande's "Astronomie", this passage of the Memoires is referred to, and Ptolemy is depicted as a very poor observer. In the second and third edition of the work, Lalande sharpens his judgment once again and transforms "Ptolemy, the poor observer" into a Ptolemy "who was actually no observer at all".[22] It remained for the most illustrious student of Lalande to historically undermine these brief and superficial comments on Ptolemy which also appeared to be the result of a reading of only secondary texts.

In 1780, while Lalande was holding lectures in Paris at the Collège de France and mentioned during one of them the Greek poet Aratus, whose didactic poem was later discussed critically by Hipparchus, the student Jean-Baptiste Delambre attracted much attention when he stood up and recited the entire passage in question from memory and was even able to comment on it in great detail. Delambre, who criticized Lalande's "Astronomie" by fastidiously writing remarks in the margins, became his assistant and later his colleague.[23]

2.3 Delambre's Investigations

After the first historical interpretations of the star catalogue by Tycho, Lalande and Laplace had been made, J. B. Delambre dedicated himself at the beginning

---

[20]There is no evidence in the Almagest supporting Lalande's assertion.

[21]Lalande, J. J. (1757), p. 421.

[22]Lalande, J. J. (1764), Traité d'astronomie, Paris. 2. ed. 1771, 3. ed. 1792. Wilson, C. (1984), p. 38.

[23]Cohen, I. B. (1971), Delambre, *Dictionary of Scientific Biography*, ed. C. C. Gillispie, New York, vol. IV, p. 14.

of the 19th century to a comprehensive investigation of the history of astronomy, concentrating especially on the astronomy of the Almagest.[24] His work gave legitimacy to the allegations that Ptolemy, in contradiction to his own claims, did not really observe the stars listed in the Almagest at all, but had assimilated them from Hipparchus. The second volume of "Histoire de l'astronomie ancienne" after an introductory chapter on Greek mathematics, treats exclusively of the astronomy of Ptolemy and comments on each book of the Almagest. The commentary on the seventh book provides the point of departure for all subsequent historical investigations of the origin of the Ptolemaic fixed star catalogue. The interpretation of the errors in longitude by Laplace could not be accepted by Delambre. In his commentary to the third book of the Almagest in which the solar theory is developed and Delambre refers to the error of the mean sun, we read:[25]

> "...but the error in the mean motions which makes the epoch [of the era Nabonassar] a bit too large, would not produce any inconvenience for the epoch at which he [Ptolemy] lived. The errors that might obtain in the solar longitudes, from which the stellar positions were to be deduced, resulted from the error in his equinox and the error in the [solar] motion over [only] a small number of years."

The question as to how the errors of the mean sun position can be reconciled with the error in equinox remains unanswered and is later labeled a "strange statement" by Dreyer.[26]

With the exception of this more favorable interpretation of Ptolemy's observations, Delambre adheres to the inductive argumentative figure of Lalande: the numerical error of individual parameters, e.g. for the position of the farthest points of the solar orbit from the earth (apogee) as well as for the observations described in the Almagest can easily be explained when they are considered as being theoretically derived and not as the contingent results of actual observations.

Ptolemy establishes the first numerical proof of the precession motion with a Hipparchan observation of the star Regulus in the constellation Leo.[27] Ptolemy reports that on 23 February +139 the last degree of Taurus culminated just at sunset (5 1/2 equinoctial hours after noon), while the moon had an elongation of $92\frac{1}{8}°$ from the sun, whose position was at 333° ecliptical longitude. Half an hour later, as the fourth part of the zodiacal sign Gemini ($67\frac{1}{2}°$) culminated, the position of the moon in the astrolabe was adjusted on the inner ring and the star Regulus on the outer ring. On the graduation on the inner ring, the longitudinal difference of Regulus from the moon could then be read off as $57\frac{1}{6}°$. At sunset, the sun stood at $333\frac{1}{20}°$, so that at this time the apparent moon stood at the longitude $65\frac{1}{6}°$ after the addition of the elongation. Within one half hour, the moon had moved 1/4° farther and the parallax had increased

[24]Delambre, J. B. J. (1817), *Histoire de l'astronomie ancienne*, 2 vols., Paris.

[25]Delambre, J. B. J. (1817), vol. II, p. 138.

[26]Dreyer, J. L. E. (1918), On the Origin of Ptolemy's Catalogue of Stars, Second Paper, *Monthly Notices of the Royal Astronomical Society* **78**, p. 347.

[27]Ptolemy, C. (1984), pp. 328ff.

by $\frac{1}{12}°$. The apparent position of the moon during the time of the second observation then amounts to $65°10' + 15' - 5' = 65°20'$. The longitude of Regulus is then given as $65°20' + 57°10' = 122°30'$: in full agreement with the value of the catalogue. Hipparchus had observed Regulus about 267 years before with the ecliptical longitude of $119°50'$, from which Ptolemy derived a precession of exactly $2°40'$.

Similarly, Ptolemy observed the star Spica and "the brightest stars close to the ecliptic". In reference to these stars he determined, according to his own testimony, the positions of the remaining stars, whose latitudes were once again approximately as large as what Hipparchus had observed, while the distances from the vernal point had increased by about $2°40'$ in the direction of the zodiacal signs. Delambre emphasizes that the peculiarity of the Regulus observation is highly significant for the interpretation of the star catalogue. Actually, Regulus would have had to move $3°40'50''$ on the ecliptic since the time of Hipparchus. Nonetheless, the observation confirms a value one degree too small. Delambre uses irony to attack the readings of authentic observations and the explanation of the errors in longitude through a deficit in the solar theory.[28]

"The conclusion from all of this is that Ptolemy compared only Regulus and Spica directly with the sun; that he took the distances of the other stars with respect either to each other, or to Regulus or Spica; and that the longitudes must have been affected by the errors in his solar longitudes. He calculated these longitudes from tables entirely in agreement with those of Hipparchus; hence the error ought to be attributed to Hipparchus, were it not that Ptolemy also assured us that he himself has observed the sun, that he has found the same values for the intervals in time between the equinoxes and the solstices, that he derived therefrom the same eccentricity, the same apogee position, and consequently the same equation and the same mean longitude. Besides which he had the same mean motion, since he assigned the same length to the year."

This interpretation would award Ptolemy a high degree of credibility and scientific precision, but would at the same time discredit the astronomical authority of Hipparchus. Delambre continues:[29]

"One could explain everything in a less favorable but all the simpler manner by denying Ptolemy the observation of the stars and the equinoxes, and by claiming that he assimilated everything from Hipparchus, using the minimal value of the latter for the precession motion."

Thus Delambre considers the longitudinal errors of the mean sun as a possible source of error for the star catalogue. It only appears implausible to Delambre that the most comprehensive astronomical observations, namely, those of the coordinates

---

[28] Delambre, J. B. J. (1817), vol. II, p. 250.
[29] Delambre, J. B. J. (1817), vol. II, p. 250.

of the star catalogue, had been carried out by Ptolemy or his students, while he copied the most important astronomical parameter of the entire system, that of the solar theory, from Hipparchus.

The significance of the preparatory work of Hipparchus for the Ptolemaic "observations" and the values handed down by Timocharis becomes clear from the calculation of the precession constant. Ptolemy lists the declinations of 18 stars from the time of Timocharis (−293), Hipparchus (−128) and his own measurements and demonstrates with six examples that the precession sums up to 2°40′ since Hipparchus. Delambre computes the precession constant of a set of declination observations with the equation:

$$p = \frac{(\delta' - \delta)\cos(\frac{\delta'+\delta}{2})}{n \sin \epsilon \cos \beta \cos(\frac{\lambda'+\lambda}{2})} \tag{2.1}$$

p = precession constant ; δ and δ' = declinations
n = time difference between two measurements; ϵ = obliquity of the ecliptic.
β = eclipitical latitude, λ and λ' = ecliptical longitudes

Delambre obtains the ecliptical longitudes of the stars from the longitudes P of the Ptolemaic star catalogue by adding one degree to compensate for the systematic error, and from that he subtracts 3°40′ for the Hipparchan longitudes and 5°40′ for the longitudes at the time of Timocharis:[30]

$$\begin{aligned} \lambda_{Ptol} &= P + 1° \\ \lambda_{Hipp} &= P - 2°40' \\ \lambda_{Timo} &= P - 4°40' \end{aligned} \tag{2.2}$$

The mean longitude $(\lambda' + \lambda)/2$ is then:

(P − 50′)	between Hipparchus and Ptolemy
(P − 3° 40′)	between Timocharis and Hipparchus
(P − 1° 50′)	between Timocharis and Ptolemy

The results are listed in the following tables. A few figures will be corrected because of changes in the new text editions of the Almagest.[31]

The mean value of all 18 values is invalidated through the stars with ecliptical longitudes close to the solstices, for the declinations are hardly changed by the precession. Consequently, the calculated precession constants for those stars involve large inaccuracies. Nevertheless, for this data set the mean value of the precession

---

[30]Delambre, J. B. J. (1817), vol. II, p. 252.

[31]The accurate tabulations can be found in table (3.13). Corrections are necessary for star η Tau, $\omega_{Hipp} = 15°10'$; α Aur, $\omega_{Ptol} = 41°10'$; α Lib, $\omega_{Ptol} = -7°10'$; β Lib, $\omega = 1°12'$. In the table only the signs of the declinations for α Lib are corrected. Name: modern star name. No. Al.: catalogue number in the Almagest; T-A: declinations from the school of Timocharis and Aristyllos; Hipp: Hipparchan declinations; Ptol: Ptolemaic declinations; T-H: Precession constant in ″/year derived from declinations T-A and Hipparchus; H-P: Precession constant derived from declinations of Hipparchus and Ptolemy; T-P: Precession constant derived from declinations of the school of Timocharis and Ptolemy; Ptolemy demonstrates the value of the precession constant with the examples assigned by a dot.

No.	Name	No. Al.	T-A	Hipp.	Ptol.	T-H	H-P	T-P	Choice
1	α Aql	288	5;48	5;48	5;50	0.00	24.46	23.77	
2	η Tau	411	14;30	15;30	16;15	68.47	27.40	43.00	•
3	α Tau	393	8;45	9;45	11;00	78.89	54.85	63.13	
4	α Aur	222	40;00	40;24	41;30	33.38	51.61	44.63	•
5	γ Ori	736	1;12	1;48	2;30	600	41.03	48.26	•
6	α Ori	735	3;50	4;20	5;15	61.22	66.37	64.45	
7	α CMa	818	-16;20	-16;00	-15;45	92.63	45.75	51.09	
8	α Gem	424	33;00	33;10	33;24	48.80	50.91	49.98	
9	β Gem	425	30;00	30;00	30;10	0.00	62.12	32;95	
10	α Leo	469	21;20	20;40	19;50	79.79	50.03	59.93	
11	α Vir	510	1;24	0;36	-0;30	49.86	37.08	41.56	•
12	η UMa	35	61;30	60;45	59;40	45.88	35.88	39.32	•
13	ζ UMa	34	67;15	66;30	65;00	46.22	50.03	48.74	
14	ε UMa	33	68;30	67;36	66;15	56.17	45.24	33.06	
15	α Boo	110	31;30	31;00	29;50	31.10	39.59	36.61	•
16	α Lib	529	-5;00	-5;36	-7;30	38.13	66.51	56.63	
17	β Lib	531	1;40	0;24	-1;00	83.50	51.11	64.11	
18	α Sco	553	-18;20	-19;00	-20;15	50.38	53.22	52.18	
	mean value:					51.39	47.45	47.41	

Table 2.1: Precession constants as function of declination variation.

constant remains essentially unchanged when the "critical" stars No. 1, 7, 8 and 9 are excluded. The three means varying from 47″ to 52″ then agree with the accurate value of 50″ per year. Apparently, Ptolemy had used only those six stars for the numerical demonstration that come closest to Hipparchus' lower limit of 36″ per year.

Naturally, Ptolemy had no access to the trigonometrical and statistical methods of calculation and evaluation which are at our disposal today. The use of a mean value as an advanced method of data evaluation was unknown at that time. For that reason he had to look at the relationship of the observations to the values theoretically obtained in a rather simple way: a series of observations verify the assumed precession value. Consequently, they are cited by Ptolemy as a proof of its correctness. The other observations whose evaluation provided no promising agreement were neglected.

Such a historical interpretation of the possible Ptolemaic strategies – acknowledging the ancient scientific practice and the conceptual and methodological instruments available – is quite alien to Delambre. What he does instead is to impose the scientific standards of his time on the Almagest and to accuse Ptolemy of making a biased selection of data:[32]

> "He (Ptolemy) begins by asserting on extremely flimsy grounds that the changes in the observed declinations agree with a precession of 36″ per year, while in actual fact there is no other value to choose than 47 to 49″."

---

[32]Delambre, J. B. J. (1817), vol. II, p. 255.

For Delambre the star catalogue is "a precious monument of the history of science."[33] The true authorship is impossible to derive from its data alone, but – so Delambre writes in a provocative assertion – the catalogue does not agree with the epoch of Ptolemy but with the epoch of Hipparchus, when the false value of the precession of 2°40′ is taken into consideration.[34] Ptolemy would have to have observed at least some of the stars of his catalogue himself, if, of course, these were visible at the more southerly latitude of the site of observation in Alexandria and not at the observation site of Hipparchus on the island of Rhodes, 5° farther north. But the Ptolemaic catalogue contains no star which could not be seen from Rhodes,[35] which Delambre understands as indirect evidence of Hipparchan authorship, or, failing that, at least it documents how strongly Ptolemy aligned himself with the Hipparchan tradition.

Delambre comments on the accuracies of the catalogued coordinates only fleetingly: the data for longitude and latitude are generally exact to 10′, that is to say, to 1/6°. Only in exceptional cases would one find 1/12° values or 5′ incorporated in the catalogue.[36] The average deviation of the star positions from the accurate positions is derived by Delambre from extensive comparisons with the accurately recalculated stellar coordinates. The relevance of his results is vitiated by the philological uncertainties pervading the authentic reconstruction of the Almagest, which is why Delambre expresses himself very cautiously with respect to the validity of the results.[37]

However, it is impossible to neglect the fact that the systematic error in longitude of one degree is fairly well distributed over all parts of the celestial sphere. When corrected by this error, the coordinates of the star catalogue have a standard deviation of 15′ to 30′.[38]

All of these investigations provide no clue to an authentic measuring of the coordinates by Ptolemy. Although no conclusive proof can be found that the stars of the Almagest were originally measured by Hipparchus, the overall evaluation of Ptolemy's accomplishments as an observer leads Delambre, as it did Lalande before, to conclude that inaccurate figures in the Almagest are not normal errors of measurement, but generally either taken over from predecessors, or they were only calculated as illustrations of a particular theory.[39]

Finally, Delambre's appraisal of the scientific quality of Ptolemy's work expressed the emotions of historians to come:[40]

> "If Ptolemy made the observations himself, then he must have compared them [with the Hipparchan ones in the usual fashion]; if those were suppressed in order to avoid discrediting his catalogue and observations, then he acted in bad faith; he did not possess that astronomical

---

[33] Delambre, J. B. J. (1817), vol. II, p. 264.

[34] Delambre, J. B. J. (1817), vol. II, p. 264.

[35] Delambre, J. B. J. (1817), vol. II, p. 284.

[36] Delambre, J. B. J. (1817), vol. II, p. 286. This statement is false, since a fraction of the degree of e.g. 15′ is not accurate to 5′, but to 1/4° .

[37] Delambre, J. B. J. (1817), vol. II, p. 290.

[38] Delambre, J. B. (1817), vol. II, pp. 287ff.

[39] Delambre, J. B. J. (1817), vol. I, p. XXV.

[40] Delambre, J. B. J. (1817), vol. I, p. XXXI.

integrity which is indispensable for the observer. To this we add that he was clumsy as well. He would have done better to have reported everything as it was instead of leaving it to the imagination of his readers to go beyond reality."

3. The Rehabilitation of Ptolemy

Thirty years were all that were needed to change Ptolemy's negative image as a counterfeiting eclectic into a positive one. During the time when Jacob Burckhardt's cultural history of Greece exerted its greatest influence, the results of the intensive, historical grappling with ancient astronomy cast doubt on the dominant interpretation of Delambre. The manuscripts that were handed down from the works of Ptolemy, Hipparchus, Geminos, Proclus, Aristarchus, Archimedes and Eratosthenes were published in critical editions, valid even today, whose quality such names as Manitius, Heath, Maass and Boll can vouch for.

Nevertheless in Oswald Spengler's *The Decline of the West*, Ptolemy is still accorded a place in the last period of the decline of Greek culture:[1]

> "... and we are now experiencing the *decrescendo* of brilliant gleaners who arrange, collect and finish off, like the Alexandrian scholars of the Roman age. Everything that does not belong to the practical side of life – to politics, technics or economics – exhibits the common symptom. After Lysippus no great sculptor, no artist as man-of-destiny, appears, and after the Impressionists no painter, and after Wagner no musician. The age of Caesarism needed neither art nor philosophy. To Eratosthenes and Archimedes, true creators, succeed Posidonius and Pliny, collectors of taste, and finally Ptolemy and Galen, mere copyists."

But after Boll's examination of an astrological manuscript managed to shake the argumentative edifice of Delambre, then Vogt, with his reconstructed stellar coordinates of the hitherto lost star register of Hipparchus, levelled it completely to the ground.

3.1 The Number of Hipparchan Stars

Various discoveries of manuscripts, along with a meticulous rereading and reappraisal of medieval astrological texts within a space of ten years at the end of the last century, combined to make a formidable argument against the dominating view which held the lost fixed-star register of Hipparchus to be genetically identical with

---

[1]Spengler, O. (1926), *The Decline of the West*, London, pp. 424f. Transl. of Spengler, O. (1923), Der Untergang des Abendlandes, München.

Ptolemy's catalogue. Some of these manuscripts contain descriptions of a fixed-star register from which the total number of stars can be derived.

In 1892 Ernst Maass edited two constellation indexes from an eighth-century codex in Basel.[2] One of the registers is attributed to Eratosthenes, the other to Hipparchus. The indexes contain neither the names of the stars nor their location, but only the names of the constellations are mentioned. The temporal and cultural variations in the names and the form of the constellations make it possible to garner an approximate chronological ordering of the catalogue merely from the names of the constellations that are used, and with that, to re-check the attested authors of the Basel text. Rehm's research shows that the text contains Hipparchan names. Beyond that, the sequence of the constellations mentioned provides further evidence for the existence of a Hipparchan register that is referred to by Ptolemy in the Almagest.[3] Alessandro Olivieri's discovery in 1898[4] and especially Franz Boll's in 1901, of two registers containing names of constellations with a supplementary list in which the number of stars in each constellation is enumerated, augmented and rerouted the knowledge of Hipparchus' register in a surprising direction.[5] In a comprehensive astrologer's hand-written text of 1550 (Parisinus 2420) under the "corrupted" title "ἐκ τῶν ὑπάρχων ἀστέρων", Boll found a register of star totals similar to the one examined by Rehm. Boll discovered that this star total is contained in a further astrological text (Parisinus 2506) with the same title, and he was able to demonstrate that this manuscript had served as a model for text 2420.[6] This brief document lists the names of the constellations and the number of stars in them.

The list is part of the following table; Ptolemy's names for the constellations are used, while the numbers of stars in Boll's manuscript are placed opposite those from the Almagest.[7]

Two additional columns list the number of the so-called external stars not directly belonging to the constellation, which Ptolemy records after the stars of the respective constellation, along with the number of stars in the Almagest whose ecliptic latitude contains a fraction of the degree of 1/4. [8]

---

[2]Maass, E. (1892), Aratea. *Philologische Untersuchungen* **12**, pp. 371ff. Reedited in Maass, E. (1898), *Commentariorum in Aratum reliquiae*, pp. 134–139.

[3]Rehm, A. (1899), Zu Hipparch und Eratosthenes. *Hermes* **34**, pp. 251–279.

[4]Cf. Rehm, A. (1899), pp. 264ff.

[5]Boll, F. (1901), Die Sternenkataloge des Hipparch und des Ptolemaios, *Bibliotheca Mathematica*, 3. Folge, Bd. 2, pp. 185ff.

[6]Boll, F. (1901), p. 186.

[7]Dreyer, J. L. E. (1917), On the Origin of Ptolemy's Catalogue of Stars, *Monthly Notices of the Royal Astronomical Society* **77**, p. 529. The numbers are updated according to Toomer's edition of the Almagest.

[8]Index of columns: Modern name of the star as identified in the Almagest; *Hipp.*: total of stars in Boll's manuscript; *Ptol.*: total of stars in the Almagest; *ext.*: total of external stars not belonging to the constellation; for constellations belonging to the northern (N), southern (S) hemisphere, or the zodiac (Z); *1/4 d.*: total of stars with a 1/4 fraction of the degree in latitude; *1/4 ext.*: total of the external stars with a 1/4 fraction of the degree in latitude.

No.	Constellation	Hipp.	Ptol.	ext.	H.	1/4 d.	1/4 ext.
1	Ursa Major	24	27	8	N	5	2
2	Ursa Minor	7	7	1	N	0	0
3	Draco	15	31	0	N	5	0
4	Bootes	19	22	1	N	2	0
5	Corona Borealis	9	8	0	N	1	0
6	Hercules	24	28	1	N	6	0
7	Ophiuchus	17	24	5	N	5	0
8	Lyra	10	10	0	N	1	0
9	Cygnus	14	17	2	N	1	0
10	Aquila	4	9	6	N	0	0
11	Sagitta	4	5	0	N	0	0
12	Delphinus	9	10	0	N	2	0
13	Pegasus	18	20	0	N	1	0
14	Cepheus	19	11	2	N	3	0
15	Cassiopia	14	13	0	N	2	0
16	Andromeda	20	23	0	N	0	0
17	Triangulum	3	4	0	N	0	0
18	Perseus	19	26	3	N	5	0
19	Auriga	8	14	0	N	1	0
20	Hydra	27	25	2	S	7	1
21	Crater	10	7	0	S	0	0
22	Corvus	7	7	0	S	1	0
23	Argo	13	45	0	S	7	0
24	Centaurus	26	37	0	S	7	0
25	Lupus	13	19	0	S	1	0
26	Ara	4	7	0	S	2	0
27	Corona Austalis		13	0	S	0	0
28	Piscis Austrinus	12	12	6	S	3	0
29	Cetus	14	22	0	S	1	0
30	Eridanus	-	34	0	S	5	0
31	Orion	18	38	0	S	8	0
32	Lepus	-	12	0	S	2	0
33	Canis Major	21	18	11	S	6	2
34	Canis Minor	3	2	0	S	0	0
35	Cancer	16	9	4	Z	2	1
36	Leo	19	27	8	Z	4	0
37	Virgo	49	26	6	Z	1	0
38	Libra	4	8	9	Z	3	2
39	Scorpio	15	21	3	Z	2	1
40	Sagittarius	16	31	0	Z	2	0
41	Capricornus	26	28	0	Z	4	0
42	Aquarius	18	22	3	Z	10	1
43	Pisces	41	34	4	Z	8	0
44	Aries	17	13	5	Z	1	0
45	Taurus	18	33	11	Z	7	2
46	Gemini	19	18	7	Z	3	1

Altogether then, the list contains 46 constellation names. Missing from the Ptolemaic constellations are Equuleus, Serpens and the partial constellation Aqua of Aquarius which Hipparchus had not yet included:

Additional Constellation	Star total Almagest	External Stars
Equuleus	4	0
Serpens	18	0
Aqua in Aquarius	20	0

Table 3.1: Additional constellations.

In the next tables are listed the numerical relationships of the number of stars in both star indexes for a series of subtotals upon which Boll grounds his estimation of the number of stars in the alleged Hipparchan register.[9]

Subset of stars	Hipp.	Ptol.
Star total of constellations 1–46:	653	877
Minus total of No. 27, 30, 32:	653	818
Minus Argo, No. 23:	640	773
Total in the zodiac:	228	270
Total of southern stars minus No. 27, 30, 32:	168	239
Total of southern stars minus No. 27, 30, 32, Argo:	155	194
Total of northern stars:	257	309
Total of Ptolemaic stars in No. 23, 27, 30, 32:		104
Total of Ptol. stars in Equuleus, Serpens, Aqua:		42
Total of stars in constellations missing in the Hipparchan list:		
Serpens, Aqua, Equuleus, Eridanus, Lepus, Argo, CrA:		146
Total 1/4-fractions minus No. 23 (Argo), 27, 30, 32:	123	
Total of external stars of all constellations:	108	
Total of stars in the Almagest:		1027

Table 3.2: Estimation of Hipparchan stars.

The ratios of the totals reflect the different sizes of the star registers:

Totals Ptol/Hipp minus No. 23, 27, 30, 32:	1.208
Ptol/Hipp in zodiac:	1.184
Ptol/Hipp south minus No. 27, 30, 32	1.423
Ptol/Hipp south minus No. 27, 30, 32, Argo:	1.252
Ptol/Hipp north:	1.202

Table 3.3: Ratio of star totals.

[9]Excluded is the double catalogue entry for star Herculis 29 = Bootis 9. Boll and Dreyer, J. L. E. (1917), p. 530 also exclude star Tauri 21 = Aurigae 11, but not the third double entry α Piscis Austrini = Aquarius 42. Here, the totals include both last double catalogue entries. This explains the increased total of 1027 Ptolemaic stars in comparison to Boll. It changes the result of his estimation from 851 to 850 stars.

Estimations of the number of stars for the missing constellations of the Hipparchan list, based on the ratios of star totals calculated above:

Totals for the missing constellations No. 23, 27, 30, 32:	121
Total of all Hipparchan stars without external stars:	761
Hipparchan external stars according the same ratio values:	89
Total of Hipparchan stars:	850

Table 3.4: Estimated size of star registers.

In a three-step argument Boll extrapolates the number of stars of the missing Hipparchan register from the somewhat obscure list in the astrological manuscript:

(i) The list of star totals in the astrological manuscript is identified as referring to a Hipparchan register of stars.

(ii) The Hipparchan register is identical to the register Ptolemy had at his disposal, whose coordinates Ptolemy is supposed to have used for the catalogue in the Almagest.

(iii) The comparison of the Hipparchan star total with the catalogue of the Almagest allows the calculation of the upper limit of the number of stars in the Hipparchan register.

For the first step Boll buttresses himself on the detailed analysis of Rehm.[10] Two aspects of the terminology turn out to be meaningful for the chronological ordering of the star registers.

The systematic sequence in which the constellations are listed indicates how strongly canonical was the figurative form of the constellations. Lists of constellations before the time of Hipparchus, for example the reconstructed Catasterisms attributed to the astronomer and geometer Eratosthenes,[11] show very simple arrangements of the constellations whose orientation is aligned according to the apparent daily rotation of the sky from east to west.

The northern constellations in the lists of the star totals examined by Boll and Rehm are arranged in three zones up to the zodiac in a rigorously systematic fashion aligned with the three polar constellations in a north-south direction. The southern constellations have the same sense of direction from north to south without the division into three zones. With the exception of Canis Major and Canis Minor these constellations extend from the zodiac to the border of visibility in the south in the context of a west to east rotation. The list concludes with the zodiacal signs in the customary sequence in which they are passed by the sun.

---

[10] Rehm, A. (1899), Zu Hipparch und Eratosthenes, *Hermes* **34**, pp. 251–279.

[11] Maass, E. (1898), *Commentariorum in Aratum reliquiae*, Berlin. Neugebauer doubts the existence of independent star totals in the Catasterisms; Neugebauer, O. (1975), pp. 577f. But clearly the star register discussed by Rehm, A. (1899), pp. 251ff, has to be dated before Hipparchus, as can be shown by the particular names of the constellations. The constellation Cygnus is named Κύκνος by Hipparchus while the older texts refer to it as Ὄρνις; Rehm, A. (1899), p. 262.

Hipparchus' early – and authentic – Commentary on Aratus makes use of a similar though simpler ordering principle: for the northern constellations one finds a division into groups with north-south orientation, too, but at some points, for instance Equus-Sagitta, one finds breaks in the sequence. Finally, the southern constellations are very arbitrarily arranged,[12] indicating that their proper form was fixed by Hipparchus or later. As further evidence Rehm mentions the constellation of Corona Austrinus,[13] which Aratus only outlines in a vague way and Hipparchus leaves out totally, while the list of star totals examined before uses the term Στέφανος νότιος which is the usual nomenclature later. Furthermore, the sense of rotation in which the zones are ordered is different from that in earlier authors. Whereas in the latter the northern constellations are arranged from East to West, following one's visual impression, Hipparchus mentions them in the Commentary in the same sense of rotation as the zodiacal constellations, from West to East, just as in the list of star totals.

Besides this sequence of the constellations, the Aratus Commentary reveals in several places Hipparchus' reform of the terminology. Whereas in the first part of the work, which is mostly polemical criticism of the astronomy of Aratus and Eudoxus, he usually employs the old terminology of his predecessors, in the second, more scientific and original part, he employs a terminology with less mythological connotations. While the constellation Bootes is called both Βοώτης and 'Αρκτοφύλαξ in the first part, only Βοώτης (which is the name customarily used later) appears in the second part of the Commentary. Just as in the second part of the Commentary, the list of star totals from the Basel manuscript contains the term Βοώτης. The same holds true for the constellation Τρίγωνον (instead of Δελτωτόν).[14] Consequently, the list can be no older than Hipparchus. There is no evidence for the existence of any systematic description of the stars between Hipparchus and Ptolemy, and the conclusion of Rehm and Boll, that we should believe the attribution in the astrological manuscript and accept Hipparchus as its originator, appears compelling.

At this point the following question deserves to be discussed: why are the totals of stars in particular constellations according to the astrological manuscript smaller than in the early text of Hipparchus' Commentary on Aratus? Based on the Commentary Rehm has derived for a series of 11 constellations a greater number of stars than indicated in the list treated by Boll. For the constellation Perseus for instance, Rehm counts 21 stars in the Aratus Commentary, whereas in the list of star totals one finds only 19 stars.[15] Similarly in the constellations Aquila (5 in the Commentary to 4 of the astrological manuscript), Heniochos (9 to 8), Thyterion (6 to 4), Cetus (14 to 13) and Libra (6 to 4), a decrease in the number of stars is to be found. However, it is not legitimate to raise an objection to the authenticity of Hipparchus' list of star totals on the basis of this indisputable reduction of the number of stars listed, for, according to Boll, it was not the intention of the Greek astronomer to compile a complete catalogue of all visible stars. So it might have been that a second check of the constellations after the writing of the Aratus Commentary resulted in smaller amount of stars being listed.

---

[12]Rehm, A. (1899), p. 256.
[13]Rehm, A. (1899), p. 272.
[14]Rehm, A. (1899), p. 255.
[15]Boll, F. (1901), p. 191.

According to a report from Pliny, it was after the appearance of a "new" star that Hipparchus decided to catalogue the positions of the stars, thereby establishing a measure of comparison for later new phenomena of the same kind.[16]

There was a further motive for compiling a new catalogue of stars. Hipparchus discovered the precession motion and with it the variability of the positions of the stars as a rotation around the pole of the ecliptic. The need for a new star catalogue, with a unified epoch for all stellar positions, could have arisen from deficiencies of the older records which might well have been cumulatively collected over a longer period of observation.

Boll adduces further reasons why some of the constellations of the new register have a smaller compass than the constellations handled in the Aratus Commentary: in the first part of the Commentary, Hipparchus refers to the traditional arrangements of the constellations and attempts, as he states in the introduction, to replace the erroneous conception of the ancients with a scientific examination of the sky. In this discussion he could very well have been referring to stars which appeared less important to him later during his own scrupulous check of the constellations. Perhaps the demarcations of the constellations changed in the course of the historical transformation of the mythological nomenclature and the astronomical terminology. Stars which had been previously counted in a constellation are dropped from the star total after the constellation was reshaped.[17]

> "Furthermore, the descriptions which we have applied to the individual stars as parts of the constellation are not in every case the same as those of our predecessors (just as their descriptions differ from their predecessors'): in many cases our descriptions are different because they seemed to be more natural and to give a better proportioned outline to the figures described."

Just how strongly the traditional form of a constellation dominates the astronomical characteristics of a star, for instance its brightness, can be seen from the constellation Bootes, whose outlines were drawn in such a way that, in the Almagest, its brightest star, Arcturus, was not counted as part of the constellation, but rather listed separately as an external star. The decrease in the number of constellations in the course of the development of the constellations can be deduced from the fact that a series of formations (Corona Borealis, Cepheus, Cassiopeia, Hydra, Crater, Canis Major, Canis Minor, Aries, Gemini, Cancer and Pisces) in the Almagest count fewer stars than the list of stars discussed by Boll. The oldest constellations, the zodiacal signs, have a disproportionately larger number of external stars, indicating that the area of the constellations was reduced in the course of time. The stars which were mentioned in the older astronomical writings handed down through tradition could not have been ignored in a comprehensive star register. Boll therefore concludes:[18]

> "However, this seeming diminution of the star totals in comparison to the predecessors is certainly nothing other than a consequence of

---

[16]The historical value of Pliny's report is disputed by Neugebauer, O. (1975), p. 289.

[17]Ptolemy, C. (1984), p. 340.

[18]Boll, F. (1901), p. 191.

a partial transformation in the shape of the constellations (...) It is, therefore, altogether conceivable that Hipparchus, in working out his new register of fixed stars, drew up somewhat narrower borders for certain constellations than those found in the older book where he had not yet completed his own view of the sky, but rather attempted to improve the astrothesis of Eudoxus and Aratus."

As a result, the list of star totals is not identical with the one in the constellations of the early Hipparchus of the Aratus Commentary: it must be a later text. Since a list of star totals handed down to us in this way must refer to an important text of Hipparchus, it appears very likely that the text in question is the star register which Ptolemy had access to and which was lost at a later date.

In the first two steps of his argument, Boll shows that the list of star totals refers to a lost Hipparchan register of stars. Its size can be calculated by a comparison with the catalogue of the Almagest, as the tables related to the reproduced star totals clearly demonstrate.

In order to determine the totals in Hipparchus' list, Boll extrapolates the missing numbers from the analogous ratios in the Almagest. The sum total of stars in that list is 653. The small number of stars in Argo (11 to the 45 of the Almagest) is corrupt, according to Boll and Dreyer.[19] Both, therefore, count only 640 correctly designated stars within 42 constellations. Furthermore, Boll adds to the missing 7 constellations[20] those external stars which lay outside of the actual formations, but nevertheless cannot be excluded from any register of stars: for instance, the bright star Arcturus as an external star of the ship of Bootes. In the 42 completely defined constellations, there is a numerical predominance of stars by the factor 1.2 in favor of the Almagest. If one divides the number of stars of the remaining 7 formations by this factor, one obtains a number of 121 supplementary stars; that is to say, a Hipparchan star total of 761 stars *within* the constellations. In the Almagest, Ptolemy catalogues 108 external stars. Using the same ratio of star totals, one obtains a total number of 90 Hipparchan external stars.

The Hipparchan register of fixed stars refers to about 851 stars according to Boll's estimation. When one considers certain limits of tolerance in the use of constant numerical ratios for the particular subsets of the catalogues of Hipparchus and Ptolemy, one is able to estimate the upper and lower limit of the number of stars in Hipparchus' register. Boll sets the lower limit at 761 or 851 stars – depending on whether the existence of supplementary external stars is accepted or not – and the upper limit at 851+20 or 30 stars "at the most".[21] The Catasterisms, which appear to be corrected in many cases according to Hipparchus' register,[22] hardly differ from the Hipparchan totals and reveal in the case of the missing constellations the same numerical ratio to the Almagest as extrapolated before.

The consequences of these estimations for the thesis of Tycho and Delambre are obvious. Boll ends his article with the following words:[23]

---

[19] Boll, F. (1901), p. 192, Dreyer, J. L. E. (1917), p. 530.

[20] Besides the six constellations completely missing the part of the constellation Aquarius called Aqua must be taken into account.

[21] Boll, F. (1901), p. 193.

[22] Boll, F. (1901), p. 193.

[23] Boll, F. (1901), p. 195.

"To all appearances, one will have to credit Ptolemy with giving an essentially richer picture of the Greek firmament after his eminent predecessors."

Boll's article marks the turning point in the interpretation of Ptolemy's star catalogue. This interpretation of the small list of star totals hidden in an astrological text must be either refuted – something which has not yet been ventured – or else the interpretation of the Ptolemaic catalogue as a simple transformation from a missing Hipparchan catalogue must be given up. The attempts at the beginning of the 20th Century to rehabilitate Ptolemy intensify, and we witness how the search for an explanation for the obvious systematic errors in the stellar longitudes gets under way.

3.2 Supplementary Catalogues

3.2.1 Björnbo's New Catalogue

The first reaction to Boll's work came from Björnbo, whose article with the title (translated) "Did Menelaos from Alexandria publish a star catalogue?" was printed, due to the whimsies of publication, immediately following Boll's article.[24] From Boll's work, Björnbo draws the conclusion that "Ptolemy's catalogue contained data about circa 170 more stars than that of Hipparchus and the task now is to determine whether the glory for this increase in the observations belongs to Ptolemy or rather to Menelaos".[25] The Islamic astronomer aṣ-Ṣūfī was the first to formulate the thesis that Ptolemy had copied stellar coordinates from Menelaus. Björnbo combines this idea with Boll's estimation of the maximal number of stars in the Hipparchan register without, however, going beyond the traditional disparaging evaluation of Ptolemy's accomplishments as an observer:[26]

> "... so the suspicion arises with good reason that his fixed star catalogue is simply an uncritical compilation of the work of several predecessors, and his calculation of the precession a result of skillful botching. I imagine it happening like this: first, Ptolemy wavered between the various speculations put forth by Hipparchus concerning the value of the constant of precession ... then, thanks to a consideration of the surely considerable but by no means commendable observations of Menelaos, he held fast to the lower limit value (36″ per year) set by Hipparchus. Then, after he accommodated himself to his preconceived hypothesis in this fashion, he succeeded, through a meticulous selection from all examples, which with the help of a few subtle tricks he is able to use as proofs, and through a careful suppression of all determinations producing different results, in stabilizing the wrong value of precession constant for a number of centuries."

Aṣ-Ṣūfī maintains that Ptolemy copied the 41 years older coordinates of Menelaus and added 25′ to the longitudes of the stars instead of the accurate value of 34′.[27] Björnbo argues that aṣ-Ṣūfī's thesis is devoid of any philological basis, for aṣ-Ṣūfī himself admits that the only ancient Greek source he knows is the Almagest.[28] Likewise, it still remains unclear whether or not astronomical reasons alone prompted aṣ-Ṣūfī to make his claim. Assuming that Ptolemy added 25′ to the longitudes of Menelaus, the resulting longitudes in the Almagest would be systematically too small by only 9′. This difference affords no reason to question the authorship of Ptolemy. Similarly, without the additional value for precession – provided that the Almagest contains for a fact the original coordinates of Menelaus – only one sixth of the actual miscalculation can be compensated. Even for this conjecture no textual evidence can be found in aṣ-Ṣūfī's statements.

---

[24]Björnbo, A. A.(1901), Hat Menelaos aus Alexandrien einen Fixsternkatalog verfaßt?, *Bibliotheca Mathematica*, 3. Folge, Bd. 2, pp. 196–212.

[25]Björnbo, A. A. (1901), p. 197.

[26]Björnbo, A. A. (1901), p. 210.

[27]Cf. Section I.2.

[28]Björnbo, A. A. (1901), p. 202.

A passage in al-Battānī's astronomy misled Björnbo into accepting aṣ-Ṣūfī's unfounded speculation. In an old Latin translation of al-Battānī made by Plato of Tivoli, Björnbo came upon a reference to a seemingly original star catalogue of Menelaus, which Ptolemy is supposed to have evaluated for the calculations of his stellar coordinates. Björnbo, however, could adduce only this single passage as supportive of his interpretation, and he himself remarks "that all of this conjecturing is based hitherto exclusively on the report of al-Battānī and is dependent on the credibility one attributes to this man; on the other hand, that is justified by the fact that in Ptolemy's 7th Book much must be amiss."[29] Later, it turned out that the passage in question had been poorly translated and the new translation once more accorded Ptolemy alone the authorship of the observations.[30]

Björnbo's efforts to bring other sources besides the Hipparchan register into the discussion as possible models for the Ptolemaic catalogue clearly fail. Nevertheless, with his hypothesis of a compilation from various other star catalogues, he points out a way of treating Boll's results without completely excluding Hipparchan sources from the Almagest. What must be explained, however, is how the missing difference of approximately 175 stars in addition to the Hipparchan register found its way into the Almagest, as well as how these stars are to be differentiated from those remaining. In 1917 Dreyer tried to identify an especially conspicuous subset of stars with the required surplus amount of stars.

3.2.2   Dreyer's 1/4 Degree Stars

In two successive articles, Dreyer offers a solution to the problems posed by Boll's work.[31] In the first article Dreyer sums up the stars of the Hipparchan register according to Boll's results and searches for a set of a special type of stars in the Almagest, which can be set off from the others and which can fill the gap of about 175 stars between the quantity of stars in the Hipparchan register and the catalogue of the Almagest.

In Boll's table of star totals, the frequency of an especially conspicuous group of stars is entered into a column; this group will be referred to from now on as *1/4*

---

[29] Björnbo, A. A. (1901), p. 211.

[30] Björnbo quotes the corrected old, but still erroneous translation, Björnbo, A. A. (1901), pp. 204f: "...ipsarum (stellarum fixarum) autem loca secundum longum et latum in Ptolomaei libro anno primo Regis Antonini, qui est annus 886 a Rege Nabuchodonosor inuenimus; in una illarum obseruationum, per quas Ptolomaeus operatus est, fuit obseruatio Menelai, qua usus est anno 845 à Nabuchodonosor Rege; dixitque stellam septentrionalem, quae inter duos Scorpionis oculos positur [sic] (β Scorpionis), velut per Lunam cum sphaera circulorum experimentatus est, illo anno in 5° 55' Scorpii existere; *ac secundum quod ipse in libro suo scripserat*, cor Leonis (i.e. Regulus) illo eodem anno in 2 gradibus et sexta (2 1/6°) Leonis esse, Leumia (Sirius) vero in 17 gradu Geminorum esse debuerat". Nallino, C. A. (1907), p. 270, corrects the translation to: "Una ex observationibus stellarum qua Ptolemaeus usus est, fuit observatio quam Menelaos narravit, ex anno 845 dicti regis Nabonassaris; et narravit stellam borealem ex iis quae sunt inter duos oculos Scorpii fuisse eo anno, cum eam sphaera armillarii metitus esset per lunam, in 5°55' Scorpii. Et necesse erat, juxta id quod descripsit in libro suo, ut esset Cor Leonis eo anno memorato in 2 1/6° Leonis ac esset Sirius in 17° Geminorum." The subject in "descripsit in libro suo" is Ptolemy and not Menelaus, as interpreted by Björnbo.

[31] Dreyer, J. L. E. (1917), On the Origin of Ptolemy's Catalogue of Stars, *Monthly Notices of the Royal Astronomical Society* 77, pp. 528–539. Dreyer, J. L. E. (1918), On the Origin of Ptolemy's Catalogue of Stars. Second Paper, *Monthly Notices of the Royal Astronomical Society* 78, pp. 343–349.

degree stars.[32] For the most part, the coordinates are given with an accuracy of 1/6 degree. In our catalogue these Ptolemaic fractions of the degrees are represented as 0', 10', 20', 30', 40', and 50'. The ecliptical longitudes of the stars are, with four exceptions – three in Virgo – exclusively of this accuracy. On the other hand, the ecliptical latitudes reveal a large set of stars recognizably distinct from the rest by their containing a fraction of a 1/4 degree. In our catalogue their latitudes are listed with the fractions 15' and 45'. Coordinates of the same accuracy, but with the remainders of 0' and 30', cannot be distinguished from the stars with an accuracy of 1/6 degree. Their number, therefore, can be only estimated. In one table, Dreyer tabulates the numbers of stars in the Hipparchan register against those of the stars in the Almagest as well as the 1/4 degree stars contained in the respective formations, and finds a surprising correspondence.[33]

For example, in the case of the first two constellations, the astrological manuscript records 7 stars in Ursa Minor and 24 stars in Ursa Major, while the Almagest contains 7 stars and no 1/4 degree stars in Ursa Minor, and 27 stars and 3 1/4 degree stars in Ursa Major. Dreyer wants to interpret the 1/4 degree stars as the ones with which Ptolemy expanded the Hipparchan star register by roughly 175 stars. Often enough, the usually larger amount of stars in Ptolemy's constellations corresponds exactly to the number of recognizable stars with an ecliptical latitude of 15' and 45'. The Almagest contains 145 stars of that type, and Dreyer concludes:[34]

> "On reading Boll's paper it struck me at once that, as about 175 stars cannot in any case have been taken from the catalogue of Hipparchus, it was not unlikely that they were represented by the 145 stars plus some others the minutes of which are 30 or 60, and may therefore have been observed either with an instrument divided to 1/6° or with one divided to 1/4°."

Dreyer assumes that Ptolemy himself observed the 1/4 degree stars and his interpretation of the passage in the Almagest where Ptolemy recounts the use of the astrolabe for measuring the coordinates, can be paraphrased in the following way:[35]

> "Hence, again using the same instrument [as we did for the moon, V 1], (because the astrolabe rings in it are constructed to rotate about the poles of the ecliptic), we observed as many stars as we could sight down to the sixth magnitude."

In this passage Dreyer discerns that "it would be impossible to affirm more distinctly that he has made a large number of observations, though Ptolemy does not say that he has observed every single star in the catalogue. Unless we are prepared to accuse that distinguished mathematician and astronomer of deliberate fraud, it is impossible to maintain that he copied the latitudes of the stars from Hipparchus and merely added 2°40' to his longitudes."[36] Ptolemy's method of

---

[32]Cf. section III.1.
[33]Dreyer, J. L. E. (1917), pp. 531ff.
[34]Dreyer, J. L. E. (1917), p. 531.
[35]Ptolemy, C. (1984), p. 339.
[36]Dreyer, J. L. E. (1917), p. 536.

measurement, as it is presented in the Almagest, could involve a series of systematic errors which distort the stellar position of the same order as the simple conversion of the Hipparchan coordinates with a false precession constant. Ptolemy writes that he first of all measured the longitudinal difference between sun and moon shortly before sunset to obtain the lunar position for the night by an easy interpolation. The position of bright reference stars, then, can be measured by their relative position to the moon.

Dreyer lists a number of systematic errors whose sums can increase up to one full degree in longitude.[37]

(i) The difference in longitude between the sun and the moon is measured at sunset. It is the time when the sun can just be seen above the horizon, though geometrically the sun has disappeared up to 30' under the horizon due to the refraction of the light. The effect of the refraction can reduce the longitude of the moon up to half a degree. All the stars whose ecliptical longitudes are determined during the night relative to the position of the moon, then, have a longitude which is up to one half degree too small.

(ii) Ptolemy had taken over the entire solar theory from Hipparchus, including the numerical values of the parameters. The position of the sun according to the observational procedure mentioned is not determined independently a second time, but rather derived from the theory. For Ptolemy's time, Tannery claimed that the location of the sun was systematically up to 22' too small. It would be a miscalculation which likewise is incorporated in the measurements of stellar longitudes.[38]

(iii) Both the measurements of the precession as well as the measurements of the spring equinox are marked by systematic errors, so that Ptolemy had no indication for the errors in the stellar longitudes of his catalogue. The observation of Regulus with which Ptolemy demonstrates the precession motion, is not based on the solar theory directly, but on the lunar theory with its even larger inaccuracies.[39] The three stars cited separately by Ptolemy as a proof for the small precession constant of one degree in one hundred years, show in similar fashion an ecliptical longitude which is too small; this time, without mentioning the measuring methods used. Together with Regulus the longitudes of these four stars are 63' too small and Dreyer, at the end of this list of possible errors, asks provocatively: "is it then necessary to believe that Ptolemy borrowed his star-places from Hipparchus?"[40]

In his first article, Dreyer summarizes the results of the historical research on the Ptolemaic star catalogue under the seven following points:[41]

---

[37]Dreyer, J. L. E. (1917), pp. 536f.

[38]Tannery, P. (1893), *Recherches sur l'histoire de l'astronomie ancienne*, Paris, p. 171. Tannery's value is too small. The correct value of about one degree had been calculated already by Laplace one hundred years before.

[39]Ptolemy, C. (1984), p. 328.

[40]Dreyer, J. L. E. (1917), p. 538.

[41]Dreyer, J. L. E. (1917), pp. 538f.

(i) The Hipparchan star register contained no more than 850 stars, which means that Ptolemy could not have taken over all of the positions from Hipparchus.

(ii) If Ptolemy had taken over these 850 stars (something for which no absolute evidence is available), then 175 stars, which are probably identical with the stars that had been measured with the aid of an instrument with a 1/4 degree graduation in the circle for the ecliptical latitude, had been observed by Ptolemy himself.

(iii) The thesis initiated by Arabic astronomers that Ptolemy copied parts of his catalogue from Menelaus is totally unjustified.

(iv) The methods with which Ptolemy determined the longitudes of the reference stars introduce formidable systematic errors into the longitudes of all other stars.

(v) The longitudes of the four stars used by Ptolemy for the computation of the precession constant are on the average too small by 63'. This error coincides with the systematic error of all the stars.

(vi) Although for that reason Ptolemy determined the vernal equinox at about one degree to small, there is no reason to doubt that

(vii) he had observed a large number of stars himself.

3.2.3 Dreyer II

In the second article, from 1918, Dreyer intensifies his efforts to obtain a satisfactory explanation of the systematic error in the stellar longitudes and he amends several interpretations proffered in the first article.

Indeed, Dreyer does not further explore the thesis that the stars missing from the Hipparchan register coincide with the 1/4 degree stars of the Almagest; rather, he carries out the examination of the possibilities of error in an authentic Ptolemaic determination of the position measurements, which allow Ptolemy to be rehabilitated.

(i) The refraction of the setting sun, in the event that Ptolemy neglected to consider it, gave rise to an error in longitude of 34.6' in the single example thoroughly demonstrated in the Almagest.[42] The larger the declination of the sun, in the northern hemisphere as well as in the southern, the greater the effect of the refraction on the error in longitude. For a declination of 20°, Dreyer calculated a longitude error of 60' on the average, in other words, exactly the amount by which the ecliptical longitudes of the stars are too small. Dreyer, however, rejects this effect as an explanation for the longitude errors, for an observation of the stars at sun-rise would not diminish but increase their longitudes. "If Ptolemy divided his observations fairly equally between sunrise and sunset, he would therefore eliminate the effect of refraction." [43]

---

[42] Dreyer, J. L. E. (1918), p. 344.
[43] Dreyer, J. L. E. (1918), p. 344.

(ii) With a similar argument Dreyer also rejects one specific systematic error from the Ptolemaic solar theory. The error in the eccentricity of the solar orbit and with it in the equation of the centre cannot account for the longitudinal errors of the reference stars, for this error varies randomly in the course of a year. If the reference stars are fairly well distributed on the celestial sphere, their longitudes would show all errors within the interval of the possible deviations of the solar theory. Its mean value is close to zero and therefore it could not be responsible for a systematic error of one degree in the stellar longitudes.[44]

(iii) Finally, according to Dreyer, the error in the mean longitude of the sun can fully account for the anomaly of the stellar positions.[45] For 10 stars Dreyer calculates the mean error of the solar theory when the sun is in opposition[46] as $59' \pm 0.9'$, which is almost the same as the mean longitudinal error of the stars.

The error in the mean sun is due to an error in the Hipparchan value of the tropical length of the year of $365\frac{1}{4} - \frac{1}{300}$ days, which is 0.00435^d too long. Ptolemy, who takes over the entire solar theory from Hipparchus, sets the beginning of the solar tables at the era Nabonassar, -746 February 26. At that time he postulates a mean solar position of $330°45'$, whereas in the Neugebauer tables the mean sun should be at $327°56'$: in other words, it should be $2°49'$ less. Since the tropical year is 0.00435 days too long, the mean motion of the sun during one Egyptian year of 365 is $15.7''$ too small. With that, the error of the mean sun changes from $-2°49'$ at the epoch of Nabonassar over the 885 years until the time of Ptolemy to

$$-2°49' + 15.7'' * 885 = +1°2'$$

This is identical with the mean error of longitude.

This explanation obviously convinced Dreyer to such an extent that his previous interpretation of the nature of the 1/4 degree stars as Ptolemy's supplement to a unknown Hipparchan register is no longer mentioned in the résumé of the later article:[47]

> "This, then, is the error of Ptolemy's equinox, and as it is (within a minute or two) equal to the average error of Ptolemy's longitude of stars, it is impossible to doubt that he really founded his catalogue on new observations of stars and the sun, taking the places of the sun from the solar tables in the third book of his Syntaxis."

In contrast to Delambre, the erroneous mean sun is, as Dreyer sees it, the reason for the much too small precession constant. The mean sun for the stellar positions measured by Timocharis in the year -288 was therefore $50'$ too large, and for the time of Menelaus (+96) $50'$ too small. The stellar longitudes which Ptolemy gives for these

---

[44]Dreyer, J. L. E. (1918), p. 344.
[45]Dreyer, J. L. E. (1918), p. 345.
[46]Meaning that, approximately, the star rises when the sun sets.
[47]Dreyer, J. L. E. (1918), p. 345.

times are 38' too large in the case of Timocharis, and 44' too small for the year of +96, and therefore correspond to the errors of the solar theory.[48] For Dreyer, then, all of this indicates that the explanation for the large errors in the determination of the precession constant found earlier by Laplace and Ideler must be correct.[49] A further confirmation of the seriousness of Ptolemy's statements according to Dreyer can be gleaned from the list of 18 stars, from which Ptolemy chooses 6 in order to support the value of the lower limit of the Hipparchan precession constant.[50] Just as Delambre calculated earlier, the average of all 18 observations of stellar declinations would not result in an average value of 36″ per year, but rather 46.9″, and would therefore lie very close to the true value. This shows that the set of declinations is independent of the solar theory (and the star catalogue) and therefore they were surely measured by observation.

For Dreyer the circle of attempts to rehabilitate Ptolemy closes. In the 18 stars of the calculation of the precession, he was able to find original Ptolemaic observations, and he could show that the striking systematic longitudinal errors of the star catalogue are without exception to be accounted for by the solar theory of the Almagest. In this light it is understandable that Dreyer, in his second article, neglects to mention his first interpretation of a large part of Hipparchan coordinates in Ptolemy's catalogue.

3.2.4   Fotheringham

One month after Dreyer's article appeared, the astronomer Fotheringham likewise published an examination in the "Monthly Notices of the Royal Astronomical Society", in which the secular acceleration of the sun is derived with the aid of ancient observations of eclipses and the Hipparchan measurements of the equinoxes.[51] After his evaluation of the Hipparchan data, Fotheringham considers the Ptolemaic measurements, "because it was the means of that false determination of the equinox which gave rise to the curious allegation that Ptolemy's star catalogue was not authentic."[52] In this quotation, the radical shift in the historical interpretation of the star catalogue is easy to recognize. Ptolemy appears no longer as forger, rather, it is the interpretation of the historian, fixed exclusively on one single aspect, that leads to the "mysterious accusations".

Fotheringham examines three of Ptolemy's observations of the equinoxes. One of these is reported in the Almagest as: [53]

> "Now that we have established that, among the first of the equinoxes observed by us, one of the most accurately determined was the autumnal equinox which occurred in the seventeenth year of Hadrian."

---

[48] Dreyer, J. L. E. (1918), p. 346.

[49] Dreyer, J. L. E. (1918), p. 347.

[50] Dreyer, J. L. E. (1918), p. 348.

[51] Fotheringham, J. K. (1918), The Secular Acceleration of the Sun as determined from Hipparchus' Equinox Observations; with a Note on Ptolemy's False Equinox. *Monthly Notices of the Royal Astronomical Society* **78**, pp. 406–423.

[52] Fotheringham, J. K. (1918), p. 419.

[53] Ptolemy, C. (1984), p. 168.

Although Ptolemy made his observations with the utmost accuracy, according to his own testimony, all three equinoxes turn out to be about one degree too small. Fotheringham computes the errors for the three observations as:[54]

Time of equinox	error in longitude
+132 Sept. $24^d23^h53^m$	$-1°17'25''$
+139 Sept. $25^d16^h50^m$	$-1°17'26''$
+140 March $21^d23^h9^m$	$-0°45'14''$

Table 3.5: Equinox observations.

According to Fotheringham the internal consistency of the results together with such a large systematic error allows only two explanations: either the number of observations is so small that the coincidence with the theoretical Hipparchan values is purely accidental, or else the observations had been selected with the aim of correspondence to the traditional values.[55] Independent of the question as to whether the observations of the equinoxes were brought into agreement with the theoretical values or not, the Ptolemaic solar theory obtains parameters which lead to a longitude of the mean sun being 1°9.7' too small at the epoch of the star catalogue.[56] From Peters-Knobel, Fotheringham takes the mean longitudinal errors of the zodiacal stars as 1°6.1'. The close agreement of both numerical values clearly illustrate in Fotheringham's opinion that:[57]

> "It is clear, therefore, that the false equinox in the catalogue reproduces to within a few minutes the false equinox as obtained five years earlier from the erroneous equinox observation from which Ptolemy professes to derive the epoch of the Sun's motion. As the star places are professedly derived from the solar places, it is curious that the fable which accuses Ptolemy of having copied his star catalogue from Hipparchus, merely adding a false precession, should ever have gained currency."

Within two decades after the appearance of Boll's examination of a long overlooked astrological fragment, the image of Ptolemy as a despicable forger was transformed into one of a serious, even if occasionally unlucky observer, and an outstanding theoretician. For the following historian, Heinrich Vogt, it is even a historical enigma that such a suspicion could have been held against Ptolemy for so long.

Up to now the investigations have shown that Ptolemy could not have simply copied from a Hipparchan star catalogue, and that the systematic error in the stellar positions can be completely and satisfactorily explained by the error in the mean sun of the solar theory. For a complete rehabilitation of Ptolemy only one piece

---

[54]Fotheringham, J. K. (1918), p. 419.
[55]Fotheringham, J. K. (1918), p. 420.
[56]Fotheringham, J. K. (1918), p. 421.
[57]Fotheringham, J. K. (1918), pp. 421f.

of evidence is lacking: The proof that the Hipparchan coordinates of the assumed lost star register were independently observed from the ones in the Almagest. Seven years after Dreyer and Fotheringham, Heinrich Vogt published an article focusing on this question.

3.3   The Reconstruction of the Hipparchan Catalogue

The question concerning the origin of the Ptolemaic star catalogue would be quickly answered if at least a part of the Hipparchan source were available to us. Failing that, the obvious move here would be to reconstruct as many stellar coordinates as possible from the Hipparchan texts which have been preserved to the present, and then to compare these positions with the corresponding data in the Almagest. Should it turn out that the stars of Hipparchus deviate significantly from the positions of the stars in the Almagest, whose ecliptical longitudes, according to Delambre's interpretation, would be decreased by 2°40′, it would prove that the positions of the stars could not have been obtained by simply taking over the older, reconstructed stellar coordinates.

The only Hipparchan manuscript to have survived in its entirety is the Commentary on Aratus and Eudoxus.[58] In this text Hipparchus criticizes in a mostly polemical manner the "Stellar Phenomena" of Aratus and Eudoxus.

The first section of Aratus's pedagogical poem "The Phenomena and Weather Prognostications", the "Phenomena", is divided into two major parts: the first considers the constellations along with the fundamental celestial circles, the second records the zodiacal signs rising and setting while the stars are being observed simultaneously.[59]

Hipparchus structures his Commentary in an analogous way and he outlines his intentions quite clearly in his introduction addressed to Aischrion as follows:[60]

> "Since my reading of Aratus reveals in most and the most important points contradictions between the data recorded there and the phenomena and the actual celestial constellations, while the other interpreters, even Attalus, seem to recognize them without hesitation as valid, I have decided, for the satisfaction of your desire for knowledge and for the general benefit of others, to discuss everything that I consider to be incorrect in a special treatise ... My intention is, more precisely, to prevent you and all others desirous of knowledge from uncritically taking over ideas which are incompatible with the scientific conception of the phenomena of the cosmos."

In his actual commentary on Aratus and Eudoxus, which makes up the first part and the main body of the work, Hipparchus compares their results with what he holds to be the "true" phenomena for the horizon of Athens; that is to say, he describes the constellations in their relation to the horizon for the geographical latitude of 37°.[61]

In the second part of his work Hipparchus describes the rising and setting of the constellations independent of the material given by Aratus and Eudoxus:[62]

---

[58] Edited by C. Manitius, *Hipparchi in Arati et Eudoxi Phaenomena Commentariorum Libri Tres*, Leipzig 1894. In the following the text is quoted as Hipparchus (1894).

[59] Hipparchus (1894), p. 291.

[60] Hipparchus (1894), p. 4.

[61] Hipparchus (1894), p. 28.

[62] Hipparchus (1894), p. 6.

> "Besides the presentation of errors made in the "Phenomena" of
> Eudoxus and Aratus as well as by those interpreters who agree with
> their declarations, I have compiled for you the simultaneous risings and
> settings of all the constellations, including the twelve zodiacal ones, as
> they occur in reality. This will enable you to check for yourself, in a
> meticulous treatment of all the details, the data of the other interpreters
> as well."

These data do not apply to the geographical latitude of Athens, but to Rhodes[63]
with a latitude of 36°, the place where Hipparchus probably carried out most of his
investigations.[64]

The two sections of the Aratus Commentary also differ from one another in
their terminology. In the critical part on Aratus and Eudoxus, for instance, the
numerical data are expressed in full degrees. The expression "$\mu o \widehat{\iota} \rho \alpha$ α'" is equivalent
to "first degree" of the sign. In the systematic second part Hipparchus' terminology
changes. The term "first degree" is missing entirely, while all other full degrees are
fairly evenly distributed over the zodiacal sign. From the beginning of the sign,
the "$\dot{\alpha} \rho \chi \acute{\eta}$", up to λ' for 30° all degree values appear approximately 6 to 7 times.
Since Hipparchus would have hardly used two different terms for the beginning of
a zodiacal sign, namely $\dot{\alpha} \rho \chi \acute{\eta}$ and λ', it is plausible to interpret the expressions of
the degrees as "from the beginning of the ... degree". The term β' then means "The
beginning of the second degree", i.e. 1° of the sign. Only then one does one obtain
consistency in the use of the Hipparchan terminology.[65]

In the systematic second part Hipparchus depicts the positions of the constellations mainly by their rising and setting phenomena:

(i) the zodiacal signs rising and setting simultaneously with the outermost stars
of the particular constellation.

(ii) the degree on the ecliptic which culminates at the moment when the constellation rises or sets.

(iii) the degree on the ecliptic which culminates simultaneously with the particular
star of the constellation.

The following example from the Manitius translation will be important for us
later. It refers to the first southern constellation, Hydra:[66]

---

[63]Hipparchus (1984), p. 184.

[64]Cf. Neugebauer, O.(1975), p. 275. Neugebauer emphasizes that Hipparchus did not make all of
his observations on Rhodes. Nonetheless, the latitude of 36° referred to in the Commentary and the
treatise "On Simultaneous Risings" provides the evidence that the writing of the Aratus Commentary,
which Manitius dates -135 as the latest (Hipparchus (1894), p. 287), Rhodes is actually also the place of
observation. There are no known Hipparchan observations of stars mentioned on a different latitude.

[65]This is the interpretation of Manitius, Hipparchus (1896), pp. 288f. But in the edition of the
Commentary Manitius does not convert the expressions into cardinal numbers. Therefore any calculation
with the Hipparchan data has to subtract 1° from the numerals in the Manitius translation in the case
of full degrees. Vogt accepts this interpretation: Vogt, H. (1925), Versuch einer Wiederherstellung von
Hipparchs Fixsternverzeichnis, *Astronomische Nachrichten* **224**, no. 5354–55, pp. 17–45.

[66]Hipparchus (1896), p. 219.

↑ 108.5° − 195.5° † : 2.5° − 97°
Rising: The northern star of those in the gaping jaws (δ Hydrae)
Setting: The one in the tip of the tail (π Hydrae)

The arrow pointing upward precedes the degree on the zodiac simultaneously
rising with the star of the constellation and the sign "†" indicates that the following
numbers are the longitudes of the zodiac culminating simultaneously with the rising
constellation. The line following "Rising:" states that δ Hydrae is the first one of
the constellation Hydra to rise over the horizon of Rhodes simultaneously with the
ecliptical degree of 108°.5. Finally, the star π Hydrae is the last star of Hydra and it
rises simultaneously with the ecliptical degree 195°.5.

Analogously, Hipparchus describes the simultaneous culmination of the ecliptic
with the setting constellation. The accuracy of the data is never higher than half a
degree, but it is not yet clear whether Hipparchus intended in every case to express
each full degree value with an accuracy of half a degree.

Assuming that the coordinates in the Almagest are founded on Hipparchan
measurements which had likewise been recorded in some form in a lost star register,
the next sensible step would seem to be to reconstruct as many stellar coordinates as
possible from the Hipparchan Aratus Commentary. If the supposition is correct that
Ptolemy had added to the Hipparchan longitudes a constant value of 2°40′ while
keeping the latitudes unchanged, then the reconstructed Hipparchan coordinates
can only deviate from the data given in the Almagest by exactly the errors which
are inherent in the method of reconstruction, while both coordinates share the
observational error of the Hipparchan measurements. In his *Histoire de l'Astronomie
Ancienne*, Delambre wrote that the attempted reconstruction must unavoidably fail
as a direct consequence of the incomplete information of the Commentary coupled
with the necessity of relying on auxiliary hypotheses.[67]

Armed only with the assumption of a particular value for the obliquity of the
ecliptic and the geographical latitude which Hipparchus had adequately established
through the value of the maximal length of the day as 36°, Heinrich Vogt, in 1925,
was successful in reconstructing a set of Hipparchan star coordinates. Since discovery
of the coordinates of a lost star register of Hipparchus, would immediately provide
more information about its relationship with the coordinates of the Almagest, the
attempt at reconstruction is of peculiar significance for the historical interpretation
of the star catalogue. Were an obvious similarity of the reconstructed Hipparchan
coordinates with the positions of the Almagest to show up, this would prove that
Ptolemy had simply copied the data. However, should it turn out that the Hipparchan
and Ptolemaic coordinates diverge significantly from each other, and should it be
furthermore be shown that the reconstructed coordinates are also identical with the
coordinates of the lost major star register of Hipparchus, this would strongly imply
the independence of the Ptolemaic measurements.

Vogt was able to find, in all, 881 numerical data of Hipparchus, including 22
entries which Ptolemy and Strabo had reported.[68] 122 stars can be reconstructed
from this set of data without any additional auxiliary hypotheses.[69]

---

[67] Delambre, J. B. J. (1817), vol. I, pp. 146–148, pp. 187–189.

[68] Vogt, H. (1925), col. 18.

[69] Only the geographical latitude of the observation place and the obliquity of the ecliptic are necessary

With the help of these reconstructed Hipparchan values, the thesis of Brahe and Delambre can be directly scrutinized: if one is to show that Ptolemy himself had not measured the coordinates for his star catalogue, but rather obtained them either directly or through a conversion of the Hipparchan material, the ecliptical latitudes of the Hipparchan stars would have to agree with the latitudes of the stars of the Almagest. This criterion can no doubt easily be checked by stars that differ significantly from their actual positions due to particular circumstances, namely by stars whose latitude was either incorrectly measured or imperfectly recorded from the observational reports. If such errors in latitude are to be found in the reconstructed latitudes of the Hipparchan stars, they would have to be contained in the Almagest, too. If Delambre's suspicion is correct, then Hipparchan and Ptolemaic errors in latitude of the same size provide strong indication of their genetic identity. Conversely, all of the Hipparchan latitudes which deviate sharply from those of the Almagest strongly indicate their genetic independence. It is therefore easier to prove that both star catalogues stem from different sources[70] than to show their dependence, because for this only the stars with an outstanding deviation from their actual positions can be used. A similar test can be carried out on the differences in the ecliptical longitude, though in this case possible common errors from the solar theory might suggest dependent coordinates even in independent measurements.

Vogt presents his results in a table which shows the number of the reconstructed Hipparchan latitudes and the corresponding latitudes from the Almagest in nine error intervals:[71]

error interval	number of stars in the interval		
	Hipp.	Ptol.	shared errors
$[-0.10°, +0.10°]$	15	32	7
$[+0.10°, +0.33°]$	17	30	5
$[-0.10°, -0.33°]$	17	32	5
$[+0.33°, +0.67°]$	15	14	3
$[-0.33°, -0.67°]$	18	11	3
$[+0.67°, +1.50°]$	12	4	0
$[-0.67°, -1.50°]$	20	5	2
$> +1.50°$	4	1	1
$< -1.50°$	4	2	0

Table 3.6: Error classes in latitude.

The error class of the interval $[+0.33°, +0.67°]$ contains 15 Hipparchan and 14 stars from the Almagest from the total set of 122 reconstructed stars; just three of these stars share the same error class. If Ptolemy had in fact taken the coordinates from the Hipparchan register, then, according to Vogt, *all* of the Hipparchan stars within an error class should be counted again in the column for the stars of the

---

parameters.

[70] All of the stars of the catalogues can provide positive evidence for this statement, independent of the deviations from the true positions.

[71] Vogt, H. (1925), col. 23.

Almagest – in this case the column of stars with a shared error would have to contain the same number as the column of Hipparchan stars.

The patent difference between those columns prompts Vogt to make the following judgement:[72]

> "Something like 2/3 of all Ptolemaic errors in latitude display such sharp deviations from the corresponding Hipparchan errors that borrowing appears to be ruled out. This makes for an invincible counter-example against Delambre's theory."

Just as he does with the errors in latitude, Vogt divides the errors in longitude into error classes leading to the same assessment.[73]

error interval	number of stars in the interval		
	Hipp.	Ptol.	shared errors
$[-0.10°, +0.10°]$	27	32	9
$[+0.10°, +0.40°]$	11	16	2
$[-0.10°, -0.40°]$	15	21	3
$[+0.40°, +0.80°]$	15	15	3
$[-0.40°, -0.80°]$	17	19	4
$[+0.80°, +1.50°]$	10	9	1
$[-0.80°, -1.50°]$	6	4	1
$[+1.50°, +2.50°]$	7	2	1
$[-1.50°, -2.50°]$	7	2	1
$> +2.50°$	1	0	0
$< -2.50°$	6	2	2

Table 3.7: Error classes in longitude.

So in this case, too, the number of shared instances in the error classes is considerably lower than the total number of the Hipparchan stars. Here Vogt draws the same conclusion as before:[74]

> "Approximately 4/5 of all errors in longitude, with a considerable absolute value, exhibit such significant differences that the derivation of the Ptolemaic longitudes from the Hipparchan seems to be sheerly impossible. Direct observation appears to be proven. In about 15 cases, slight differences by small absolute errors make both explanations applicable."

For a small number of stars, the differences in positions from the actual locations are very large, which would indicate an error in either the measurement or the copying process. At the same time, the deviating coordinates of the reconstructed Hipparchan numbers differ hardly at all from the values of the Almagest, providing

---

[72] Vogt, H. (1925), cols. 23f.
[73] Vogt, H. (1925), col. 26.
[74] Vogt, H. (1925), col. 26.

an unmistakable trace of a common origin. For his part, Vogt holds the errors in latitude of seven stars to be sufficiently correspondent that in these cases copying seems to be possible:[75]

Name	error in latitude: $\beta_{-127} - \beta_{Hipp/Ptol}$	
	Hipparch	Almagest
β Boo	+0.47	+0.42
β Vir	+0.62	+0.48
δ Aqr	−0.57	−0.58
θ Eri	−0.63	−0.42
ζ Cas	−0.86	−0.74
α Car	−1.07	−1.08
π Hya	+4.47	+4.85

Table 3.8: Possibly copied longitudes.

In particular for the star π Hydrae, all doubt that the coordinates of the Almagest stem from early observational material of Hipparchus can be cast aside. The reconstructed ecliptical latitude of the Hipparchan star deviates from the accurate position by the extreme value of $4°47$ – even for ancient astronomers this is an outrageous error and it would just have to attract attention at each renewed check – and the location of the star in the Almagest reveals a similar error of $+4°85$. Shared errors of such size could not have come about merely by chance. Vogt proceeds:[76]

> "In these cases of the coincidence of large errors – there are a total of about ten of these – it would be rather difficult to believe that all of this was due to chance. Indeed, probably the only way to explain this coincidence would be to argue that in these special cases, either Ptolemy or one of his collaborators has borrowed the stellar latitudes from Hipparchus."

Similarly, the errors in longitude make it possible to recognize stars that one must assume were borrowed. In Vogt's table, the deviations of the longitudes from the accurate positions of the Hipparchan coordinates are juxtaposed to those of the Almagest (table 3.3).[77]

For these nine stars Vogt admits transmission according to Delambre's interpretation. However, as Vogt reads the data, only similar errors in both coordinates, namely the ecliptical latitude and longitude together, provide *sufficient* evidence for their genetic identity. Altogether, then, a Ptolemaic loan is sufficiently proven for only 5 stars (τ Ari, α Car, υ Boo, π Hya, θ Eri).[78]

---

[75]Vogt, H. (1925), col. 24.
[76]Vogt, H. (1925), col. 24.
[77]Vogt, H. (1925), col. 26.
[78]Vogt, H. (1925), col. 26.

Name	error in longitude: $\lambda_{-127} - \lambda_{Hipp/Ptol}$	
	Hipparchus	Almagest
τ Ari	−0.51	−0.50
τ Psc	−0.51	−0.46
ζ Cyg	−0.58	−0.57
υ Boo	+0.56	+0.62
α Car	+0.76	+0.88
θ Gem	−2.96	−2.71
π Hya	−2.36	−2.02
θ Eri	−3.82	−4.32

Table 3.9: Possibly copied longitudes.

"Considering the complete correspondence of the huge errors in latitude and the absolute value of the errors in longitude of θ Eridani, it may be correct, in spite of a difference of half a degree, to classify this star in this group (of the copied stars) as well. Assuming this to be the case, in only about 5 instances is there sufficient evidence for the borrowing of both Ptolemaic coordinates from Hipparchus."

Therefore, in only 5 of 122 cases does Vogt succeed in finding convincing evidence for Ptolemaic borrowing of the coordinates from Hipparchus. As Vogt maintains, the small number of errors common to the reconstructed Hipparchan coordinates and the values of the Almagest in the error classes, indicates in a persuasive way their genetic independence. At the end of his article, Vogt sums up:[79]

"Boll's discovery has taught us that Ptolemy relied on his own observations at least for the non-Hipparchan stars. The direct comparison of many Ptolemaic coordinates with those of Hipparchus now proves that Ptolemy's latitudes equal their Hipparchan counterparts only in the mean, not in the individual values. And it is only in the mean value that his longitudes exceed the Hipparchan ones by $2°40'$. This allows us to consider the fixed star catalogue as of his own making, just as Ptolemy himself vigorously states."

The reconstruction of a number of Hipparchan stars permits an analysis of the genetic relationship between the stars of the Almagest and the known data of Hipparchus. For Vogt the analysis demonstrates an overriding independence of both observations, with the exception of 5 (out of 122) stars, in the case of which it has been proven by means of the reconstruction that they were copied.

Next, Vogt tackles a series of arguments which are supposed to support the thesis of the Ptolemaic borrowing from Hipparchus.

---

[79] Vogt, H. (1925), col. 43.

3.3.1   The Determination of the Precession

Delambre airs the suspicion that Ptolemy, in the third chapter of the seventh book
in which he furnishes proof for the precession of one degree per hundred years,
selected from the total of 18 observations mentioned just six with which he could
demonstrate the lower limit of the motion estimated by Hipparchus.

Be that as it may, the correctness of this suspicion has no immediate influence on
the thesis of the originality of the Ptolemaic star catalogue. However, it does throw
light on Ptolemy's attitude towards the weight to be assigned to those observations
of his which are in conflict with Hipparchan measurements, as well as towards the
priority of theoretical considerations in general. In the event that Ptolemy had in
fact merely picked out the useful observations that would best support traditional
Hipparchan insights, the interpretation of Ptolemy as a mere copyist would become
appreciably more convincing, as opposed to the view that Ptolemy had compiled
his catalogue from his own observations. Yet, in having himself made observations,
Ptolemy had ignored the recognized authority of Hipparchus and, lastly, he nowhere
mentions a direct comparison with Hipparchan coordinates with deviations or even
a comparison of accuracy in relation to Hipparchus' register.

Hence the methodological treatment of observations for the determination of
precession characterizes the relationship between the Ptolemaic method of measuring
and the results of theoretical deduction. It allows an analogous inference about the
status of Ptolemaic observations in the star catalogue.

Starting from the 18 declination measurements in the Almagest for the time of
Timocharis, Hipparchus and Ptolemy, Delambre calculated the mean value and obtained a precession constant of 47.45″ per year. Proceeding from the transformation
of the ecliptical to the equatorial coordinate system,

$$\sin \delta = \cos \epsilon \sin \beta + \sin \epsilon \cos \beta \sin \lambda \tag{3.1}$$

one obtains through differentiation the motion of precession as

$$p = \frac{(\delta_2 - \delta_1) \cos(\frac{1}{2}(\delta_2 + \delta_1))}{n \sin \epsilon \cos \beta \cos(\frac{1}{2}(\lambda_2 + \lambda_1))} \tag{3.2}$$

with p =   precession constant

        ϵ =   obliquity of the ecliptic, for Ptolemy $23.86°$

        n =   years between Hipparchus and Ptolemy ($n = 265^y$)

   λ_1, λ_2 =   ecliptical longitudes

        β =   ecliptical latitudes

   δ_1, δ_2 =   declinations, taken from the Almagest

On the right side of the equation Delambre was not able to substitute the
Hipparchan ecliptical longitude of a star, for the Hipparchan register is not available
and Ptolemy does not quote any sources. For this reason, Delambre assumed that all
star coordinates had been computed by simply adding 2°40′ to the longitudes of the
Hipparchan longitudes and, subsequently, he calculated the Hipparchan ecliptical
longitudes by subtracting 2°40′ from the longitudes of the Almagest. Now, the
approximately true ecliptical longitudes of the stars at the epoch of Ptolemy can be

gained by adding one degree to the catalogued longitudes. By this the systematic error of Ptolemy's star catalogue is corrected.

All the longitudes are calculated from the ecliptical longitude l of the Almagest as:

$$
\begin{aligned}
\lambda_2 &= l + 1° & (3.3) \\
\lambda_1 &= \lambda_2 - 3°40' \\
&= l - 2°40' \\
\tfrac{1}{2}(\lambda_2 + \lambda_1) &= \tfrac{1}{2}(2l - 1°40') = l - 50'
\end{aligned}
$$

The longitude $\frac{1}{2}(\lambda_2+\lambda_1)$ for the calculation of the precession in (3.2) presupposes, according to Vogt, that Ptolemy calculated the ecliptical longitudes for his catalogue through the addition of 2°40' to Hipparchan longitudes, and not on the basis of actual observations. From that Vogt concludes:[80]

"Now we see on the right side of the equation for p nothing which is unknown, and so Delambre can determine the precession that, with a value of 47.45″, wonderfully approaches the true precession of 50″. It is unhappily the case, however, that in the results it is not so much the truth that emerges, but rather the hypothesis which had already been introduced as its very premise. In logic, this is known as petitio principii."

Next, Vogt works out the values of the precession constant once more from the observations of the declinations and reaches a totally different result, as can be seen from the following table (table 3.10).[81]

The mean of all 16 positive values of Vogt's tabulation corresponds, therefore, exactly to the erroneous Hipparchan value of one degree every hundred years, which means that Ptolemy must have selected a representative set in choosing the stars used for the demonstration of the precession constant. On the other hand, according to Delambre's calculation of the precession constant the Ptolemaic subset (designated in the table with points) takes on a highly selective flavour. Vogt insists that Ptolemy " ... had, without calculating exactly, selected 6 stars suitable for the purpose of illustration due to their proximity to the equinoxes".[82]

However, both Vogt's method of calculation and his critique are unintelligible. The precession constant can be calculated from equation (3.2), if all variables on the right side are known, especially the Hipparchan and Ptolemaic longitudes. If Vogt had substituted his reconstructed Hipparchan longitudes and the Ptolemaic longitudes from the Almagest into the equation, it could have been solved without the benefit of any supplementary suppositions; the problem in that case is that the

[80]Vogt, H. (1925), col. 35.
[81]Vogt, H. (1925), col. 36.
[82]Vogt, H. (1925), col. 36.

star	Vogt's precession [''/y]	Delambre's precession
ε UMa	55.36	45.24
α Tau	53.94	54.85
η UMa •	50.81	35.88
ζ UMa	43.34	50.03
α Boo •	43.20	39.59
α CMa	40.75	45.75
α Vir •	38.05	37.08
β Lib	36.28	51.11
η Tau •	34.92	27.40
α Leo	33.96	50.03
α Lib	33.56	66.51
α Sco	30.30	53.22
α Aur •	27.45	51.61
β Gem	24.18	62.12
γ Ori •	20.38	41.03
α Gem	1.09	50.91
α Ori	-5.98	66.37
α Aql	-119.10	24.46
means (positive values):	35.47	47.45

Table 3.10: Vogt's precession table.

results are very close to Delambre's values and cannot be reconciled with those of Vogt.[83] Vogt describes his calculation as follows:[84]

> "As the problem cannot be solved without supplementary supposi-
> tions, if one posits the Ptolemaic latitudes as unchanging and known, a
> further supposition is not necessary. However, from simply solving the
> astronomical triangle PES with given ϵ, δ, β, the resulting values for λ
> and the precession have nothing in common with those of Delambre."

3.3.2 Dreyer's 1/4 degree stars

In 1917 Dreyer speculated that the stars whose latitudes are catalogued with an
accuracy of one sixth of a degree (with a fraction of a degree of 0′, 10′, 20′, 30′, 40′ or
50′), originally stem from Hipparchus, while the less accurate 1/4 degree latitudes
(these are identifiable by the minute values 15′ and 45′) were measured either by
Ptolemy himself or else by one of his contemporaries. Vogt uses two convincing
arguments against this line of reasoning:[85]

---

[83]There is hardly any impact on the result when the Ptolemaic longitudes are corrected by their
systematic error. The problem in Vogt's argument will be discussed in section 4.2.

[84]Vogt, H. (1925), col. 35. The spherical triangle PES is defined by the equatorial pole P, the ecliptical
Pole E and the star S.

[85]Vogt, H. (1925), col. 39. The numbers are quoted from Vogt. They vary slightly because of changes
in the new edition of the Almagest by Toomer.

(i) In the Almagest there are 95 latitudes with the fraction of 15′ and 47 entries with 45′; taken together, then, 142 directly recognizable 1/4 degree stars. In his Aratus Commentary Hipparchus mentions 374 stars with coordinates which, according to Vogt, must have made up part of a lost star register. 47 of these are catalogued in the Almagest as recognizable 1/4 degree stars.

The Almagest contains 441 stars with a recognizable fraction of 1/6 degree (i.e. 10′, 20′, 40′, 50′) and 131 of them appear in the Aratus Commentary. It follows that about a third of the stars in Hipparchus' book are catalogued in the Almagest as 1/4 degree stars and it is impossible to understand why precisely this large set of stars should not be copied but observed by Ptolemy himself.

(ii) Dreyer underestimated the total number of 1/4 degree stars that are not immediately recognizable as such. His argument depended on Boll's analysis, according to which the extent of the Hipparchan register cannot possibly exceed a total of 850 stars, and thus Ptolemy, as a logical consequence, must have himself observed about 170 stars at the very least. In the Almagest, 142 stars are directly identifiable as 1/4 degree stars, and Dreyer estimated that the 30 which are missing, along with a few more, are to be drawn from the 1/4 degree stars with a catalogue fraction of 0′ and 30′. Obviously, this is clearly underestimated.

Vogt tallies the number of 1/4 degree stars with the fraction of 0′ and 30′ in the following fashion: taken in sum, there are 142 stars with 15′ and 45′ in the Almagest. The average frequency of 1/4 degree stars is 71 stars. Moreover, 441 stars are directly identifiable as 1/6 degree stars, which amounts to an average frequency of 110 stars. But there are 210 stars with 30′ latitude in the Almagest – 30 stars more than the combined total of 71+110 stars for both types of fractional accuracy, and 49 more thans the combined frequencies for the full degrees, of which the Almagest lists 230. Assuming that the graduation of the observational instrument did not indicate fractions of 1/6 degree, but was carried out more crudelye, it would naturally follow that the values which were read off from a mark on the scale, that is to say, the full and half degrees, would occur more frequently than the estimated values residing in between. When this surplus is added proportionally to the two classes of accuracy, the final sum is 315 stars with an accuracy of 1/4 degree and 709 stars with an accuracy of 1/6 degree.

Accordingly, not only are some stars to be added to the readily recognizable 1/4 degree stars, but rather, a total of 173 extra stars with either a full or half degree fraction in latitude. The total of 1/6 degree stars is much too small to cover the extent of the Hipparchan register as estimated by Boll.

The historical cause of two different accuracies of the stellar latitudes in the Almagest remains tantalizingly open.

3.3.3 Peters' Hypotheses of two Observation Instruments

At the meeting of the Astronomische Gesellschaft in 1887, Peters proposed to explain the two different accuracies through the use of two different observation instruments, one equipped with a 1/4 degree graduation, the other with one of 1/6 degree.[86] Vogt points out that the 18 declinations in the Almagest contain both accuracies, too. From the 18 values there are 3 full degrees, 2 half degrees, 1 third, 5 fourth, 1 fifth and 6 sixth degrees. The values of 1/4 and 1/6 degree are therefore fairly equally balanced. For Vogt it is difficult to believe that Ptolemy employed two different observation instruments for measuring the longitudes, latitudes and declinations respectively. Vogt proposes to look for an explanation of the two graduations requiring as few instruments as possible.[87]

3.3.4 Graduation of the Astrolabe

According to the information he gives in the Almagest, Ptolemy claims to have observed the star positions with an astrolabe, but he mentions neither its size nor its graduations. Vogt now reverts for a reconstruction of plausible graduations to the commentary of Pappus, according to which the outer meridian ring possessed a diameter of one cubit and a thickness of 1/60 cubits, the size by which the diameter of the inner rings have to become progressively smaller.[88] The circle of the ecliptic from which the longitudes are read off, then, has a radius of 28/60 cubit and the circle for the latitudes a radius of 27/60 cubit. Taking the measure of an Egyptian cubit of 525 mm as the base, the degree graduation marks on the longitude circle would be 4.28 mm apart, and those on the latitude circle 4.28 mm. If one reckons with Roman cubits (443.6 mm), the degree marks on the longitude circle would be 3.61 mm apart and on the latitude circle 3.48 mm. With an average distance of the degree marks of 4 mm, the marks of half a degree would have a distance of 2 mm and, correspondingly, the 1/4 and 1/6 degree marks a distance of 1 mm and 2/3 mm respectively.

These distances are very small. From the Almagest we know about the size of the finest graduations in the case of the parallactic instrument. According to Ptolemy's description the size of the measuring rods should not be smaller than 4 cubits. First, Ptolemy divides the rod in 60 parts, and from these he proceeds "subdividing each section into as many subdivisions as possible".[89] This graduation enables him to read an angle of 35', that is, 7/12 degree, which means that the scale has to be divided at least in intervals of 1/12 degrees.[90]

With the total minimal length of 4 cubits for the whole rod, a graduation of 1/12 degree corresponds to a distance between the marks of 2.5 mm to 2.9 mm. If, in fact, the distance between the graduation marks on the parallactic instrument is the smallest possible for Ptolemy, then one should have the same minimal distance in the case of the astrolabe. In that case the astrolabe can only have a graduation of half degrees.

---

[86]Peters, C. H. F. (1887), Mitteilungen, *Vierteljahresschrift der Astronomischen Gesellschaft* **22**, p. 270.

[87]Vogt, H. (1925), col. 40.

[88]Vogt, H. (1925), col. 41.

[89]Ptolemy, C. (1984), p. 244.

[90]Ptolemy, C. (1984), p. 247.

Vogt seeks support for his estimation of the instrumental accuracy from an indirect remark of Pappus concerning the scale of the astrolabe. His argument has to depend on a number of uncertain assumptions. Therefore, Vogt endeavors to fortify his considerations through an analysis of the frequency of the degree fractions in longitude and latitude.

The 1/6 degrees of longitude are not distributed equally:[91]

fraction:	1/6	2/6	3/6	4/6	5/6	6/6
number:	174	183	97	240	101	222

Table 3.11: Fractions of a degree in the longitudes.

The mean frequency for each 1/6 degree amounts to 170 stars, which is only approached by the number of stars with 10′ fraction. The numbers for 30′ and 50′ are considerably smaller, while the others lie well above it. In toto 645 stars come on the even sixth degree (with 0′, 20′, and 40′), which is much more than the tripled mean amount of 510=3*170 stars. In contrast the odd sixth degree longitudes amount to only 372 instead of 510. Vogt explains this significant difference by supposing a graduation of 1/3 degree. The values lying between the marks, the uneven sixths, are then estimated by the observer when the outer ring is adjusted roughly between two marks. According to Vogt, this is the only way of explaining the astonishing small number of half degrees in the longitudes.

Whereas the 1/4 degree graduation appears only four times in the longitudes, it is frequent in the latitudes. Here the 1/4 degree fractions, with 95 instances in all, are represented twice as frequently as the 3/4 degree fractions. This lopsided distribution is, as Vogt sees it, incompatible with a 1/4 degree graduation of the rings, for in this case both counts ought to be about equal. Vogt believes that he can detect here a graduation of half degrees. Since the 1/6 degree fractions in latitude force Vogt to assume a graduation of thirds of a degree, one had to abandon the idea of one instrument for all observations – a consequence Vogt does not follow any further.[92]

3.3.5   The Epoch of Observation for the Hipparchan Coordinates

In order to obtain sufficient data for the reconstruction of Hipparchan coordinates, Vogt's argument is based on the assumption that the underlying data reproduce in a large measure the data of the otherwise unknown so-called Hipparchan star register. Pliny writes that Hipparchus assembled a star catalogue at a later time than the writing of the Aratus Commentary, when a new star appeared, in order to ensure a better comparison of the celestial phenomena. Most probably it was only after he wrote his Commentary that Hipparchus discovered the motion of precession, and it is therefore possible that he renewed his observeations of a series of star coordinates. Vogt, for his part, wants to pursue this question and to gather

---

[91] These are Vogt's figures. They are corrected in section V.3.
[92] Vogt, H. (1925), col. 42.

evidence for chronological differences in his reconstructed coordinates. The date of a coordinate is determined by the best fit with the calculated accurate position of the star in the sky. After excluding 10 cases with an error of more the 65 years from the average, Vogt obtains as a mean observation epoch the year -150 with a "mean error" of ± 3.77 years using 77 stars in the zodiac and around the equator.[93]

From 18 declinations of the Commentary Vogt excludes 7 cases with an error of more than 84 years and computes as mean observation epoch the year -156 ± 10. From the 16 Hipparchan declinations mentioned by Ptolemy in the Almagest, Vogt derives after the exclusion of 4 data an epoch of -130 ± 6.46. He similarly calculates a later period for the right ascensions of the star clock described at the end of the Aratus Commentary.

This result supports the interpretation that the data which the Almagest attributes to Hipparchus actually stem from a later period than the coordinates of the Aratus Commentary.[94] Still, one cannot exclude the possibility that an essential part of the Hipparchan star register had been observed later. Vogt's discovery that certain stars of the Aratus Commentary such as θ Eridani and π Hydrae had been taken over in the Almagest serves to argue against the compilation of a new and complete star catalogue by Hipparchus.

After the motion of precession had been discovered, it turns out to be practical to enter the star coordinates in an ecliptical coordinate system. The coordinates could be reduced to any other time by a simple addition of the precession to the ecliptical longitudes. Delambre and Vogt suspect that the coordinates measured by Hipparchus are equatorial and were converted into the ecliptical system afterwards.[95]

> "If this is the case we should not conceive of the revision of the Hipparchan star catalogue, which was achieved after the discovery of precession, as a sharp break, but rather as a transition from one form to another. Hipparchus seems to have used the new observations with the old ones, the new device and the old one ... If it is furthermore true that it is only possible to make a formal, but not a genetic, division between the old equatorial and the new ecliptical coordinates, it follows that my reconstruction, which can rely for the main part only on older data and only in a small part on later entries, may claim to offer a reliable representation of the Hipparchan ecliptical star catalogue in general, even if each and every detail is not done justice."

In particular, four results of Vogt's work have succeeded in expunging in the eyes of many the pervasive uncertainty over the origin of the Ptolemaic star catalogue.

(i) An attempt at a partial reconstruction of the lost Hipparchan star register has succeeded. Although the reconstruction is based on the earlier data of the Aratus Commentary and it is reported that Hipparchus, after the appearance of a new star and the discovery of the precession, carried out at least some

---

[93] Vogt, H. (1925), cols. 31f.
[94] Because of Pliny's report.
[95] Vogt, H. (1925), col. 32.

new observations, Vogt nevertheless understands his reconstruction as a representative subset of a later star catalogue which Ptolemy, according to Brahe and Delambre, is supposed to have copied for the Almagest.

(ii) The reconstructed coordinates display errors which do not completely coincide with the errors of the coordinates in the Almagest. In the eyes of Vogt it demonstrates that the two catalogues are of different origin. According to Vogt, in only 5 cases out of 122 is there sufficient reason to assume a simple transcription of Hipparchan data.

(iii) Vogt defuses Delambre's accusations that Ptolemy had manipulated the derivation of the precession constant. Dreyer's suggestion that only the stars with an accuracy of 1/4 degree in longitude had been observed by Ptolemy does not tally with Boll's projection of the maximal size of a Hipparchan star register.

(iv) Vogt maintains that a series of systematic errors, e.g. an error in the geographical latitude of the observation site, the neglected influence of refraction, an inadequate solar theory and the use of an erroneous lunar theory can satisfactorily explain the significant systematic longitudinal errors of the stars in the Almagest.[96]

The reaction of the historians is univocal. In his commentary on the Almagest, Pedersen remarks on Vogt's arguments:[97]

> "Taken together they seem, however, to offer convincing evidence for the conclusion that Ptolemy was entitled to present his catalogue of the fixed stars as a result of his own observational work. In return we must acknowledge that he was perhaps not as good an observer as Hipparchus, and that his erroneous equinox may account for many of the errors. But the fact that one was possibly a more diligent observer does not make the other a scientific fraud."

---

[96]Vogt cites Tannery for a maximal error of the solar theory of 22'. This is the reason Vogt provides a whole list of errors instead of concentrating on the error of the mean sun.

[97]Pedersen, O. (1974), *A Survey of the Almagest*, Odense, p. 258.

3.4   Gundel's List of Hipparchan Stars

After Vogt had succeeded in reconstructing a significant proportion of the Hipparchan star register and in presenting strong evidence of its genetic independence from the coordinates of the Almagest, a small list of stars from an astrological manuscript once again introduced slight discrepancies in the newly gained rehabilitation of Ptolemy, though without having the revolutionary consequences of Boll's work.

In the year 1936 Wilhelm Gundel published an early astrological text in which a number of Hipparchan and even older ecliptical longitudes of stars is contained.[98]

The astrological manuscript, probably written in 1431[99] is entitled "Liber Hermetis Trismegisti". Although the text is structured only weakly through chapter titles, this still allows us to recognize a total of 37 separate sections whose lengths, contents and significance vary considerably.[100] For the questions that concern us, the third chapter on the bright stars is of special interest. The Hermes Trismegistos treatise is neither a copy nor a Latin translation of a coherent original text, but rather, in all probability, a conglomeration of older treatises that had relevance for astrology.[101] Furthermore, it is in no way merely a compendium of an older text, for it occasionally contains allusions to other astrological doctrines or texts which are not to be found in "Hermes".[102] After the first book of "Hermes", which contains a treatise on the decans of the zodiacal signs,[103] and the second, according to Gundel not very informative chapter on the "constructive forms of the male and female degrees", an interesting chapter follows entitled "De stellis lucidis et qualitatibus signorum".[104]

This section offers a listing of 68 bright stars, distributed over the entire sky, with coordinates which are as a rule ecliptical longitudes. For an astrological text, it is difficult to accept the number 68 as the total of stars under consideration, and Gundel speculates that 72 stars were contained in the original text as a doubling of the number of decans. Gundel suggests locating the four missing stars in (i) the constellation of Sagittarius, where the text mentions only 2 instead of 4 stars, (ii) in Orion where the middle star of the belt is missing and (iii) in the head of Aries where indeed three bright stars are expressly mentioned, yet only 2 are specified with regard to their position.[105] Before further discussion, the following information is gathered in the table: Gundel's list of 68 stars with their modern name, their catalogue number $No.$ in the Almagest, the true longitude λ_{-128}, the longitude in "Hermes" λ_T, the corresponding longitude in the Almagest λ_A, the

---

[98]Gundel, W. (1936), Neue astrologische Texte des Hermes Trismegistos. *Abhandlungen der Bayerischen Akad. d. Wissenschaften*, Philos.-hist. Abteilung, Neue Folge, Heft 12, München.

[99]Gundel, W. (1936), p. 3.

[100]Gundel, W. (1936), p. 3.

[101]Gundel, W. (1936), p. 4.

[102]Gundel, W. (1936), p. 4.

[103]Decans become visible each ten days in their heliacal rising, i.e. they become visible for the first time shortly before the sun rises on the eastern horizon. Each zodiacal sign is therefore divided into three decans, which amounts to a total of 36 decans.

[104]Gundel, W. (1936), p. 123.

[105]Gundel, W. (1936), p. 126. Neugebauer criticizes this interpretation in Neugebauer, O. (1975), p. 286, n. 17: "His assumption (p. 135, p. 142, n. 1) that the original number of stars must have been 72 seems to me unfounded".

magnitude according to the Almagest and the difference in longitudes of "Hermes" and the Almagest:[106] The columns with the catalogue numbers of the Almagest along with the accurate longitudes are added to Gundel's list. Apparent errors in Gundel's edition have been corrected.[107]

	Name	No.	λ_{-128}	λ_T	λ_A	magn.	$\lambda_T - \lambda_A$
1	γ Cnc	452	97;59	98	100;20	4,3	-2;20
2	δ Cnc	453	99;06	99	101;20	4,3	-2;20
3	μ Leo	464	111;54	111	114;20	3	-3;20
4	ε Leo	465	111;06	112	114;10	3,2	-2;10
5	ζ Leo	466	117;55	117	120;10	3	-3;10
6	η Leo	468	118;18	118	120;10	3	-2;40
7	γ Leo	467	119;47	119	120;40	2	-3;10
8	α Leo	469	120;23	118 (98)	122;30	1	-4;30
9	i Leo	477	124;53	124	127;00	6	-3;00
10	δ Leo	481	131;33	131	134;10	2,3	-3;10
11	θ Leo	483	133;48	133	136;20	3	-3;20
12	ι Leo	484	137;50	137	140;20	3	-3;20
13	β Vir	501	147;06	148	149;00	3	-1;00
14	δ Vir	506	162;04	164	164;10	3	-0;20
		or			162;10	3,2	+1;50
15	α Vir	510	174;17	172	176;40	1	-4;40
16	α Lib	529	195;32	194	188;00	2	-4;00
17	μ Lib	530	194;36	195	197;00	5	-2;00
18	α Boo	110	174;38	170(?)	177;00	1	-7;00
19	σ Sco	552	218;13	218	220;40	3	-2;40
20	α Sco	553	220;11	220	222;40	2	-2;40
21	τ Sco	554	221;53	221	224;30	3	-3;30
22	A Oph	247	230;38	230	233;00	4	-3;00
23	ζ Oph	252	219;37	219	222;10	3	-3;10
24	σ Sgt	575	252;47	252	255;20	3	-3;20
25	ζ Sgt	591	254;03	253	256;20	3	-3;20
26	α Cap	601	274;11	274	277;20	3	-3;20
27	β Cap	603	279;27	274	277;20	3	-3;20
28	γ Cap	623	292;05	292	294;50	3	-2;50
29	δ Cap	624	293;51	293	296;20	3	-3;20
30	β Aqr	632	293;50	294	296;30	3	-2;30
31	unknown			295			
32	Hipp., ris.			327			
33	Hipp., ris			329			

---

[106] Gundel, W. (1936), p. 148.

[107] Gundel's identification of star 11, 22, 23 are mistaken. The table shows the corrected stars. Furthermore, in identifying star 67 as β Col, Gundel shifted the Almagest longitude of the latter by a full zodiacal sign; the description of it as "in extremitate caudae" cannot refer to a star in the leg, but must mean the star η CMa, "The star on the tail". In this case a large difference in longitudes of 8°10′ still remains, which could indicate an error in copying. The coordinates are given in degree and minutes. The star 31 could not be identified, and for 32, 33 and 41 one finds a Hipparchan simultaneous phenomena instead of a longitude.

	Name	No.	λ_{-128}	λ_T	λ_A	magn.	$\lambda_T - \lambda_A$
34	ω Psc	681	332;58	333	336;00	4	-3;00
35	ι Cet	732	331;18	332	334;20	3,4	-2;20
36	β Cet	733	332;45	333	335;40	3	-2;40
37	η Cet	727	342;00	342	345;00	3	-3;00
38	θ Cet	726	346;39	347	349;40	3	-2;40
39	δ And	335	352;17	352	355;20	3	-3;20
40	η Psc	695	357;14	358	00;40	3	-2;40
41	Hipp., merid.			348			
42	γ Ari	362	3;36	4	6;40	3,4	-2;40
43	β Ari	363	4;23	5	7;40	3	-2;40
44	η Cas	180	10;22	11	13;00	4	-2;
45	φ Per	350	15;08	15	17;10	4,5	-2;10
46	β Per	202	26;38	27	29;40	2	-2;40
47	α Tau	393	40;10	39	42;40	1	-3;40
48	β Tau	400	52;59	53	55;40	3	-2;40
49	ζ Tau	398	55;12	55	57;40	3	-2;40
50	o Per	215	31;35	31	34;10	3,4	-3;10
51	λ Per	207	40;14	40	43;00	4	-3;
52	α Aur	222	52;16	52	55;00	1	-3;
53	β Ori	768	47;12	47	49;50	1	-2;50
54	ι Aur	229	47;04	47	49;50	3,4	-2;50
55	δ Ori	759	52;45	52	55;20	2	-3;20
56	ϵ Ori	760	53;52	54	57;20	2	-3;20
57	α Ori	735	59;09	59	62;00	1,2	-3;00
58	ϵ Gem	433	70;21	70	73;00	3	-3;00
59	τ Gem	427	75;51	75	78;40	4	-3;40
60	θ Gem	426	71;31	72	76;40	4	-4;40
61	α Gem	424	80;43	80	83;20	2	-3;20
62	β Gem	425	83;59	84	86;40	2	-2;40
63	β Aur	223	60;21	59;30	62;50	2	-3;20
64	α CMa	818	74;52	74	77;40	1	-3;40
65	ϵ CMa	832	81;18	81	83;40	3	-2;40
66	δ CMa	831	83;56	83	86;40	3,4	-3;40
67	η CMa	835	90;07	84	92;10	3,4	-8;10
68	α CMa	848	86;33	88	89;10	1	-1;10

Table 3.12: Longitudes of Hermes.

The table of stellar longitudes had a particular significance for astrological use because the physical energies connected with the bright stars exerted along with the planetary constellations a characteristic influence on the events of daily life.[108] For the scientific astronomers after Hipparchus, the astrological connotations of the stars lost their significance. Instead, their research concentrated on the measurable

---

[108]Gundel, W. (1936), p. 124.

quantities like the rising and setting times, or the magnitudes of the stars. It is therefore understandable that astrological texts from a later period revert to ancient compendia in which the author had not yet worked his way free of the astrological terminology. In the introduction to a Greek astrological manuscript by an anonymous author from the year +379, particular attention is called to the fact that the idiosyncrasies of the stars had been examined by only a few of the older astronomers. The author adds that, for practical reasons, the stellar longitudes of the stars are corrected for the year +379 according to the law of the divine Ptolemy, that is to say, one degree per hundred years is added to the longitude.[109] We know from these remarks that older lists of stellar longitudes were circulating in astrological circles.

As a rule, the longitudes of the stars in "Hermes" are smaller then those of the Almagest, distinctly indicating their older origin. The description of the stars within the constellations confirms this interpretation. The "Hermes" terminology does not follow Ptolemy's, but rather that of still older astronomers. Gundel provides several examples as evidence:[110]

(i) The ten stars of Leo are arranged in such a way that four stars are placed *before* the mane and *in* the breast of Leo, while Ptolemy distributes them in the neck. Star 10, δ Leonis, is described in Hermes as "in supercilio" (in the eye brow), Ptolemy places it on the rump of the lion. The lion's head is therefore depicted as much larger that in the Almagest, in full agreement with the older tradition.

(ii) Arcturus, the brightest star in Bootes, is located in the belt. For Eudoxus and Aratus, Arcturus is part of the belt, too, whereas Eratosthenes, Vitruvius and Ptolemy count it as part of the knee, or locate it between the thighs, and, furthermore, Ptolemy sees it as an external star not belonging to the constellation at all.

(iii) The three stars of Scorpius lie in its breast. This arrangement is known only from Hipparchus. Eratosthenes places them in the back, Ptolemy on the body.

(iv) Both stars of Sagittarius are similarly described in Hipparchan terminology. There, σ Sagittarii and ζ Sagittarii are determined in their position by a rectangle of fainter stars. The Almagest sets them on the left shoulder and under the armpit respectively. In the first chapter of the seventh book Ptolemy describes the Hipparchan alignments of the stars and compares them with his own observations for demonstrating the fixed relative position of the stars to each other. Here, both stars of Sagittarius are described by a rectangle as it was done by Hipparchus.[111]

In this case "Hermes" clearly employs Hipparchan terminology, a fact that is also reflected in the accounts of the brightness of the stars. The Almagest is the first surviving document classifying the brightness of stars into six magnitude classes. In his edition of the Aratus Commentary Manitius compares the

---

[109]Gundel, W. (1936), pp. 124f.

[110]Gundel, W. (1936), pp. 125ff.

[111]Gundel, W. (1936), p. 128.

Hipparchan descriptions of brightness with those from the Almagest and he finds that in general the first three magnitudes are called "bright/luminous" (λαμπροί), and those of the fourth and fifth magnitude "small" (μικροί) with a further differentiation by the predicate "strong" (ἐκφανής, ὀξύς) and "weak" (ἀμαυρότερος).[112] In "Hermes" both stars of Scorpius, are labelled "luminous", as they are by Hipparchus, though they are catalogued in the Almagest as stars with a magnitude of three and four.[113]

(v) Aquarius is sketched in "Hermes" following the older terminology. The second star, which Ptolemy assigns to the water, can be found in the right foot. Ptolemy does not mention a foot of Aquarius at all, while Eratosthenes placed stars in it. This terminology was also known by Eudoxus and Aratus. In the Almagest these are called external stars. Again Ptolemy's constellations are smaller than those of his predecessors.

(vi) The constellation Cetus in "Hermes" might be the replacement for the ancient Egyptian Crocodile, for one still finds in it the partial constellation "magnum rostrum", whereby most likely the throat of the Crocodile is referred to.[114]

In addition, the constellations Aries, Libra, Virgo, Taurus and Canis Major refer in an unmistakable way to a Hipparchan or pre-Hipparchan terminology. Only in the case of Ophiuchus and Capricornus are elements of the later Ptolemaic grouping discernible, but in the overwhelming majority of constellations"Hermes" follows the ancient arrangements.

This result is confirmed by the longitudes of the stars. The list contains 22 stars having a longitude from 2°20' to 2°50' smaller than those of the Almagest. As Gundel sees it, the conversion of the longitudes with the Ptolemaic precession constant leads us back to "the time of Hipparchus or his students. Among those we have to think in the first place of the astrologer Serapion, who has become a discernible figure for us through the recently edited texts from astrological manuscripts".[115]

An even larger number of stars (31) displays a difference in longitude from 3° to 3°40' and Gundel holds them to be coordinates from the school of the astronomers Timocharis and Aristyllos of the third century B.C.E.[116]

Another group of stars has coordinates which at first sight cannot accord with ecliptical longitudes in any way. Star 41 at the end of Ursa Minor has a longitude of 348° compared with the Almagest's 60°10'; it is impossible to interpret this as an ecliptical longitude. In the Aratus Commentary Hipparchus notes: "It would be more accurate to say that Perseus and Cassiopeia are in the region of the end of the tail of Ursa Minor; for the most outer and brightest star (α) of Ursa Minor lies on the meridian of Pisces 18° (348°) or, as Eudoxus divides the zodiac, on that of Aries 3°(3°)". [117]

The degree value in "Hermes" is therefore identical with the point of the ecliptic culminating simultaneously. Its longitude is defined, not as in the case of ecliptical

---

[112]Hipparchus (1896), p. 294.
[113]Gundel, W. (1936), p. 128.
[114]Gundel, W. (1936), p. 129.
[115]Gundel, W. (1936), p. 131.
[116]Gundel, W. (1936), p. 131.
[117]Hipparchus (1896), p. 56.

longitudes, by the perpendicular to the ecliptic, but by the degree of the intersection of the meridian circle and the ecliptic. One finds a similar confusion of coordinates at the stars 14 and 32 (here it is the longitude of the point of the ecliptic rising simultaneously), and star 33.[118] Gundel interprets the origin of the stellar coordinates in the Hermes Trismegistos text as follows:[119]

> "As a result of our survey of this new star catalogue we are able to obtain from the forms of the constellations, the positions of the stars in them, and from their longitudes, the insight that we have before us a compilation of various astronomical observations from different periods ... The numerous similarities with Hipparchus' statements – in addition to the longitudes of the stars and their function in the constellation we find the determination of one rising and three meridian phenomena – leads to the insight that a friend or student of Hipparchus has compiled this catalogue ... Up to now it has been seen as fairly certain that Hipparchus was the first to use this type of position measurement for his new star catalogue. This, of course, is undermined by this new star catalogue and its pre-Hipparchan coordinates."

Should Gundel's statement turn out to be correct, then, along with the reconstructions of Vogt, more authentic Hipparchan coordinates would be available to us for a comparison with the Almagest. With one sole exception "Hermes" notes the stellar coordinates in full degrees. We can be assured that the Hipparchan coordinates, just as in the Aratus Commentary, were given with a higher accuracy and had been rounded to full degrees, as so often happened in medieval texts.[120] In the event that the corresponding coordinates from the Almagest were obtained through the addition of $2°40'$ to the Hipparchan coordinates, then, according to Gundel, the stars of "Hermes" with a difference of $2°$ to $2°50'$ are those which had been rounded from the precise Hipparchan positions of the lost star register.

Gundel's dating of the coordinates with the help of the too small Hipparchan-Ptolemaic precession constant does not depend on one's position with regard to the origin of the Ptolemaic star catalogue, because genuine Hipparchan coordinates are in both cases roughly $2°40'$ smaller than those in the Almagest. It is dubious, however, whether Gundel is correct in his identification of an older star register than that of Hipparchus. Neugebauer compared all the longitudes in "Hermes" with the true positions[121] and discovered that out of 59 stars suitable for a test, just one of them has a longitude of less than $1°$ relative to the stellar position of the epoch -130. All in all, 96.5% of the stars can be dated to the time from -130 to -60.[122]

Thus the Hermes Trismegistos text appears to contain a list of about 60 Hipparchan longitudes which are available for further examination of Vogt's reconstructed coordinates and the pursuit of the question of the authorship of the Ptolemaic star catalogue.

---

[118]Gundel, W. (1936), p. 132.

[119]Gundel, W. (1936), p. 134.

[120]Neugebauer, O. (1975), p. 286.

[121]taken from Peters-Knobel.

[122]Neugebauer, O. (1957), *The Exact Sciences in Antiquity*, 2nd, New York, pp. 68f.

3.5 Precession and Solar Theory

3.5.1 Pannekoek's Calculation of Precession

Exact knowledge of the motion of precession and the sun's motion makes it possible to re-examine the statements of Delambre and Vogt on the Ptolemaic demonstration of the Hipparchan precession constant and the explanation of the longitudinal errors suggested by Laplace.[123]

In the second chapter of the seventh book, the Almagest quotes Hipparchus from a now lost work entitled "On the Displacement of the Solstitial and Equinoctial Points" where he describes the phenomenon of precession and estimates its value. According to Ptolemy's account, Hipparchus compared the data of earlier lunar eclipses from the time of Timocharis (ca. -280) with his own observations and found that the bright star Spica, which is often used as a reference star for the definition of the ecliptical coordinate system, had increased its longitude by $2°$.[124] Total lunar eclipses are particularly suitable for the measuring of coordinates because the sun at the moment of centrality is located exactly $180°$ apart in opposition to the moon on the ecliptic, and the position of the moon, after the parallax is considered, is immediately given through the position of the sun. A systematic error in the position due to a deficient lunar theory cannot enter into any measurement involving the moon. It seems surprising that the corresponding value for the precession constant amounts to $50''$ per year, with an increase of longitude of $2°$ in the time between Timocharis (ca. -280) and Hipparchus (-140).[125]

Ptolemy recounts that Hipparchus could only approximate the value for the precession constant, since the time difference between him and Timocharis was not sufficiently large enough and older, reliable observations were not available to Hipparchus.[126] Furthermore, Ptolemy goes on to cite a solstice evaluation of Hipparchus in the Almagest: "For if the solstices and equinoxes were moving, from that cause, not less than $\frac{1}{100}$th of a degree in advance [i.e. in the reverse order] of the signs, in the 300 years they should have moved not less than $3°$".[127] Obviously, Hipparchus does not commit himself to the exact value of $1°$ per hundred years for the precession; rather, he posits just a minimal value which does not contradict the modern value for the precession of $1°$ in 72 years. As it was Hipparchus who discovered the precession, a cautious estimation of its value seems reasonable. After that, the Hipparchan lower limit was considered as the representative value of precession constant. This shows that no astronomer between Hipparchus and Ptolemy carried out a serious astronomical analysis of the precession motion.

In the Almagest this is precisely what Ptolemy sets out to do. Unlike Hipparchus, though, he does not make use of the lunar eclipses, which allow a fairly exact determination of the stellar positions, but rather of the stellar declination. The declination measurement is a relatively simply method easy to handle for ancient

---

[123]Pannekoek, A. (1955), Ptolemy's Precession, *Vistas in Astronomy* 1, pp. 60–66. Petersen, V. M., Schmidt, O. (1968), The Determination of the Longitude of the Apogee of the Orbit of the Sun according to Hipparchus and Ptolemy, *Centaurus* 12, pp. 73–96.

[124]Ptolemy, C. (1984), p. 327.

[125]Pannekoek, A. (1955), p. 61.

[126]Ptolemy, C. (1984), p. 329.

[127]Ptolemy, C. (1984), p. 328.

astronomers: at the moment of culmination, i.e. the highest position of the star over the horizon on the meridian, the altitude h_c is measured. With the geographical latitude φ of the observation site, the declination is obtained immediately as:

$$\delta = h_c + \varphi - 90° \tag{3.4}$$

Ptolemy cites 18 declination measurements from the time of Timocharis, Hipparchus, and himself for the demonstration of the precession constant.[128] Pannekoek, for his part, tabulates the measurements compared with the accurate declinations.[129]

Name	No.	HR	observ. Decl.			calc. Decl.				Differences		
			T.-A.	Hipp.	Ptol.	P.-K.	-288	-128	137	T.-A.	Hipp	Ptol
α Aql	288	7557	5.80	5.80	5.83	5.20	5.67	5.68	5.78	0.13	0.12	0.05
η Tau	411	1178	14.50	15.17	16.25	16.13	14.52	15.33	16.63	-0.02	-0.17	-0.38
α Tau	393	1457	8.75	9.75	11.00	10.37	9.00	9.70	10.80	-0.25	0.05	0.20
α Aur	222	1708	40.00	40.40	41.17	40.36	39.77	40.43	41.45	0.23	-0.03	-0.28
γ Ori	736	1790	1.20	1.80	2.50	2.00	1.25	1.78	2.65	-0.05	0.02	-0.15
α Ori	735	2061	3.83	4.33	5.25	4.05	3.80	4.25	4.95	0.03	0.08	0.30
α CMa	818	2491	-16.33	-16.00	-15.75	-15.37	-16.20	-16.07	-15.88	-0.13	0.07	0.13
α Gem	424	2891	33.00	33.17	33.40	33.08	33.07	33.25	33.47	-0.07	-0.08	-0.07
β Gem	425	2990	30.00	30.00	30.17	29.37	29.97	30.08	30.18	0.03	-0.08	-0.01
α Leo	469	3982	21.33	20.67	19.83	19.40	21.13	20.67	19.83	0.20	0.00	0.00
α Vir	510	5056	1.40	0.60	-0.50	-0.20	1.43	0.55	-0.93	-0.03	0.05	0.43
η UMa	35	5191	61.50	60.75	59.67	59.10	61.58	60.68	59.20	-0.08	0.07	0.47
ζ UMa	34	5054	67.25	66.50	65.00	65.08	67.50	66.62	65.10	-0.25	-0.12	-0.10
ε UMa	33	4905	68.50	67.60	66.25	65.38	68.62	67.75	66.30	-0.12	-0.15	-0.05
α Boo	110	5340	31.50	31.00	29.83	29.30	32.27	31.30	29.72	-0.77	-0.30	0.11
α Lib	529	5531	-5.00	-5.60	-7.16	-6.22	-4.77	-5.63	-7.07	-0.23	0.03	-0.09
β Lib	531	5685	1.20	0.40	-1.00	-0.22	1.08	0.25	-1.12	0.12	0.15	0.12
α Sco	553	6134	-18.33	-19.00	-20.25	-19.25	-18.37	-19.10	-20.28	0.04	0.10	0.03
									$\mu =$	-0.07	-0.01	0.04
									$\sigma =$	0.02	0.01	0.02

Table 3.13: Precession and declination measurements.

With the differences in declination and time, Pannekoek calculates the precession constant according to:[130]

$$p = \frac{\Delta\delta}{\Delta t \sin\epsilon \cos\alpha} \tag{3.5}$$

For the mean values of the first half of the data (stars 1–9), Pannekoek reckons the precession constant as 46.4″ per year and for the second group as 46.0″,[131] which is very close to the accurate value.

---

[128] Ptolemy, C. (1984), pp. 331f.

[129] Pannekoek, A. (1955), p. 64. Name: modern star name; No: running star number in the Almagest; HR: star number in the Bright Star Catalogue; the observed declinations for Timocharis/Aristyllos (T.-A.), Hipparchus (Hipp) and Ptolemy (Ptol); the calculated declinations on the basis of the ecliptical coordinates of the Almagest (P.-K.); the calculated true declinations for the years -288, -128 and 137; the errors of the observed declinations at the time of Timocharis/Aristyllos (T.-A.), Hipparchus (Hipp) and Ptolemy (Ptol).

[130] Pannekoek, A. (1955), p. 63. The equation is an approximation of the equation (3.2) with the simplification:

$$\cos\alpha = \frac{\cos\beta\cos\lambda}{\cos\delta}$$

[131] Pannekoek, A. (1955), p. 63.

In the methodology of the Almagest the calculations could not be performed with the help of modern trigonometrical equations. In place of this Ptolemy used the table of sphaera recta (Alm. II, 8) which provides the corresponding declination of the ecliptic for each ecliptical longitude. In an approximate fashion, the subsequent motion in the ecliptical longitude is obtained with the ecliptical longitude of a star and the declination difference.[132].

In Ptolemy's time, the statistical notion of a mean value of a set of values was not yet known to science, and as a result of this it could not have made sense to Ptolemy to derive the mean value of the precession out of the sample of 18 stars. The values, each one taken in isolation, represent a confirmation or refutation of a hypothesis. The Almagest describes no evaluation of all examples, but rather a selective demonstration on the basis of 6 stars. These instances confirm the lower limit of the Hipparchan precession constant of one degree every century.[133]

As Delambre had already noted, the Hipparchan value can only be confirmed with this special set of data, the mean of which Pannekoek fixes at $38''$ per year.[134] In his view, the result confirms Delambre's interpretation. "There can be no doubt that Ptolemy selected these six stars because they were favourable to his assumed value of the precession and could be quoted as confirmations, and that other stars were omitted because they did not confirm his assumption."[135]

Pannekoek has no comments on Vogt's different analysis. So, the question is still open whether Vogt's harsh criticism of Delambre's procedures, and with it Pannekoek's, is founded.

From his inquiry Pannekoek draws yet another, more important conclusion. It is obvious that Ptolemy, too, reported observations in the Almagest which, if they are evaluated, do not support the Hipparchan precession constant. One trait of early scientific research could be that at first a whole set of measurements is gathered from which a satisfactory confirmation or a hypothesis is sought in particular instances. Only at the beginning of the 17th century did it become customary to interpret observational data by taking arithmetic means.[136] In the context of the interpretive strategy of the defenders of Ptolemy's reputation whose arguments rely basically on an explanation of the errors in longitude by the deficient solar theory, it has to be interesting whether the declinations in the Almagest are likewise distorted. To be sure, the method of measuring declinations is not dependent on the solar theory, but the uncomplicated transformation from the ecliptical coordinates of the stars in the catalogue to their declinations must make a direct comparison with the measurements appear virtually inevitable.

Pannekoek's table includes a further column in which the declinations which are calculated from the coordinates in Ptolemy's star catalogue are set next to the declinations measured by Ptolemy. They agree very poorly. One must therefore assume that either Ptolemy himself did not carry out the control calculation – at least with the fundamental stars of the catalogue – or else he is silent about the inconsistency.

---

[132]Ptolemy, C. (1984), pp. 99ff.
[133]The selected stars are No. 2, 4, 5, 11, 12, 15.
[134]Pannekoek, A. (1955), p. 64.
[135]Pannekoek, A. (1955), p. 64.
[136]Pannekoek, A. (1955), p. 65.

According to Pannekoek's reading, the errors in declination distribute normally, so that even in the case of the 18 Ptolemaic stars one finds a set of data not distorted by later corrections or systematic errors, such as an inadequate lunar theory or a mistaken position of the equinox.[137] The measurements of Hipparchus are the most exact values, those of Ptolemy are in accuracy roughly equal to the older observations of Timocharis and his school.

3.5.2   The Hipparchan Solar Theory

After the presentation of the mathematical foundations of the coordinate systems and the transformations from one to another in the first two books of the Almagest, Ptolemy develops the solar theory as a basic theory for all further astronomical theories. In the case of this central theory Ptolemy relies entirely on the work of Hipparchus. He reports on Hipparchus' analysis of the length of the year:[138]

> "The ancients were in disagreement and confusion in their pro-
> nouncements on this topic, as can be seen from their treatises, especially
> those of Hipparchus, who was both industrious and a lover of truth. ...
> Hence Hipparchus comes to the idea that the sphere of the fixed stars
> too has a very slow motion, which, just like that of the planets, is towards
> the rear with respect to the revolution producing the first [daily] motion,
> which is that of a [great] circle drawn through the poles of both equator
> and ecliptic.
>   As for us, we shall show this is indeed the case, and how it takes
> place, in our discussion of the fixed stars (the theory of the fixed stars,
> too, cannot be thoroughly investigated without previously establishing
> the theory of the sun and the moon)."

The uncertainty that Ptolemy discovers in Hipparchan texts in no way hinders him from either confirming or just copying the Hipparchan elements and parameters of the solar theory.[139] In their joint article "The Apogee of the Orbit of the Sun" Petersen and Schmidt compare in detail the accuracy of the solar theory of Hipparchus with its confirmation by Ptolemy. They are able to show that the high degree of accuracy of the Hipparchan parameters for the eccentricity and the longitude of the apogee are merely coincidental; and also that the confirmations by Ptolemy are within the expected limits of tolerance.[140]

The duration of spring (J_1) and summer (J_2), the length of the year (J), the eccentricity of the sun (e) and the longitude of the apogee (α) at the time of Hipparchus and Ptolemy are compared with the true values (table 3.5.2).

Ptolemy, as one sees, employs the same values as Hipparchus, although his longitude of the apogee deviates more than 5 degrees from the true position which Hipparchus for his time surprisingly approaches. Since both the eccentricity as well as the apogee are calculated exclusively from the lengths of each quarter year (J1

---

[137]Pannekoek, A. (1955), p. 63.

[138]Ptolemy, C. (1984), p. 131.

[139]An account of the solar theory can be found in Neugebauer, O. (1975), pp. 53ff.

[140]Petersen, V. M. und Schmidt, O. (1968), pp. 73ff.

	J_1	J_2	J	e	α
Hipparchus	94.5	92.5	365.2467	0.0417	65.5
calculated −145	94.01	92.34	365.2423	0.0351	66.23
Ptolemy	94.5	92.5	365.2467	0.0417	65.5
calculated +140	93.90	92.56	365.2423	0.0349	71.09

Table 3.14: Derivation of solar apogee.

and J2), and their values remain within the limits of tolerance both for the time of Hipparchus as well as that of Ptolemy, one cannot find fault with Ptolemy because of his more inaccurate value for α, as Manitius indeed does in the commentary to his translation of the Almagest:[141]

> "The fact that such a significant difference in the position of the apogee could remain unknown to him casts no favourable light on this observational talent. Indeed, one can go on even further and doubt whether he had observed the summer solstice at all. Observations he mentions only once, while at three other occasions he assures us that he had "exactly calculated" this."

Petersen and Schmidt demonstrate that the first reproach of Manitius is illegitimate. According to the rules of error propagation, the values for the longitude of the apogee vary within a margin of 14°, hence it was due solely to coincidence that Hipparchus was able to determine such an accurate value.

Since the work of Fotheringham, the comparisons of the Ptolemaic solar theory with the actual positions of the sun show that at the time of the stellar observations as described in the Almagest the mean solar position was slightly more than 1° too small, just like the stellar longitudes. Petersen and Schmidt buttress their new calculation with the improved tables by Tuckerman and obtain exactly the same result.[142] In figure (3.1) is a plot of the errors of the Hipparchan-Ptolemaic solar theory for the mean sun.[143]

The diagram shows two important aspects for the interpretation of the genetic origin of Ptolemy's star catalogue.

(i) The position of the mean sun at the time +137, the epoch of the Ptolemaic star catalogue, is too small by about 1.1°. This error alone could explain the systematic error of the stellar longitudes.

(ii) At the time of Hipparchus the solar theory is void of errors in the position of the mean sun. Just so, if Hipparchus had fixed his position measurements with the help of his solar theory, the star positions would have contained no large systematic error.

---

[141] Ptolemy, C. (1963), vol. I, p. 428.

[142] Tuckerman, B. (1962/64), Planetary, lunar and solar positions 601 BC to A.D.1 at five-day and ten-day intervals. *The American Philosophical Society*, Philadelphia, 1962. Planetary, lunar and solar positions A.D.2 to A.D. 1649 at five-day intervals. *The American Philosophical Society*, Philadelphia, 1964.

[143] Petersen, V. M., Schmidt, O. (1968), p. 89.

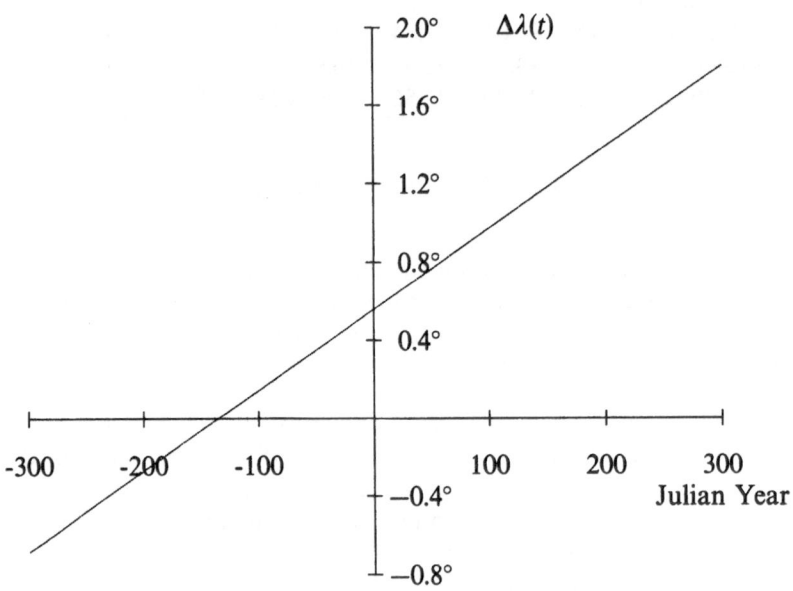

Figure 3.1: Error of the mean sun.

The development of the arguments in this century have until now not brought further support for the thesis of Brahe and Delambre. Only Gundel's list of Hipparchan star coordinates has expanded the materials needed for comparing the two catalogues without any further consequences so far.

On the other hand, those who defend the reputation of Ptolemy have been able to bring into play powerful arguments in his behalf. In the first place, the longitudinal error of the stellar positions is completely derivable from a simple systematic error in the solar theory and, for that reason, it is in complete agreement with the description of the reports of observations in the Almagest. To be sure, this piece of evidence is by no means sufficient to rule out the possibility of a contending theory, namely, that Ptolemy had taken over the coordinates from Hipparchus – in whatever form they had been written down originally. But the work done by Boll indicates that a Hipparchan list after the Aratus Commentary did not have the same number of stars as the catalogue of the Almagest, and Vogt collects material for his point that the reconstructed Hipparchan coordinates are genetically incompatible with the Ptolemaic ones.

3.6   Accusations

In the foreword to his book "The Crime of Claudius Ptolemy", R. R. Newton defines
the direction of his interpretation of the Almagest:[144]

> "This is the story of a scientific crime. By this, I do not mean a crime
> planned with the care and thoroughness that scientists like to think of
> as a characteristic of their profession, nor do I mean a crime carried
> out with the aid of technological gadgetry like hidden microphones
> and coded messages on microdots. I mean a crime committed by a
> scientist against his fellow scientists and scholars, a betrayal of the ethics
> and integrity of his profession that has forever deprived mankind of
> fundamental information about an important area of astronomy and
> history."

Newton operates in an argumentative context which Lalande and Delambre
established 200 years ago: the multitude of inaccuracies of the empirical data in
the Almagest can be best explained if Ptolemy's alleged observations are interpreted
as theoretical constructions. The proof of this thesis can, however, no longer be
maintained by an argumentation which proposes an interpretation and declares that
the resulting consequences are consistent with the main features of the historically
documented scientific activities: besides that, one must bring forward sufficient
evidence to exclude alternative interpretations.

In the case of the Ptolemaic star catalogue, the serious objections to Delambre's
interpretation developed by Boll and Vogt must, in addition, be invalidated. In his
book, Newton attempts to justify the allegation of forgery in every part of the
Almagest. Swerdlow has forcefully demonstrated how many of Newton's statements
contain fallacies and unjustified, non-historical assumptions.[145] However, his strong
assertions and the growing public interest in forgery and other moral traps of
scientific practice have stirred up the historians' attention.

The following presentation restricts itself to the central arguments of Newton
against the authenticity of the Ptolemaic star catalogue. With one exception, these
arguments had already been formulated by Delambre.

The major problems for the evaluation of the Ptolemaic star coordinates focus
on the formulation of a criterion to decide whether the coordinates are a result of
observations during the time of Ptolemy, or whether they are calculated from the
older observations of Hipparchus.

Newton tries to show by means of statistical arguments that the coordinates
cannot be a result of Ptolemaic observations. In order to determine whether a
catalogued coordinate deviates randomly from the recalculated accurate position
of the star at that time, one can utilize the concept of a standard deviation of a
distribution of data. From the data of the Almagest and the Aratus Commentary,
Newton computes the following standard deviations:[146]

---

[144]Newton, R. R. (1977), *The Crime of Claudius Ptolemy*, Baltimore, p. XIII.

[145]Swerdlow, N. M. (1979), Ptolemy on Trial, *The American Scholar* **48**, pp. 523–531.

[146]Newton, R. R. (1977), p. 216. The figures are not entirely reliable. Especially the estimates for the
ecliptical coordinates could not be reproduced.

Observer	Source	coordinate	standard deviation
Hipparchus	Ptolemy	ecliptical longitude	22.3′
Hipparchus	Ptolemy	ecliptical latitude	20.8′
Hipparchus	Hipparchus	declination	12.3′
Timocharis	Ptolemy	declination	8.8′
Hipparchus	Ptolemy	declination	6.6′
Ptolemy	Ptolemy	declination	7.2′

Table 3.15: Newton's standard deviations of measurements.

Newton himself anticipates the outcome of his analysis already here and names Hipparchus as the observer of the ecliptical coordinates of the Almagest, whose standard deviation – reduced to the epoch of Hipparchus – Newton estimates as 22′ in ecliptical longitude and 21′ in latitude. Only the stars of the zodiacal constellations are used for the error estimation, because, according to Newton, small observational errors with stars of large latitude will lead to a large error in longitude, and secondly, "there are more than 1000 stars in the catalogue, and using all of them in estimating the errors would be highly laborious."[147]

The standard deviation is a measure for the random dispersion of a data set around a mean value. A standard deviation of 22′ in the ecliptical longitudes states that approximately 2/3 of all instances appearing in the sample fall within an interval of 22′ around the mean value. If the size of the interval around the mean value is increased to the double amount of 2σ, one can find about 95% of all values within the limits. A stellar coordinate with a difference of more than 3σ to the mean value of randomly distorted observations can be found only in 0.3% of all cases.

3.6.1   The Observation of Regulus and Spica

Newton intends to show with the aid of these standard deviations that the observations of Regulus and Spica, which Ptolemy claims to have made and which are preferred as proof of the precession motion, are in all probability fabricated.[148] From the observation of Regulus, whose measured ecliptical longitude is identical to the longitude in the catalogue, Ptolemy obtains a precession constant of exactly one degree per century. According to the Almagest, Ptolemy measured the difference in longitude between the moon and the sun just before sunset with the astrolabe. Just after sunset he repeated the measurement, this time for the difference in longitude between the moon and Regulus. After the corrections for the parallax Ptolemy then derived the difference in longitude between the sun and Regulus and with it the longitude of the star.

The positions of the sun and the moon, which Ptolemy, as Newton sees it, claims to have observed, do indeed agree with the theoretical positions. These theoretical predictions differ by one degree in the longitude from the accurate mean sun and the error of the lunar theory oscillates with a standard deviation of $0°.5$

---

[147] Newton, R. R. (1977), p. 217.
[148] Newton, R. R.(1977), pp. 217f.

around the accurate value. Newton concludes here the impossibility of the Ptolemaic measurements, because, according to his calculations, the probability of an accidental coincidence of an observation with the theoretical value for the longitude of Regulus is about 1 in 1,000,000.

Swerdlow already emphasized the absurdity of this type of argument.[149] It assumes that (i) if Ptolemy did observe, his measurements would be free of any systematic distortion – especially those of the solar theory – and (ii) that the errors in the measurements are independent of each other. Only then can the total probability of the event be calculated as the multiplication of the probability of observing the same position of the sun as theoretically derived and the probability of measuring the theoretical lunar position. In the case of the Regulus observation, Ptolemy did not observe the position of the sun but derived it from his theory.[150] This fact already violates assumption (i) in Newton's calculation of probabilities. Any Ptolemaic observation of stellar longitude is connected to the errors of the solar theory and the average of a whole series of observations cannot come close to the recalculated accurate values. So only the improbability of observing the theoretical lunar position remains. It might well be that Ptolemy calculated the empirical data from the theory, but it is also possible that he measured it. The odds for such a procedure are far from the fictitious numbers Newton mentions. Hence, it could well be that Ptolemy did construct the numerical examples; only, Newton's methodological instruments provide no basis to gain any insights in that respect.

The same arguments apply for the observation of Spica. Once again, the allegedly measured longitudes coincide exactly with the values of the catalogue, though these differ from the recalculated correct values by $1°.28$. With a standard deviation in the ecliptical longitude of 22.3′, the measured longitude deviates as much as 3.5 times of the standard variation from its mean value. Newton figures the probability of a statistical error to be about 1:2000. The same criticism as before applies. In fact Newton shows only that Ptolemy's observations cannot be measurements free of systematic errors, something no one ever seriously asserted. This result is far from the conclusion Newton draws:[151]

> "Thus, with odds of about 2000 to 1, the longitude of Spica given in Ptolemy's table was fabricated instead of being observed. The odds that the longitude of Regulus was fabricated have just been estimated at more than 1000000 to 1. With enormously high probability, the only two longitudes that we can test directly were both fabricated, contrary to Ptolemy's claim that he observed them."

3.6.2 The Measurements of Declination

Like Delambre and Pannekoek before him, Newton evaluates those 18 Ptolemaic declination measurements, from which 6 stars are selected to demonstrate the precession constant of one degree per century. Similarly to Delambre, Newton notes that for the 6 chosen stars the mean value of the precession constant, $38°.1$, accords

---

[149]Newton, R. R. (1977), p. 218, Swerdlow, N. M. (1979), p. 530.
[150]Ptolemy, C. (1984), p. 328.
[151]Newton, R. R. (1977), p. 218.

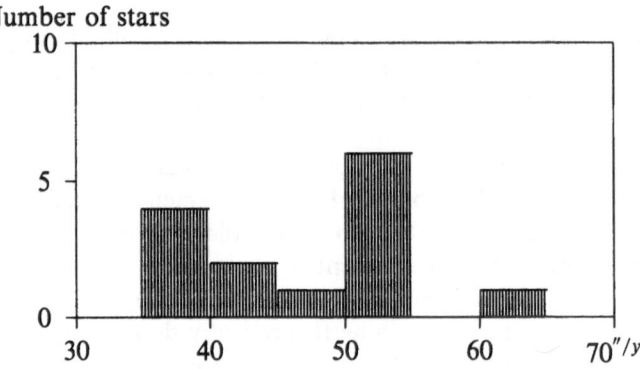

Figure 3.2: Distribution of precession values.

well with the Hipparchan lower limit. Four of the 18 stars have a position close to the solstices. Here the ecliptic is perpendicular to the declination circle and consequently a small variation in longitude does not change the declination. Conversely, measuring the declinations of such stars has no significance for the determination of precession motion. The other declination measurements which are indeed suitable for an evaluation but which Ptolemy does not make use of yield a mean precession motion of 52.8″ per year. This value is compatible with the accurate value of 49.8″ per year and in contradiction to the 1 degree per century proven by Ptolemy.

Newton objects to Pannekoek's conclusion that Ptolemy had only selected favourable data out of a real data set of observations.[152] The standard deviation of a single calculated precession constant amounts to 5.5″ per year.[153] The values which Ptolemy used to demonstrate his precession constant lie about 2.7 times the standard deviation from the mean value, which according to Newton's reckoning can happen with statistically dispersed measurements only with a frequency of 1:290. If the six evaluated observations had been selected from a total of genuine measurements, as Pannekoek insists, the total number of observations must have been very large indeed in order to interpret the 6 small values of the precession constant as the lower end of a normal distribution. According to Newton, a large series of observations is improbable, because the conversions of the declination variations without the aid of modern trigonometry would be difficult and extensive.[154]

A histogram of the resulting precession constants seems to contradict normal distributed values (fig. 3.2), as one expects it in cases of genuine observation without large systematic distortions.

Here, two explanations are suggested by Newton: either the values must have been very selectively chosen out of an extensive set of observations, or the six precession constants calculated by Ptolemy are not grounded on actual observations. "He deliberately decided to 'prove' a false value of the precession by the use of spurious data. In order to conceal what he was doing, he mixed the spurious data

---

[152]Newton, R. R. (1977), p. 222.
[153]Newton, R. R. (1977), p. 222.
[154]Newton, R. R. (1977), p. 222.

with some genuine data, so that he could pretend that he was using typical data."[155]
Again Newton's statements are unjustified in several points:

(i) The number of data is too small to prove statistically that the distribution of resulting precession constants is not a normal distribution.[156]

(ii) The underlying assumption that Ptolemaic observation would generate unbiased data cannot be justified.

(iii) It is unintelligible how Ptolemy could deliberately prove a false precession constant. Such a procedure would assume that Ptolemy knew the accurate value and how it is statistically related to the whole set of measurements. Both presumptions lack any historical understanding.

3.6.3 Stellar Positions from Occultations by the Moon

Along with the investigation of the changes in declination, Ptolemy evaluates conjunctions of the moon with the Pleiades and the stars Spica and β Scorpii.[157] For Newton, these observations are fabricated, too. In a series of cases, Ptolemy's descriptions of celestial phenomena are not correct when gauged by today's standards of accuracy. Ptolemy, for instance, tells us that Timocharis saw on 29 January -282, how the southern half of the moon covered exactly either the rearmost third or the rearmost half of the Pleiades. In contrast to this Ptolemy proceeds in his calculations as if the northern part of the moon had covered the eastern part of the Pleiades[158] as it would be calculated from the theories of the Almagest. Whether in this case a copying error is at work, whereby a "southern" resulted out of the "northern" in the report of Timocharis, or whether Ptolemy adjusted the observations to suit his theoretical values, cannot be determined. Along with a series of similar cases Newton points with special vigour to two conspicuous demonstrations in the Almagest:
Ptolemy reports of the observation of Timocharis in November -282:[159]

> "In the 48th year of the same [First Kallippic] Circle, he says that on the sixth day from the end of the last third of Pyanepsion, which is Thoth 7, when as much as half an hour of the tenth hour had gone by, and the moon had risen above the horizon, Spica appeared exactly touching the northern point on the moon.
>
> This moment is in the 466th year from Nabonassar, Thoth 7/8 in the Egyptian calendar [-282 Nov. 8/9]; [the hour is], according to Timocharis himself, $3\frac{1}{2}$ seasonal hours after midnight, or approximately $3\frac{1}{8}$ equinoctial hours, since the sun was near the middle of Scorpius; but, according to logical reasoning, [it must have been] $2\frac{1}{2}$ hours after midnight. For that is the time when 82°30′ is culminating, and 172°30′ (approximately) is rising: and that was the longitude of the moon at that moment when, as he says, it was rising."

---

[155] Newton, R. R. (1977), p. 226.
[156] Cf. chapter IV.2.
[157] Ptolemy, C. (1984), pp. 333ff.
[158] Ptolemy, C. (1984), p. 334.
[159] Ptolemy, C. (1984), p. 336.

The interesting aspect about the citation is that obviously Ptolemy had corrected the time of observation, as given down to him, in accord with his theoretical predictions and in turn calculated with his correction the small precession constant. Newton determines the rising time of Spica and finds that the star did not rise, as Ptolemy calculated, 2 1/2 hours after midnight, but rather 2 hours and 47 minutes after midnight.[160] Ptolemy no doubt trusted the data he had gained from his theory more than the traditional observation reports. In Newton's opinion, Ptolemy changed the time of observation by one hour without checking whether Spica actually (that is to say, theoretically) had risen or not. "However, he had to have 2 1/2 hours for the time in order to get the position of the moon that he needed. Hence he simply stated this time without checking to see whether Spica had yet risen."[161]

With this observation reported in the Almagest, Ptolemy had obtained an astonishing result: "So in the 12 years between the two observations [of Spica] it moved about $\frac{1}{6}^{\circ}$ towards the rear from the summer solstice".[162] Ptolemy calculates with the corrected time of the second observation that the longitude of Spica had increased by 10', equivalent to the precession constant of one degree per century.

Newton tabulates all the position measurements of the Pleiades, Spica and β Scorpii and compares them with the data of the catalogue.[163] There is agreement in all cases and Newton remarks: "The table gives incontrovertible proof that the conjunctions and occultations have been fabricated".

3.6.4   Fraction of the Degrees

Up to now, Newton's interpretations have not really diverged from those of Lalande and Delambre, namely, that the inaccuracies of the Ptolemaic observations would be more easily explained when understood as theoretically constructed demonstrations, not as authentic observations.

Newton's study of the degree fractions of the coordinates does lead to a further evaluation of both competing interpretations.[164] Delambre and Dreyer had already noted that the latitudes of the stars in the Almagest have a graduation of 1/6 degree (10', 20', 30', 40', 50') as well as coordinates with a graduation of 15' and 45'. They seem to be, as Dreyer had interpreted them, coordinates with an accuracy of a 1/4 degree, mixing at 0' and 30' with the fractions of the 1/6 degree coordinates.

If all 1/4 degree fractions were measured with approximately the same frequency, the sum of the 1/4 degree stars with 0' and 30' fractions in latitude would be as large as the amount of stars whose latitude is catalogued with 15' and 45' and which are easy to count. The ecliptical longitudes of the stars in the Almagest show, with the exception of 4 stars, no 1/4 fractions of the degree, but a highly vacillating frequency in the fractions. Newton has collected them in a table.[165]

Newton, for his part, proposes a different interpretation than Dreyer:

---

[160]Newton, R. R. (1977), p. 236.

[161]Newton, R. R. (1977), p. 236.

[162]Ptolemy, C. (1984), p. 336.

[163]Newton, R. R. (1977), p. 230.

[164]Newton, R. R. (1977), pp. 245ff.

[165]Newton, R. R. (1977), p. 245. Obviously the figures cannot be correct. Newton also counts the double entries in the catalogue and thus arrives at the sum of 1027. The correct numbers are tabulated later.

fraction of degree	number of stars		
	longitude	latitude	theoretical
0	226	236	171
10	182	106	128
15	4	88	86
20	179	112	128
30	88	198	171
40	246	129	128
45	0	50	86
50	102	107	128
totals	1027	1026	1026

Table 3.16: Newton's fractions of the degree.

The instrument is graduated in half degrees. An observer enters every full or half degree in his logbook, when the lines of demarcation lie exactly on the dividing marks or very near them. If the line of demarcation lies exactly between the dividing marks, 1/4 degree values will be recorded (e.g. 15′ or 45′); should the line lie slightly before or behind the middle between the division marks, the coordinates will then be recorded with 1/6 degrees accuracy (e.g. 10′, 20′, 40′, 50′).

If for instance the accurate coordinate corresponds to fractions between 55′ and 5′, the observer would write down a 0′, for coordinates between 42.5′ and 47.5′ the degree fraction would be 45′, and if the value lies in the interval 35′ to 42.5′, 40′ would then be noted down. Newton calculates the theoretical number of 40′ values expected in the catalogue of the Almagest according to this procedure from the relative size of the intervals between the rounded catalogue fractions, as summarized in the column "theoretical" of the table (3.16).

The number of fractions in latitude are close to the proposed theoretical distribution. Newton wants to examine whether the actual frequency distribution of the fractions is compatible with the reading procedure he proposes in that he assigns to a frequency n of one type of fractions a standard deviation $\sigma = \sqrt{n}$ as "allowed" variation. Newton finds only a moderate agreement.[166] In particular, the small frequency of 45′ (Newton counts 50 cases) in comparison to the 15′ fractions (86 stars) can be explained only by mistakes in the reading procedures. Because in Greek the 45′ fraction is expressed by the sum of 1/2+1/4, the 15′ by the fraction 1/4, a mistake by copying could result in converting a 45′ fraction into a 15′.

Newton is not able to provide evidence for the required large amount of very special copying errors, and he even refers to an additional argument which speaks against it. In the event that copying errors to this extent had really occurred and shown up in the revisions of the Almagest, it would follow that the degree fraction 50′ (1/2+1/3) likewise would be less frequent than the degree fraction 20′ (1/3). This is, however, not the case and Newton has to concede: "The number of 45′

[166]Newton, R. R.(1977), pp. 246f. "If the total number of cases is N, the standard deviation of the difference is $\sqrt{N}$". The appropriate statistical test would compare the two distributions in general, e.g. by a χ^2-test. Newton's method has hardly any significance.

has been depleted by some process that we cannot explain satisfactorily".[167] The hypothesis of the two measuring devices with different graduations falls victim to the same difficulties.

By contrast, one finds – with four exceptions of 15' – no 1/4 degree fractions in the longitudes. Three of these stars lie in the constellation Virgo. Newton assumes that originally no longitude was catalogued in a fraction of 1/4 degree and that copying errors are the cause of the peculiarity.

In average there are 124 longitudes with a fraction of 10', 30' or 50', and 217 stars are catalogued with a fraction of 0', 20', or 40'. Such discrepancies cannot be explained as the outcome of the measurement with only one observational device with a graduation of 1/6 degree. Vogt interpreted the distribution as a result of measuring with a 1/3 degree graduation, whereby there are fewer values lying in between the marks.

Against Vogt's explanation Newton raises the objection that the respective number of fractions, either the fractions on the mark or the estimated values, should be about the same.[168] In fact, there are far more 40' fractions than others. The difference between the 40' and 20' values amounts to 67 stars. That is more than three times as much as the standard deviation and, as Newton avers, a highly significant difference. The same goes for the other degree fractions as well, which makes it necessary to assume that the longitudes had not been read with a measuring device having a division of 1/3 degree.

Newton can derive the distribution of the fractions of the longitudes on the hypothesis that the longitudes all stem from the Hipparchan source to which Ptolemy adds a constant of 2°40'. Out of a Hipparchan longitude, for example, one with a fraction of the degree of 0', a Ptolemaic longitude with 40' evolves after the addition of 2°40'. Newton supposes that the scaling of the observation instrument was similarly structured for the ecliptical longitudes and latitudes and that for this reason the frequency distribution of the degree fractions of the longitudes, as they were originally noted down by Hipparchus, resembles the distribution of the latitude fractions.

Initially, there were according to this interpretation 236 longitudes with 0', 106 with 10' and 88 with 15' and so on, just the same as in the latitudes. After the addition of 2°40', the 236 longitudes with 0' transform into values with 40'. It is still unclear how the 1/4 degree values 15' and 45' were rounded off. It is unreasonable to catalogue beyond the 1/6 degree accuracy, which would result from the addition 40' to a 1/4 degree longitude. The original 15' would convert into 55' = 15' + 40' and the 45' would turn into 25'. In those cases the conversions have to be rounded either to the next higher or lower standard fraction.

Newton does not try to solve this problem by recourse to the customary practice of rounding in Greek mathematics nor that of the Almagest, rather he discusses what kind of rounding could be best reconciled with the desired similarity of the theoretical distribution of the degree fractions in longitude with those of the Almagest.[169]

---

[167]Newton, R. R. (1977), p. 248.
[168]Newton, R. R. (1977), p. 249.
[169]Newton, R. R. (1977), p. 251.

> "We could try to speculate about whether he would choose to change a 15 (which became 55 after addition of 40) into a 50 or a 0, and whether he would change a 45 into a 20 or a 30. Speculation of this sort would be inconclusive and, luckily, it is not necessary."

The speculation cannot be as superfluous as Newton would have us believe, for without it the argument comes close to a *petitio principii*: on the one hand, Newton wants to prove on the basis of the frequency distribution that the Ptolemaic star catalogue coincides genetically with the Hipparchan register; on the other, he defines the practice of rounding the transformed 1/4 degree fractions in a way that the frequency distribution of the documented fractions can be matched: the number of instances of 0' without the rounded coordinates is less in Newton's theoretical distribution than in the Almagest. Accordingly, Ptolemy's 55' values must have been rounded to the full degree. Likewise, the theoretical frequency of 20' is too small if the 25' fractions are rounded to the half degree. Consequently, Ptolemy had to round in one case the fractions to the full degree and in the other to the third degree to derive a well fitting theoretical distribution of fractions to the empirical one based on countings of the star catalogue.[170]

fraction	Number of cases			
	original,	with 40' added,	with 40', reassigning	distribution
of degree	(latitudes)	without 15' & 45'	15' & 45'	Almagest
0	236	112	200	226
10	106	198	198	182
15	88	-	0	4
20	112	129	179	179
30	198	107	107	88
40	129	236	236	246
45	50	-	0	0
50	107	106	106	102

Table 3.17: Theoretical distribution of degree fractions (Newton).

Expressed in two diagrams (fig. 3.3), the similarity between the frequency distributions becomes obvious, if one assumes a derivation of the coordinates according to Newton's interpretation.

Once again, we see how Newton's argument, as Delambre's long before him, is set up abductively: a large number of peculiarities in the Almagest can be explained on the hypothesis that Ptolemy constructed all so-called observations theoretically. Whether or not any alternative interpretation can be excluded, is not considered any further. Even in Newton's examination of the distribution of the degree fractions, the asserted outcome is included in the construction of the appropriate rounding procedure. The argument would be more convincing if the assumed rounding practice could be detected in other Ptolemaic calculations, too. In addition the strength of

---

[170]Newton, R. R. (1977), p. 250.

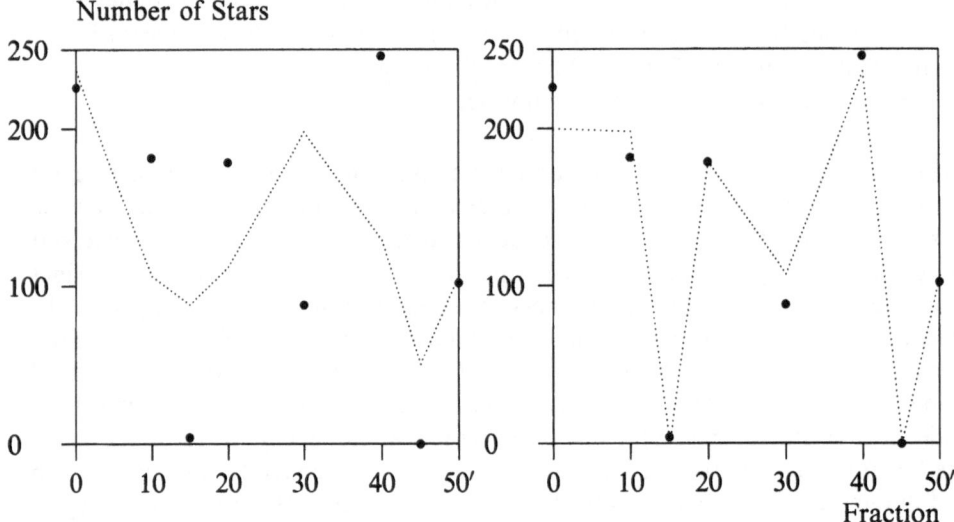

Figure 3.3: Distribution of fractions of a degree.

the argument suffers from the lack of proof that no other observational procedure can generate a similar distribution of the fractions.

Newton summarizes the result of his investigations in three points:[171]

(i) The observer employed an astrolabe scaled in ecliptical longitudes and latitudes in full degrees. The fractions are estimated up to an accuracy of 1/6 degree.[172]

(ii) The Hipparchan latitudes are copied in the Almagest as they were.

(iii) A 40′ fraction is added to the longitudes; the resulting fractions of 55′ were rounded to the full degree, and those of 25′ were rounded to 20′.

Now Newton must contend with the dominating counter-argument of Boll and Vogt. He refers only briefly to their investigations and rejects them with superficial objections:[173]

> "Here, unfortunately, is a flourishing specimen of the species *error immortalis*. The studies of Boll and Vogt do not tell us anything at all about Hipparchus's star catalogue and its relation to Ptolemy's catalogue. They cannot, because Hipparchus's catalogue is no longer available for study. Instead, the studies of Boll and Vogt actually deal with Hipparchus's *Commentary on Aratus and Eudoxus* and, to a much lesser extent, with some minor sources possibly connected with Hipparchus. The conclusions of Boll and Vogt rest upon the unlikely assumption that

---

[171]Newton, R. R. (1977), p. 252.

[172]There is an internal inconsistency in Newton's argument. Before, the assigned graduation was half a degree. Since an estimation of 1/6 degree by a graduation of only full degrees is very unlikely, we consider Newton's statement in the summary as a mere slip.

[173]Newton, R. R. (1977), p. 240.

the *Commentary* and the other sources are rigorously consistent with Hipparchus's lost catalogue. Hopefully a scholar, as he advances in his career, acquires more understanding of his subject and a greater mastery of the data."

The criticism of Boll is based on two objections:

(i) Newton believes that Boll based his identifications of the list of star totals on the fact that Hipparchus is named as the author of the medieval manuscript. Here Newton objects that Hipparchus' name is mentioned outright only in the medieval text, not in the earlier Greek writings, so that it could be an inaccuracy or a later addition of the medieval scribe.[174] Newton fails to see that Boll did not pass his judgement with the help of the title of the manuscript, rather he investigates the philological characteristics of the manuscript. Boll supports his analysis, for example, with Rehm's study of the names of the constellations. Newton, for his part, mentions not a word of this.

(ii) Boll's calculation of the extent of the Hipparchan register is allegedly defective, according to Newton. The 42 constellations of the list of star totals yield 640 stars. Since the Almagest counts up 772 stars in these constellations, Boll reasons that the Hipparchan register could not have included, altogether, more than about 852 stars. Newton points out that the constellations in the Aratus Commentary occasionally contain more stars than the Hipparchan list of star totals.

Six constellations with 61 stars are named in the Aratus Commentary, for which the list of star totals mentions just 52 stars. According to Newton it is improbable that Hipparchus should have left stars out of his later register which he had already incorporated in his Commentary (an argument which Boll had discussed and refuted before). For this reason the total of the Hipparchan stars of the list must be multiplied by the ratio (61/52) in order to calculate the extent of the lost Hipparchan star register. The number of stars referred to in the list, 640, thus increases to 751, corresponding roughly to the number in the catalogue in the Almagest. Should the list of star totals really turn out to be of Hipparchan origin, this shows according to Newton that Hipparchus' star register is just as comprehensive as Ptolemy's catalogue.[175]

It is not easy to follow Newton in increasing the star totals by just this factor, for what is summed up in the list of star totals is clearly a total of all the stars to be counted in the Hipparchan constellations. If in fact the list of star totals refers to a Hipparchan star catalogue, and taking into account that the Aratus Commentary is an early text of Hipparchus, then this catalogue must have the number of stars mentioned in the star total. One still could object that there are different sources for the star totals and the supposed Hipparchan star register; here, however, Newton offers no ground for a discussion.

Newton's arguments against Vogt's article shows his small understanding of it.

---

[174] Newton, R. R. (1977), p. 240.
[175] Newton, R. R. (1977), p. 240.

Vogt analyzed 881 data of 374 different stars from the Aratus Commentary along with some other sources. Newton doubts that the stellar phenomena in the Aratus Commentary are deduced from the coordinates of a supposedly later star register.[176] Therefore the coordinates reconstructed by Vogt cannot be those of the register.

At this point Newton contradicts his own argument against Boll. He resorts to the Aratus Commentary in order to increase the number of stars against Boll's results, and Newton objects against Vogt that the Aratus Commentary is from such an early date that it could not possibly have been derived from the star register which was the model for the Ptolemaic star catalogue. It appears that Newton seeks out exactly the aspects which support his thesis without paying attention to their internal consistency.

More than this, Newton ignores, or does not know, that Vogt can prove in at least five cases the genetic identity of the reconstructed stars with those of the Almagest and, consequently, has provided a good argument that Hipparchus did not base a later star register entirely on new observations.

Newton explains Vogt's result by noting that Vogt uses declination data from the Aratus Commentary for his reconstruction, which had possibly not been based on the Hipparchan star register, because it is usually easy to measure the declinations with a meridian instrument instead of calculating them from other coordinates of a star catalogue.[177] The comparison of the declination measurements in the Almagest with the Ptolemaic coordinates reveals a discrepancy from which one should infer, when one uses Vogt's arguments, that the coordinates of the star catalogue cannot be equivalent to the declination measurements.[178]

Newton, moreover, does not investigate to what extent the declination measurements are relevant for the reconstructed coordinates. As Vogt takes the mean value of the data in several steps of his reconstruction procedures, and most of the input data are the simultaneous longitudes of the stellar phenomena on the ecliptic, the declinations cannot be essential for the reconstructed coordinates.

Newton did not manage to refute the objections to Delambre's thesis by such excursions into the work of Boll and Vogt. His remarks seem to be mainly aimed at explaining the inaccuracies in the Almagest by attacking the scientific integrity of Ptolemy.

Newton's superficial handling of the arguments prompted Toomer to write in the introduction to his translation of the Almagest:[179]

"I hope that this will shed some light on the problem of Ptolemy's manipulation of his material (both computational and observational) in order to present an appearance of rigor in his theoretical treatment which he could never have found in his actual experience. The problem is an interesting one, which deserves an informed and critical discussion. Unfortunately, the recent book on this subject by R. R. Newton provides

---

[176]Newton, R. R. (1977), p. 240.
[177]Newton, R. R. (1977), p. 242.
[178]Newton, R. R. (1977), pp. 242ff.
[179]Toomer, G. J. (1984), p. viii.

nothing of the kind, but rather tends to bring the whole topic into disrepute."

4. The Analysis of the Star Catalogue

4.1 The Catalogue in the Almagest

The search for the origin of the Ptolemaic star catalogue requires a meticulous edition of the text and a correct identification of the ancient positional description with the stars known by their modern names. All of the investigations of this century have relied on the editions of Heiberg and Manitius and the coordinate recalculation of Peters/Knobel, published in 1915.

In 1876 C.H.F. Peters compared the older editions of the star catalogue by Flamsteed, Lalande, Bode, and Baily with the criticism of aṣ-Ṣūfī and recognized a multitude of errors.[1] Spurred by these findings, Peters examined until his death in 1890 a large number of Almagest copies during his travels over much of Europe. After his death, Edward Knobel received Peters' manuscripts from the trustees and revised the material up to the publication of 1915.

Since there is no copy of the Almagest from the time of Ptolemy available to us, its original contents must be reconstructed by means of comparing the set of later copies, not one of which is complete and without defects. Peters/Knobel had recourse to 21 Greek, 8 Latin and 4 Arabic copies of the star catalogue and they were able to eliminate many errors, in particular the copying mistakes, occasionally noted as early as Ibn aṣ-Ṣalāḥ. The critical Greek edition of Heiberg[2] and its German translation by Manitius[3] provide a reliable edition of the Almagest, which lately has been superseded by Toomer's English translation.[4] Of special interest for the Ptolemaic star catalogue is the edition and German translation of the Arabic translation of the star catalogue by P. Kunitzsch.[5]

Both of the later editions should be consulted for any analysis of the star catalogue.

---

[1]Peters, C. H. F., Knobel, E. B. (1915), Ptolemy's Catalogue of Stars. A Revision of the Almagest, *The Carnegie Institution of Washington*, Washington, p. 7.

[2]Ptolemy, C. (1898/1903), *Claudii Ptolemaei opera quae extant omnia*, ed. J. L. Heiberg, Leipzig.

[3]Ptolemy, C. (1963), *Handbuch der Astronomie*, German trans. and annot. by K. Manitius, introduction and corr. by O. Neugebauer, 2 vols., Leipzig.

[4]Ptolemy, C. (1984).

[5]Kunitzsch, P. (1974), *Der Almagest. Die Syntaxis Mathematica des Claudius Ptolemäus in arabisch-lateinischer Überlieferung*, Wiesbaden; Kunitzsch, P. (1975), *Ibn aṣ Ṣalāḥ. Zur Kritik der Koordinatenüberlieferung im Sternenkatalog des Almagest*, Göttingen; Ptolemy, C. (1986), *Der Sternkatalog des Almagest, die arabisch-mittelalterliche Tradition*, ed. and annot. by P. Kunitzsch, Wiesbaden.

A historical evaluation of the Ptolemaic text demands far more than a critical edition of the coordinates. Only a comparison of the coordinates with the accurate position of the star at the historical epoch allows conclusions regarding possible sources of errors in the Ptolemaic measurements. Hardly any historical insight can be obtained from the numerical data as they are presented in the Almagest. Typically, all inquiries of the genesis of the catalogue start with an analysis of the errors understood as deviations from the accurate procedures.

Only after two further steps can the catalogue data be compared with the accurate stellar positions. First of all, the star descriptions, coordinates and magnitude recordings of the Almagest must allow an identification of the star. One has to know which star in the sky was actually catalogued by Ptolemy. In the case of the brighter stars, one is hardly ever plagued by difficulties. The modern star names tabulated in the editions of Peters/Knobel, Manitius, Toomer or Kunitzsch undoubtedly refer to the same stars as the Almagest. With respect to the fainter stars, it is sometimes ambiguous which star appropriately matches both in position and magnitude the description of the catalogue. Identification procedures tend to harmonize possible manuscript errors by correlating the Ptolemaic data to another star with a better correspondence, or the historian tends to correct the manuscript. In general the identifications of Peters/Knobel exaggerate the revisions of the documented coordinates. They assume scribal errors in the manuscript merely on the basis of large deviations from the accurate position and change them accordingly. The editions of Toomer and Kunitzsch preserve the more authentic version of Ptolemy's star catalogue. A fair number of alternative identifications are tabulated by Peters/Knobel.[6] Out of a total of 1025 stars in the Almagest they report 252 cases with uncertain identifications.[7]

Only after a global comparison of the Ptolemaic coordinates with the accurate positions, recalculated for the epoch of Ptolemy, taking into consideration the general systematic errors of the stars in the constellation, can one solve most of these doubtful identifications.

The last recalculation of the positions for the stars of the Almagest was made by Peters and Knobel, for which Peters employed the star catalogues of Piazzi and Danckwart as initial data. Later Knobel improved the calculations for the brighter stars with the help of P.V. Neugebauer's tables.[8] The mixture of various catalogues from different epochs, paired with the inexact knowledge of the proper motions of the stars, sometimes results in significant deviations from the positions as they are recalculated today. Therefore the subsequent argumentation is based on a new reduction of the stellar positions. The recalculations of 18 selected bright stars was checked independently to ensure a reliable reduction.[9]

---

[6]Peters, C. H. F. u. Knobel, E. B. (1915), pp. 114ff.

[7]There are 3 double entries in the total set of 1028 stars in Ptolemy's catalogue.

[8]Peters, C. H. F. , Knobel, E. B. (1915), p. 51.

[9]Many thanks for the efforts of H. Schwan, Astronomisches Rechen-Institut Heidelberg, and H. W. Duerbeck, Observatorium Hoher List der Universitätssternwarte Bonn. Furthermore the recalculations can be compared with Hawkins, G. S., Rosenthal, S. K. (1967), 5000 and 10000-Year Star Catalogs, *Smithsonian Contribution to Astrophysics* 10, no. 2, Washington.

4.1.1   Critical Edition of the Catalogue

Toomer's new translation and Kunitzsch's edition of the Arabic translations are the most precise reconstructions of the star catalogue available to date. There remain only a few cases with still possible variations.

Every copy of the Almagest contains copying mistakes which express themselves in a characteristic way in each respective language. Greek numerical data, for instance, are denoted by letters, such as α for 1 or β for 2. Fractions of the degree are represented with letters together with a special symbol, e.g. " γ' " for 1/3 degree or 20 minutes of arc. This style of writing almost invites copying mistakes. A missing fraction mark can easily transform a value of 20 minutes of arc into one of 3 degrees. Peters and Knobel compiled a list of the most frequent errors:[10]

from	$\rightarrow$	to
$\varsigma=6°$		$\varsigma'=10'$
$\Gamma=3°$		$\Gamma'=20'$
$\Delta=4°$		$\Delta'=15'$
$\epsilon=5°$		$\Theta=9°$
$A=1°$		$\Lambda=30°$
$\Lambda=30°$		$\Delta=4°$
$A=1°$		$\Delta=4°$

Table 4.1: Possible copying errors.

A comparison of the numerous text variants can turn up deviations that are typical for this type of copying errors. From the epoch at which the copies are supposed to have been written and the fact that they were models for subsequent copies, most of the scribal mistakes are discoverable.[11] Besides the Byzantine tradition of Greek manuscripts of the Almagest, on which the Heiberg edition (and consequently the edition of G. Toomer) is largely based, P. Kunitzsch examined a series of Arabic translations and made it possible to draw a more detailed picture.[12]

Kunitzsch mentions several striking examples of the occasional philological troubles that can arise during the attempt to reconstruct the original coordinates.[13] The star 11 Herculis (No. 129 in the catalogue), for instance, reveals variations in longitude and latitude. As for the longitudes, all Arabic manuscripts, with one sole exception, record the value 3°50'. For the calculation of his catalogue, al-Battānī employed a value equivalent to 3°40'. In the collection of manuscripts worked on by Peters/Knobel, 5 further variants are noted: the manuscript Ven. 303 contains the value 6°30', MS. Bod. 3374 the value 6°50', MS. B.M. 7475 2°50, MS. Vienna Trap. 24 the value 4°40' and several other catalogues, among which MS. par. 2389 records the longitude as 6°40'. Due to possible transcription errors, the chronological

---

[10]Peters, C. H. F. , Knobel, E. B. (1915), p. 24.

[11]Cf. Peters, C. H. F. , Knobel, E. B. (1915), pp. 114ff.

[12]Kunitzsch, P. (1974); Ptolemy, C. (1986); Kunitzsch, P. (1975).

[13]Kunitzsch, P. (1974), pp. 152ff.

ordering of the manuscripts and the position of the star in the constellation, the last
value mentioned may well be the correct one.

Our investigation utilizes Toomer's edition of the catalogue which is reproduced
in the catalogue of Appendix B. Uncertain coordinates or identifications are marked
in an extra column.

4.1.2   Recalculation of the Coordinates for the time of Hipparchus

Because of the systematic error in longitude along with the possible older origin
of the Almagest coordinates, the positions of the stars are not recalculated for the
epoch of Ptolemy (+137), but rather for the epoch -128.[14] The fourth revised edition
of the Bright Star Catalogue of Dorrit Hoffleit supplied the accurate star data.[15]
The catalogue data are transferred from a magnetic tape and tested on plausibility;
therefore, they should not contain transcription mistakes. The coordinates are recalculated according to the rigorous formulas of Newcombe and the constants are
those that were used for the conversion of the fourth Fundamentalkatalog (FK4).[16]
According to today's knowledge of the constants of motion of the coordinate system,
and the positions and proper motions of the stars, the accuracy of the results should
generally be better than one minute of arc.[17]

4.1.3   Identification of Stars

After the transformation of the accurate star positions to the time of Hipparchus
the stars of the Almagest can be identified. If the Almagest were free of errors,
then the correlated stars whose positions had been calculated back in time would
have to have the same ecliptical latitude and an ecliptical longitude which is too
small by the value of the precession from the year -128 up to Ptolemy's epoch of
+137. In fact, most of the stars are catalogued with a significant error, most notably
a systematic error of about one degree, so that on the average the longitudes in
the Almagest are only 2°40′ larger than the accurate coordinates for the epoch
-128. Any identification technique searches for the best fit to the accurate position,
magnitude, and position within a constellation, taking into account possible errors in
the Ptolemaic star catalogue. Therefore, any identification routine has to consider all
possible distortions in the catalogue in order to assign a Ptolemaic star description to
a particular star. Besides the systematic error of the longitudes, one has to consider
possible errors in transcription and measurement. The proper identification requires
an approximate knowledge of the original measurement errors. A suitable technique
for the identification of a stars assumes a comparable measurement error for the
surrounding stars.

For the re-examination of the star identifications in the literature, every star
of the Bright Star Catalogue has been tested for its correlation to each set of

---

[14]-128/27 is the year of many Hipparchan observations reported in the Almagest.

[15]Hoffleit, D. (1982), *The Bright Star Catalogue*, Fourth Revised Edition, New Haven. The catalogue
covers nearly all visible stars and the stars of the Ptolemaic star catalogue should be a proper subset of
it. For that reason the identifications in Appendix B tabulate the catalogue numbering of the Bright Star
Catalogue, abbreviated as HR.

[16]Cf. Appendix B.

[17]Estimation communicated to me by Dr. Schwan, Astronomisches Recheninstitut Heidelberg.

coordinates and magnitude of the Almagest. Consideration was also given the fact that individual groups of stars show varying systematic errors in longitude and latitude. Possible transcription errors were taken into account as well. After that step, about 5% of the stars still correlated to another identification than Toomer's. After Ptolemy's description of the position of the star within the constellation had been reconsidered, almost all traditional identifications could be confirmed. Only some few cases, about 15, were either undecided or suggested a different identification.

One of these is star No. 836 (δ Mon) in the constellation Canis Major. In the following chart all stars of the constellation are plotted in ecliptical coordinates with their respective catalogue number.[18] The chart shows all Ptolemaic stars of the constellation and all stars of the Bright Star Catalogue, related to the epoch -128. The latter are shown by a dot of a size which represents the visual magnitude of the star. The correspondences between the two catalogued are indicated by a dotted line between the two positions. Due to the precession motion, the recalculated stellar positions of the earlier Hipparchan epoch are displaced by about 2°40′ in longitude from the positions of the Almagest.

Normally, the correspondence lines should only be oriented horizontally with a length equivalent to the precession motion minus the systematic error of longitude, for the ecliptical latitudes undergo negligible changes within these periods of time. The catalogue stars and the identifications in the configuration are recorded according to Toomer's edition of the Almagest.

In the case of the most northern star in Canis Major, a rather conspicuous deviation in latitude is noticeable for the star No. 836, indicating a faulty identification. Toomer follows the identification of Peters/Knobel, though he notes the variant of Manitius who identified the coordinates with the star 19 Monocerotis.[19] Comparing it with the other errors of the surrounding stars, Manitius' identification has to offer stronger evidence and the star reference is changed accordingly in Appendix B. The corrections to Toomer's identifications are made as conservatively as possible and they are restricted to only the clear alternatives.

For further corrections and possible improvements in the text editions of the Almagest, a set of charts replete with all the stars of the Almagest and the surrounding stars of the Bright Star Catalogue is printed in Appendix A. With these aids, almost all stars in the constellations can easily be identified.

---

[18]We chose a Cartesian coordinate system which is easier to evaluate for identification problems, though it does not properly reflect the visual impression of the sky.

[19]Ptolemy, C. (1984), p. 387. This star is plotted as a star of magnitude 4 on the right below the catalogue star.

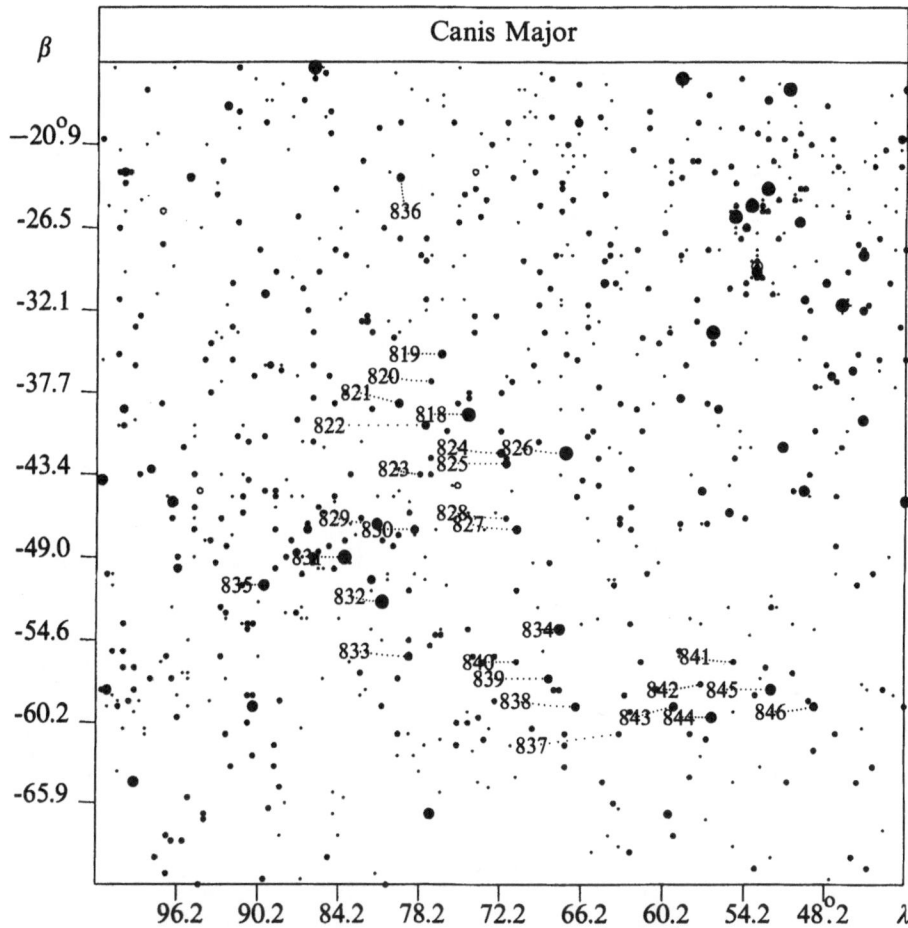

Figure 4.1: Identification Chart for the Constellation Canis Minor

4.1.4 Errors in the Almagest

After these corrections have been made there still remains a small group of stars whose longitude, latitude, magnitude, or positional description do not properly accord with any accurate data within the narrow limits of tolerance. Important cases are the stars π Hydrae and θ Eridani. The coordinates of both stars deviate considerably from the calculated positions, π Hydrae by $-4°.86$ in latitude and $4°.47$ in longitude (in reference to the epoch -128) and θ Eridani by $0°.37$ in latitude and $6°.81$ in longitude. In neither case are text variants known to us, nor are correlations with other recalculated positions available.

In such cases one has to assume an error made either in the original manuscript or during the measuring which had deformed the value to such an extent. These peculiar cases of stars with significant errors play an important rôle in the historical interpretation of the catalogue.

4.2 Criticism of Vogt

Vogt's work on the reconstruction of the Hipparchan star register stands in the center of the arguments against the allegation of fabrication. More specifically, it is the only contention arguing that a major portion of the star catalogue of the Almagest had been authentically observed by Ptolemy himself.

At the very most Boll's general estimation of the size of the Hipparchan register supports the thesis that at least one fifth of the Ptolemaic catalogue cannot possibly be attributed to a Hipparchan catalogue later than the Aratus Commentary. Only Vogt's analysis manages to make a broader claim. Quite understandably, Vogt's word is repeatedly cited in the attempt to establish the Ptolemaic origin of the star catalogue in the Almagest. The general view of Ptolemy's star catalogue was summarized as: "a critical trans. of Ptolemy's star catalogue containing detailed comparisons with modern computed positions was made by C.H.F. Peters und E.B. Knobel ... This adopts Delambre's erroneous conclusion about Ptolemy's dependence on the (hypothetical) star catalogue of Hipparchus, which was refuted by H. Vogt".[20]

Newton's remarks about Vogt's work are superficial and restrict themselves to gross speculations. A careful perusal of Vogt's article, however, exposes crucial defects in it.

4.2.1 Vogt's Interpretation of Delambre's Precession Table

Vogt accuses Delambre of committing an unwarranted fallacy in his calculation of the Ptolemaic precession constant proving a selection of data by Ptolemy. To calculate the precession constant from the data of the Almagest, Delambre employs the equation

$$p = \frac{(\delta_2 - \delta_1)\cos(\frac{\delta_2+\delta_1}{2})}{n\sin\epsilon\cos\beta\cos(\frac{\lambda_2+\lambda_1}{2})}. \tag{4.1}$$

With the sole exception of the ecliptical longitude of the star λ_1 at the time of Hipparchus, all other data can be taken from the Almagest. To calculate the precession p, Delambre had to extrapolate the ecliptical longitude of the star for the Hipparchan epoch from the Ptolemaic data. He obtained the longitude by subtracting 2°40′ from the catalogued longitude of the star, in other words, by exactly the amount with which Ptolemy increased the Hipparchan longitudes in the view of Delambre.

For Vogt, this step involves a fallacy with *petitio principii* and to that he adds rather mournfully: "... it is really a shame that what emerges is not the truth but rather the hypothesis that had been introduced into the presupposition of the argument".[21] Finally, Vogt celebrates himself as an enlightened corrector of old errors: "It was necessary to refute these century-old errors, for they are alive and continue to be reproduced today. Even such a researcher as Manitius, who had done so much for the recognition and understanding of Hipparchus and Ptolemy,

---

[20]Toomer, G. J. (1975), p. 204.
[21]Vogt, H. (1925), col. 35.

thought that Delambre had successfully answered the question".[22] In no way do the arguments against Delambre justify Vogt's arrogance.

The first problem to arise is that it is not at all clear where Delambre is guilty of a *petitio principii*. To be sure, he assumes that the ecliptical longitude for the time of Hipparchus is calculated from the Almagest according to the method by which he holds Ptolemy to have constructed his catalogue. However, this assumption has no influence on the numerical result of the calculation.

(i) The best way of approximating the Hipparchan longitudes, taking into consideration the longitudinal errors of the coordinates in the Almagest, would be to subtract the value of precession minus the systematic error of the Almagest from the Ptolemaic longitudes. Even assuming that Ptolemy himself had observed the stars of his catalogue and had thereby obtained ecliptical longitudes which were too small because of the error in the solar theory, one can approximate the Hipparchan longitudes by subtracting $2°40'$. Delambre's procedure is independent of any interpretation of the origin of the Almagest.

(ii) It is not important whether the actual Hipparchan longitudes are exactly $2°40'$ smaller than the longitudes of the Almagest. Here Vogt manages to cover up the lack of a detailed error analysis.

In the equation (4.1), the Hipparchan longitudes λ_1 are entered on the right side, but only as mean value with the Ptolemaic longitude as argument of the cosine function. The argument $(\lambda_2 + \lambda_1)/2$ is the mean longitude of the star over the period between the declination measurements. The calculation of the precession is in general not sensible to a variation in the ecliptical longitude.

The variation in the precession Δp dependent on a error of longitude $\Delta\lambda_1$ is calculated according to

$$\Delta p = \Delta\lambda_1 \frac{dp}{d\lambda_1}, \tag{4.2}$$

i.e. the relative change of p is according to (4.1) and (4.2) with λ_1 in degrees[23], calculated as

$$\frac{\Delta p}{p} = \frac{\pi}{360°}\Delta\lambda_1 \tan(\frac{\lambda_2 + \lambda_1}{2}). \tag{4.3}$$

It becomes clear from the form of the tangent function that the calculation of the precession constant for longitudes close to the solstices – where the declinations hardly change – is plagued with inaccuracies. Otherwise it is relatively insensitive in the face of errors in longitude.

As an example of this we shall now turn to the error in Delambre's evaluation of the observations of η Ursae Majoris. The Ptolemaic longitude is $\lambda_2 = 149°.83$, consequently the assumed Hipparchan longitude is $2°40'$ smaller. If this value is incorrect by one degree, then the resulting precession constant would change only

---

[22]Vogt, H. (1925), col. 36.
[23]The factor $\pi/180$ is required for longitudes given in degrees.

by the very small amount of 0.5%. Therefore, the correctness of the Hipparchan longitudes is not so important.

It is all the more astounding, though, that Vogt sets off against Delambre's results his own calculations, to which he gives the heading "precession according to the correct calculations".[24] The article does not tell us, however, how the calculation in detail was made.[25] Since the equations (4.1-4.3) are deduced from the astronomical triangle PES, it is entirely unclear how Vogt calculates his precession constants.

According to the error estimation (4.3), it is insignificant whether Vogt in (4.1) enters the ecliptical longitudes according to Delambre's procedure or according to the values of the Hipparchan longitudes he had reconstructed. The Almagest notes for the star η UMa a longitude of $\lambda_1 = 149°.83$. While Delambre inserts the Hipparchan longitude $\lambda_2 = 147°.16$ in equation (4.1), Vogt could take as his point of departure the value of the reconstructed longitude $\lambda_2 = 145°.67$. Delambre obtains a precession constant of $35.88''/y$ and Vogt would have had to calculate a precession of $36.46''/y$ with his reconstructed ecliptical longitude.[26] Instead, he tabulates a "correct" value of $50.81''/y$, of unknown origin. Relying on the results for the remaining stars which were obtained in the same way, Vogt contends that Ptolemy truly presents an exemplary set for the demonstration of the faulty constant of one degree per century. Had Ptolemy evaluated all the data with modern methods, he would, according to Vogt, have obtained just a confirmation of the previous result. Therefore it was legitimate for Ptolemy to select the 6 examples as representative demonstrations. Just like Vogt's calculation in the individual cases, the overall argument is without foundation.

The precession values which emerge from the declination differences of the Almagest are represented in the last three columns of the table (4.2).[27]

Newton stated that the resulting values are not distributed normally on the basis of the declination differences between Hipparchus and Ptolemy and he concluded that Ptolemy had fabricated the declinations of the six stars selected, instead of selecting them out of whole series of observations.[28]

The histogram in (fig. 4.2) shows the frequency distribution of the derived precession constants from the declination differences between Hipparchus and Ptolemy

---

[24]Vogt, H. (1925), col. 36.

[25]"If one assumes with Delambre, and without auxiliary assumptions one cannot solve the problem, that Ptolemy's latitude did not change and that they are known, then one needs no further assumption. From a simple conversion of the astronomical triangle PES with given ϵ, δ, β, one now obtains values for λ and p, which have nothing in common with Delambre's values. I juxtapose these calculations to the Delambre's results, ordered according the size of the true values". Vogt, H. (1925), col. 35. Vogt refers to the spherical triangle PES, defined by the equatorial and ecliptical poles P and E and the star S.

[26]Apparently Delambre uses Ptolemy's obliquity of the ecliptic $\epsilon = 23°.86$. With the accurate value of $\epsilon = 23°.71$ one obtains p=$36.17''/y$.

[27]The precession is calculated according Delambre's formula with Ptolemy's obliquity of the ecliptic $23°.856$ instead of the accurate $23°.71$. The time differences between the observations are 144 years (Timocharis–Hipparchus), 265 years (Hipparchus–Ptolemy), and 409 years (Timocharis–Ptolemy). λ, β are the ecliptical coordinates of the Almagest; δ declinations; T-H, H-P, T-P are the columns with the resulting precession constants (in seconds per year) for the declination differences between Timocharis (T), Hipparchus (H) and Ptolemy (P). For calculating the mean values the stars close to the solstices are excluded. Their corresponding precession constants show large errors (The stars No. 288, 818, 424, 425). The six stars selected by Ptolemy for the demonstration (indicated by bullets) are No. 411, 222, 736, 510, 35, 110.

[28]Viz. section 3.6.2.

Name	No.	HR	δ_{T-A}	δ_{Hipp}	δ_{Ptol}	λ	β	T-H	H-P	T-P
				Evaluation of the Declination Differences						
α Aql	288	7557	5.80	5.80	5.83	273.83	29.17	0.00	22.16	21.75
η Tau•	411	1178	14.50	15.17	16.25	33.67	3.67	46.32	41.64	43.30
α Tau	393	1457	8.75	9.75	11.00	42.67	-5.17	78.83	55.64	63.99
α Aur•	222	1708	40.00	40.40	41.17	55.00	22.50	32.72	36.18	34.89
γ Ori•	736	1790	1.20	1.80	2.50	54.00	-17.50	60.90	41.07	48.30
α Ori	735	2061	3.83	4.33	5.25	62.00	-17.00	61.40	66.72	64.68
α CMa	818	2491	-16.33	-16.00	-15.75	77.67	-39.17	91.68	45.66	63.79
α Gem	424	2891	33.00	33.17	33.40	83.33	9.50	49.74	49.97	49.74
β Gem	425	2990	30.00	30.00	30.17	86.67	6.25	0.00	68.03	35.53
α Leo	469	3982	21.33	20.67	19.83	122.50	0.17	78.99	50.46	59.99
α Vir•	510	5056	1.40	0.60	-0.50	176.67	-2.00	49.85	37.07	41.55
λ UMa•	35	5191	61.50	60.75	59.67	149.83	54.00	45.86	35.78	39.26
ζ UMa	34	5054	67.25	66.50	65.00	138.00	55.67	46.20	50.06	48.75
ε UMa	33	4905	68.50	67.60	66.25	132.17	53.50	56.16	45.26	48.96
α Boo•	110	5340	31.50	31.00	29.83	177.00	31.50	31.20	39.84	36.80
α Lib	529	5531	-5.00	-5.60	-7.16	198.00	0.67	38.12	54.50	48.6
β Lib	531	5685	1.20	0.40	-1.00	202.17	8.83	52.77	51.08	51.66
α Sco	553	6134	-18.33	-19.00	-20.25	222.67	-4.00	50.61	53.19	52.23
						μ	(14)	52.14	47.04	48.79
						σ		14.52	9.01	9.36
						μ	(Ptol)	44.47	38.60	40.68
						σ		11.12	2.57	4.82

Table 4.2: Determination of the precession constant.

(H-P).[29]

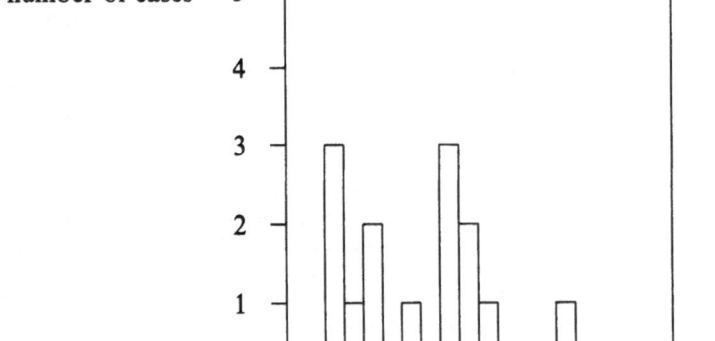

Figure 4.2: Precession H-P.

Two clusters can be recognized, indicating a possible selection of data. What the distribution does not show, however, is that Ptolemy had fabricated the declination

---

[29]The histogram covers 14 values between 30° and 80° and is divided into 20 intervals. The mean value is $\mu = 47°04$ and the standard deviation $\sigma = 9°01$.

differences of the chosen stars. Quite the contrary, it is just as possible that Ptolemy, or Hipparchus, had added other observations to the six declination differences chosen in order to get a sample of observations distributed over the whole celestial sphere. For this reason, stars like α Aql and α Gem are included in the list of declination observations, though they are not suitable for the demonstration of the precession constant. They do serve to demonstrate the effect of precession on the declinations of the stars in the various parts of the sky and to prove that the precession motion takes place around the pole of the ecliptic and not around e.g. the equatorial pole. The Almagest makes clear that this question had been raised by Hipparchus and that for this purpose he had analyzed the declination changes of the stars from all regions of the sky.[30]

> "Now Hipparchus agrees with [the idea of] the motion taking place about the poles of the ecliptic. For in 'On the displacement of the solsticial and equinoctial points' he deduces from the observations of Timocharis and himself that Spica (again) has maintained the same distance in latitude, not with respect to the equator but with respect to the ecliptic, being 2° south of the ecliptic both earlier and later periods. That is why in 'On the length of the year' he assumes only the motion which takes place about the poles of the ecliptic, although he is still dubious, as he himself declares, both because the observations of the school of Timocharis are not trustworthy, having been made very crudely, and because the difference in time between [Timocharis and himself] is not sufficient to provide a secure result. We, however, find the [latitudinal distances with respect to the ecliptic] preserved over the much longer interval [down to our times], and that for practically all fixed stars.

In the following passage Ptolemy describes the various changes of the declinations.[31]

> "In order to illustrate this point for a few easily recognizable stars we will set out, for each of the two hemispheres mentioned, their vertical distances from the equator, as measured along the great circle through the poles of the equator, as recorded by the school of Timocharis, as recorded by Hipparchus, and also as determined in the same fashion by ourselves".

Ptolemy resorted to the now lost Hipparchan work "On the displacement of the solsticial and equinoctial points" as a source for the data of Hipparchus and Timocharis and added, he tells us, his own declination measurements to them. Hence the list of declination variations should not be interpreted as a record of equally significant observations from which the individual precession constants can be evaluated. Just as Hipparchus recorded the whole set of stars and the observations, so he could also select the data for the demonstration of the precession constant. The

---

[30]Ptolemy, C. (1984), p. 329.
[31]Ptolemy, C. (1984), p. 330.

histogram of the observations between Hipparchus and the school of Timocharis reveals a distribution of the resulting precession constants similar to that between Hipparchus and Ptolemy (fig. 4.3), though this time Ptolemy had no reason to "tune" the declinations into a certain direction.

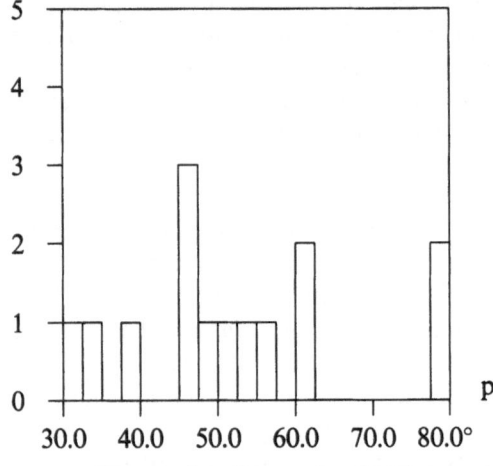

Figure 4.3: Precession T-H.

Newton's allegation that Ptolemy had mixed actually measured data with fabricated ones is groundless.[32] In that case, both the Hipparchan data and those of Timocharis must have been similarly fabricated: a claim for which no proof or plausible motive exists. Quite the contrary, the evaluations are confirmed through the Hipparchan statement that the precession has a value of *at least* one degree per century. Therefore one can conclude that Hipparchus must have likewise calculated values of the precession which come close to $36''/y$.

The distribution of all the declination differences (fig. 4.4) suit the accurate value for the precession constant of about $50''/y$. Ptolemy, for his appraisal, had simply relied on what he considered the most trustworthy observations, namely those of Hipparchus and those of his own; from these he chose the six stars that best corresponded to his hypothesis and that received their validity from this fact.[33]

4.2.2   Reconstruction of Coordinates

Vogt reports that he collected from all the available sources 881 Hipparchan numerical data for 374 stars from which he can derive both coordinates for 122 stars "without a supplementary hypothesis."[34] This reconstruction, however, is decidedly not completely free from supplementary hypotheses, as he would have us believe. The geographical latitude of the observation site as well as the time of the observations are, among other things, incorporated in the equations.

---

[32]Newton, R. R. (1977), p. 225.

[33]The histogram covers 42 values between 30° and 80° in 20 intervals. Their mean value is $\mu = 49°.32$ and the standard deviation is $\sigma = 11°.18$.

[34]Vogt, H. (1925), col. 18.

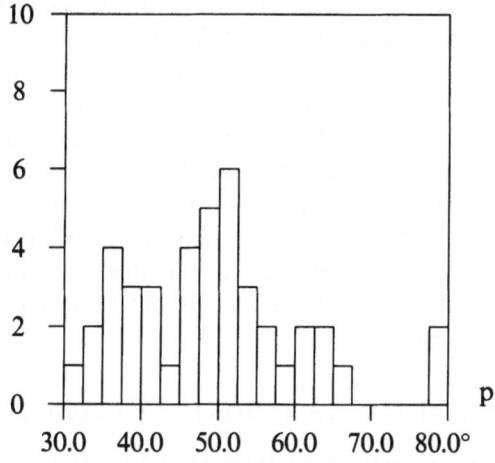

Figure 4.4: Precession, all declinations.

Vogt recognizes a chronological layer in the coordinates. For the 77 stars of the "actual" Aratus Commentary, after leaving out the cases at the extremes, he derives the year −150 ± 3.77.[35] From 18 stars of the star clock at the end of the appendix, Vogt obtains a mean time of observation of −130 ± 7.45, after eliminating 10 stars with more extensive errors. These chronological differences within the Aratus Commentary would correspond to a displacement of the reconstructed ecliptical longitude of about 15′. In the table containing the reconstructed longitudes at the end of his article, Vogt fails to supply any information about the data used in figuring the individual coordinates. In most of the cases the data of the Aratus Commentary are overdetermining for the reconstruction of the coordinates, i.e. not all information is required to derive the coordinates of the star.

From the case of π Hydrae, which is meant to serve as an example, we can see that Vogt takes mean values of intermediate parameters for the reconstruction of the ecliptical coordinates. If for example the complete set of data lead to two differing arcs of the ecliptic above the horizon, dependent on the selection of input data in Vogt's reconstruction procedure, then the mean of both values serves as parameter in the further calculation.[36] Vogt's article informs us neither about the extent to which data from various epochs are combined in the reconstructions, nor at which point mean values are taken for overdetermining data. The fact that data are combined in such a way has the consequence that the resulting coordinates would at best only approximate to any possibly existing Hipparchan coordinate.

Besides the uncertainty concerning the epoch of observation, one has to take into account the variation in the related geographical latitude. Hipparchus relates

---

[35]Vogt, H. (1925), col. 31. The "actual" Aratus Commentary in the view of Vogt seems to refer to the first major part of the book, where Hipparchus criticizes Aratus and Eudoxus: "Hipparchus' Commentary on the Phenomena of Aratus and Eudoxus is preserved and with it many data of stellar positions which we took out of this treatise. But the richest and most important parts for us are the two appendices. The first is a treatise on 42 constellations referring to Rhodes ($\varphi = 36°$); the second a star clock from hour to hour with culminations of 43 stars connected to the full and half signs of the zodiac". Vogt, H. (1925), col. 17.

[36]Vogt, H. (1925), cols. 18f. Vogt's reconstruction is also described in Neugebauer, O. (1975), pp. 281f.

the data of the main part of the Commentary on Aratus and Eudoxus to the geographical latitude of Athens ($\varphi = 37°$), while all of the entries in the second part hold true for the latitude of Rhodes with $\varphi = 36°$. In the calculation he gives as an example, Vogt works with the latitude of Rhodes, but he never mentions whether the data from both major parts of the Commentary are used in the calculation and therefore introduce a supplementary error into the reconstruction.

It is difficult to estimate the accuracy of Vogt's reconstructions, in particular with respect to the two potential sources of error just mentioned. Vogt himself does not provide any error estimation for his coordinates, something which is highly significant for assessing the question whether or not these are genetically identical with the stars of the Almagest.

4.2.3   The Accuracy of the Reconstructed Coordinates

Vogt tabulates the reconstructed coordinates with an accuracy of 1/100 degree. In contrast, the initial numbers of the Aratus Commentary are only exact to a full or half degree. It frequently happens e.g. that Hipparchus assigns slightly differing values for simultaneous culminations to the same star at different places in the Commentary.

For star β Cnc one finds three longitudes differing about $1°.5$, and even still larger differences can be found in other places.[37] To judge by the frequency with which diverging entries for the same phenomenon in the Aratus Commentary are found, the Hipparchan values cannot possibly be the product of a rigorous derivation from a fixed star register. One is compelled to the view that Hipparchus had determined the data in the Commentary with the help of a globe, in the parts where he disputes against Aratus and Eudoxus with no high standards of accuracy. Besides the aid of a globe Hipparchus must have had more exact sources for the original second part of the Commentary, on which Vogt's reconstructed coordinates are essentially based. Hipparchus introduces the second part with the words:[38]

> "In addition we will accurately describe for every constellation:
>
>   (i) The degree of the zodiacal sign standing in the meridian.
>
>   (ii) For every constellation, the fixed stars which culminate simultaneously with the rising and setting of the beginning and end of the constellation.
>
>   (iii) In how many equinoctial hours each constellation rises or sets.
>
>   We will describe each of these points approximately up to the point of an insignificant difference."

Hipparchus' additional remark that the numbers are to be given "approximately up to the point of an insignificant difference" could be understood as a reference to the limited accuracies of the figures compared to the later standard of the Almagest. A careful astronomer like Hipparchus could never have failed to realize when the same phenomenon, such as the simultaneous culmination of a star, is recorded

---

[37] Hipparchus, (1896), p. 187, p. 233, p. 267.
[38] Hipparchus, (1896), p. 184.

in different places in the book with different values – unless he used a globe which is not capable of precise reproduction of the phenomena. Accordingly, Vogt's reconstructed stellar coordinates are not obtained from data which are derived in a rigorous calculation with the highest level of accuracy. Rather, they are founded upon values which have only an indirect relationship to the Hipparchan fixed star register.

Furthermore, Vogt needs for his reconstruction more than just one of these data for either the rising, setting or culmination longitudes. He but rather has to combine at the very least two different types of phenomena.[39] Since the accuracy of the coordinates does not increase in the process, it must be assumed that the resulting coordinates can easily deviate from the original positions used by Hipparchus for the compilation of the Commentary. Taking this into consideration, one can only be astonished at Vogt's "proof" of the independence of the two star registers.

4.2.4 Vogt's Proof of Independent Observations

If Ptolemy had based his catalogue on a Hipparchan fixed star register with ecliptical coordinates, the ecliptical latitudes of the stars from the Almagest would have to be identical with those of the Hipparchan register and the longitudes increased by 2°40′ for the precession. In that case, the Hipparchan coordinates would have to be catalogued with the same accuracy of 1/6 or 1/4 degree as in the Almagest. If the Hipparchan fixed star register had recorded the stellar positions in another coordinate system, e.g. in declinations and culmination times, then Ptolemy would have had to transform them. But even then the Hipparchan coordinates should be about as accurate as the coordinates of the Almagest.

Had Hipparchus really used these coordinates in order to reckon the positions of the stars in the Aratus Commentary, or used the globe to fix them, they would have been rounded off to one or a half degree, by some process of which the mathematical or graphical method and the implied inaccuracies are unknown. Over and above the inaccuracies inherent in the original data, the uncertainty in Vogt's reconstruction procedure itself generate a further dispersion of the reconstructed coordinates. Assuming that Ptolemy did take over the coordinates from Hipparchus, it still would not be assured that Vogt's reconstructions would coincide with the values of the Almagest.

In spite of these scatterings, a number of similar deviations would have to become manifest as a result of the differences of the reconstructed coordinates and the Ptolemaic coordinates from the accurate positions of the stars. Should, for example, a star have been wrongly measured or recorded incorrectly by Hipparchus, then these coordinates as well as their corresponding ones in the Almagest would have to show similar deviations from the accurate value. Statistically speaking, the discrepancies of the coordinates of Hipparchus and Ptolemy should show correlations.

Vogt realizes this crucial test for the genetic origin of the Ptolemaic coordinates, and he introduces the correlation table in his article with the words:[40]

"Had Ptolemy then, as Delambre wants to see it, taken over the

---

[39]Cf. details of the calculation in Vogt or Neugebauer, O. (1975), pp. 281f.
[40]Vogt, H. (1925), col. 23.

Hipparchan latitudes without change, had he set $b_2 - b_1 = 0$, and had he increased the Hipparchan longitudes by $2°.67$, namely $l_2 - l_1 = 2°.67$, then it must be correct to depict the results as

(i) $(\beta_2 - b_2) \approx (\beta_1 - b_1)$.
(ii) $(\lambda_2 - l_2) \approx (\lambda_1 - l_1) + 1°$."

If the sign "$\approx$" stands for the correlation between the differences, then both (i) and (ii) are the criteria for the reconstructed Hipparchan coordinates being the original source for the coordinates of the Almagest.

In what follows, Vogt speaks of the "equality" of the differences. In doing so he uses a stricter criterion without any legitimate justification.[41]

"In fact, the Ptolemaic errors in latitude are not equal to those of Hipparchus and the Ptolemaic errors in longitude are not larger than the Hipparchan errors by even one unit: the Tables I and II teach us this...

When we divide the errors into intervals of 9 groups, we obtain the following:

error interval	number of stars in the interval		
	Hipp.	Ptol.	shared errors
$[-0.10°, +0.10°]$	15	32	7
$[+0.10°, +0.33°]$	17	30	5
$[-0.10°, -0.33°]$	17	32	5
$[+0.33°, +0.67°]$	15	14	3
$[-0.33°, -0.67°]$	18	11	3
$[+0.67°, +1.50°]$	12	4	0
$[-0.67°, -1.50°]$	20	5	2
$> +1.50°$	4	1	1
$< -1.50°$	4	2	0

Table 4.3: Common error intervals, latitude.

In the narrow limits of error of $\pm0°.33$, the Ptolemaic errors in latitude predominate with an amount of $85 = 70\%$, while only $49 = 40\%$ of all Hipparchan stars belong here. On the other hand, the Hipparchan stars are far more strongly represented in the intervals of large error. About 2/3 of all Ptolemaic errors in latitude deviate so sharply from the corresponding Hipparchan values that copying seems to be ruled out. All of this, then, makes up the invincible counter evidence against the theory of Delambre."

Although Vogt discovered in the case of seven stars a similarly large deviation in latitude by both Hipparchus and Ptolemy, indicating that the latter had bor-

---

[41] Vogt, H. (1925), col. 23.

rowed them from the former, these instances, in his opinion, are actually only the exceptions:[42]

> "In these cases of the agreement in rather large errors – there are perhaps ten of them – it would be difficult to believe solely in the influence of accident. Most likely, the agreement can be explained in no other way than that Ptolemy or his colleagues had really borrowed the stellar latitudes from Hipparchus."

This admission found no resonance in the later reviews of his article, but Vogt's declaration that the deviations present an invincible counterargument to Delambre's theory has been received with open arms. It has been overlooked that his argumentation collapses under closer scrutiny.

The arrangement of the errors in groups, by means of which the independence of the coordinates is supposed to be proven, is absolutely unsuitable for a correlation test. Vogt tries to impress us with differences that are not statistically significant; his overall method obscures the situation to the reader.

Vogt's statement that the striking differences rule out borrowing in the case of 2/3 of all errors in latitude can only refer to the totals of stars which he finds in the chosen error intervals. There, only 26 out of 122 stars have an error in latitude which is classified in the same group for the Hipparchan and for the Ptolemaic deviation. As Vogt construes this, the possibility of copying can exist only when e.g. a Hipparchan latitude shows an error of $+0\overset{\circ}{.}20$ relative to the accurate latitude, and a Ptolemaic latitude deviates from reality by $+0\overset{\circ}{.}11$. In this case, they would be counted as elements of the same error interval in the third column of the table. On the other hand, should the Ptolemaic latitude deviate from the accurate position by $0\overset{\circ}{.}09$, then the coordinates fall into different classes and are no longer considered as genetically identical.

Here the absurdity of the method becomes quite obvious. Even a minimal difference of $0\overset{\circ}{.}01$ between the latitude differences suffices, to argue with Vogt, to cause the "error in latitudes to show such large deviations that a borrowing appears to be ruled out". Now, had Vogt chosen to keep the division into groups small enough, then there would be no Hipparchan star which was borrowed by Ptolemy if we accept Vogt's line of reasoning. The star α Persei, for example, exhibits a Hipparchan latitude difference of $-0\overset{\circ}{.}15$ and a Ptolemaic difference of $-0\overset{\circ}{.}08$. Consequently it is counted in different error intervals, which is statistically an insignificant difference. Representative for a whole group of stars is τ Psc, which shows with $\Delta\beta_1 = -1\overset{\circ}{.}70$ and $\Delta\beta_2 = -1\overset{\circ}{.}02$ similarly large differences, indicating that Ptolemy had taken them over, although their absolute difference in error is about $0\overset{\circ}{.}7$. The star π Hydrae has the largest latitude differences of $\Delta\beta_1 = 4\overset{\circ}{.}47$ and $\Delta\beta_2 = 4\overset{\circ}{.}85$. Its coordinates are rightfully classified by Vogt as genetically equal, although the respective errors differ about $0\overset{\circ}{.}4$. Vogt does not seem to notice that these differences lead his classification scheme *ad absurdum*.

Not only is Vogt's classification procedure untrustworthy, even his estimation of the number of independent coordinates in relation to the genetically identical stars is false. Vogt contends that the shared differences of over half a degree in latitude

---

[42]Vogt, H. (1925), col. 24.

for seven stars are so large that one can exclude an independent observation of the coordinates.[43] These seven stars do not even include the previously mentioned case of the star τ Psc. As far as Ptolemy's stars go, only 24 of them have an error in latitude which is greater than $0°.5$. In relation to this set of stars, then, the identity of one third of the stars is proven by applying Vogt's strict criterion of identity. It is therefore all the more astonishing when Vogt later diminishes the number of seven copied stars even further by the argument that for some of them no similarly detectable large and mutual error in longitude could be found.[44]

After that, Vogt reasons: "So sufficient testimony can be found in only five cases for the borrowing of both Ptolemaic coordinates from Hipparchus", though he says nothing about the size of the set of stars by which independence is proven on the basis of the same strict criterion. It does seem as if Vogt were merely looking for indications to show the autonomy of the two registers while, in the process, he curiously neglects to make any neutral evaluation of the data.

4.2.5   Statistical Test for Independent Data

A Simulation

A statistical test can analyze Vogt's claim of independent errors in the Hipparchan and Ptolemaic coordinates.

In a simulation the next figure (4.5) shows graphically how independent coordinates with the same standard deviation in their errors distribute around zero point.

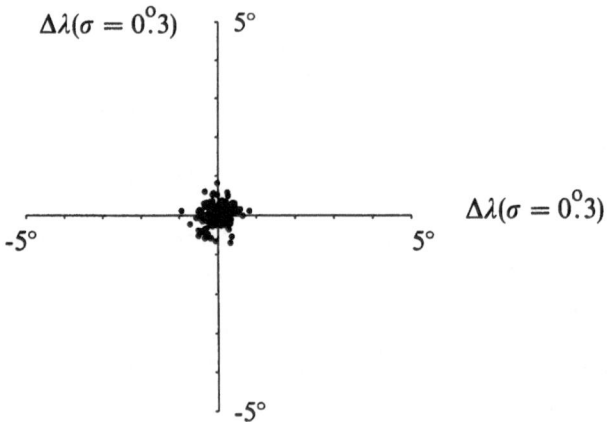

Figure 4.5: Errors of coordinates, uncorrelated (simulation).

Each point stands for a simulated coordinate error of a star, whereby the randomly generated numbers represent the differences of the Hipparchan and Ptolemaic latitude from the accurate position of the star. The points are evenly distributed over

---

[43] Vogt, H. (1925), col. 24.
[44] Vogt, H. (1925), col. 26.

all four quadrants of the coordinate system. Such a distribution is significant for independent errors of the coordinates, hence a proof for two different observational sources. This also holds true when one of the coordinate differences is dispersed to a larger extent around the mean value than the other. In this case (fig. 4.6), the points are on the average closer to one axis. Nevertheless, there are still approximately the same number of stars in all quadrants.

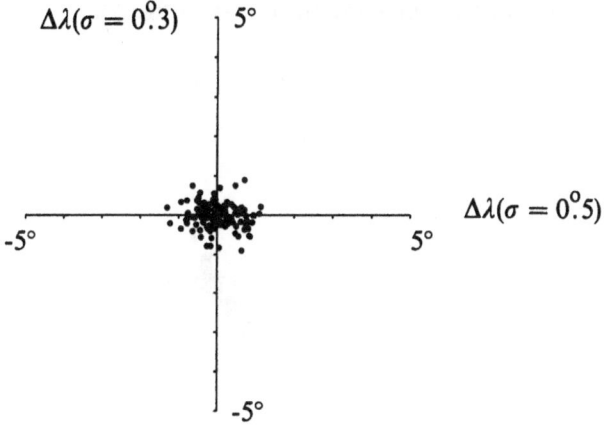

Figure 4.6: Uncorrelated coordinate errors with different standard deviations.

The distribution of these points would take on another flavor were the original Hipparchan catalogue to be found and if the Hipparchan latitudes coincide with the latitudes of the Almagest. In this case (fig. 4.7), all the related latitude differences are the same and all points lie on a straight line.

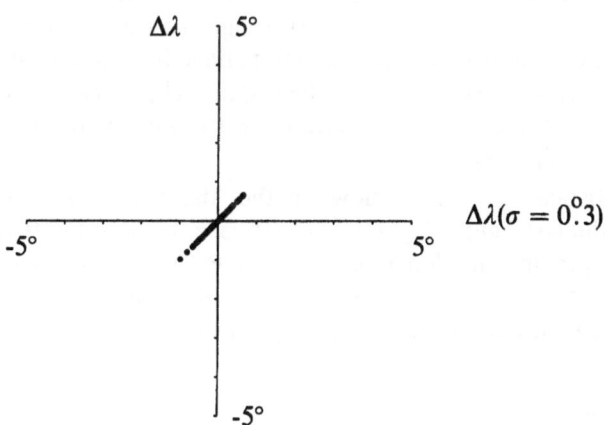

Figure 4.7: Strong correlation.

For the independent differences, one can find no straight line along which the distribution of points is aligned. In the second case all of the values lie on a straight line and show a positive correlation of 1.

Due to the Hipparchan calculation procedures in the Aratus Commentary and the uncertainty in Vogt's reconstruction procedures, one cannot assume that the reconstructed coordinate values are those possibly taken over by Ptolemy. If Hipparchus' initial data are equivalent to Ptolemy's, but the reconstructions cannot properly reproduce them as a result of the reconstruction techniques and the simplification in the Commentary, then they are scattered around the original values and the errors only align to the straight line as in fig. 4.8.

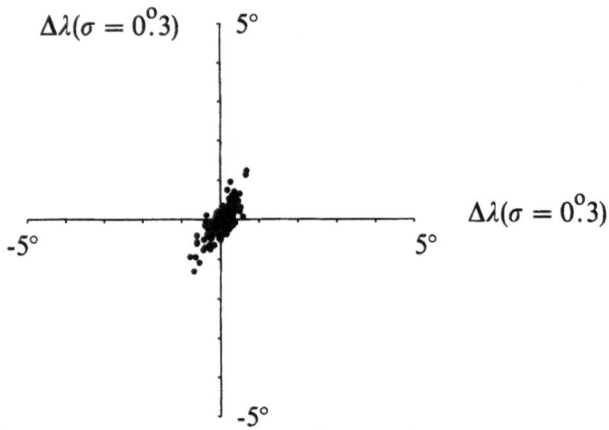

Figure 4.8: Correlating coordinate errors (Simulation).

The more the reconstructed Hipparchan coordinates differ from the original data, the more difficult it becomes to establish a significant correlation to the errors of the Ptolemaic coordinates. Then the correlation coefficient is less than 1. This situation is simulated in the next diagram. Although the distribution of points appears to be almost a random distribution, the points have been obtained by treating the random Ptolemaic errors as equivalent to the original Hipparchan ones and, proceeding from these values, superposing the random scattering for the reconstructed Hipparchan coordinates (fig. 4.9).

A statistical test has to show whether the distribution of the errors corresponds more to the last figure (fig. 4.9). In that case the coordinate differences would correlate, thus proving the dependence of the Ptolemaic values on those of Hipparchus. Otherwise, as represented in figure 4.6, the distribution is even in the four quadrants and thereby an autonomous origin could be considered probable.

Vogt's data

In the analysis of Vogt's data, the errors of the ecliptical longitudes are tested for correlation (fig. 4.10).

Contrary to Vogt's claims, the points show a significant correlation (r=0.37). The alignment of the points to the diagonal is suggestive. In this representation, the constant of $1°.19$ is added to the Ptolemaic longitudes to account for the systematic

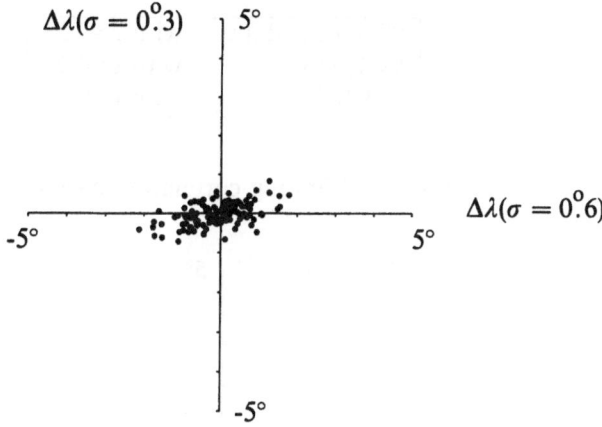

Figure 4.9: Correlating coordinate errors, with high standard deviation (simulation).

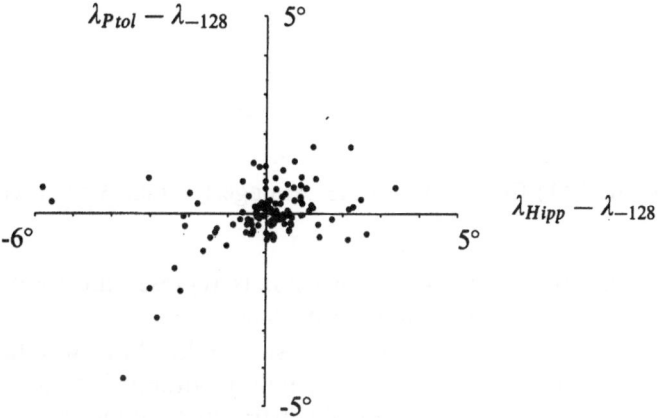

Figure 4.10: Errors in longitude Almagest versus Vogt's reconstruction.

error.

The correlation coefficient provides a measure for dependence among the errors. The numerical value of the correlation coefficient for a specific level of significance depends on the size of the sample. For 120 points a significant correlation is shown on the level of 95% probability when the correlation coefficient is greater than r=0.18. The mandatory lower limit increases with a higher level of significance:

Consequently the distribution of the differences in longitude are correlated with 99.9% significance.

A similar result is obtained for Vogt's errors in the ecliptical latitudes (fig. 4.11).

The computation yields a positive correlation of r=0.47 and here we must also assume a dependence with 99.9% significance. In the upper right-hand corner of the plot one finds the point for the star π Hydrae, which Vogt concedes that Ptolemy had taken over. Once more, one can recognize an orientation of the points along

99% probability	with r>0.23
99 1/3%	with r>0.27
99.9%	with r>0.30

Table 4.4: Limiting correlation coefficients.

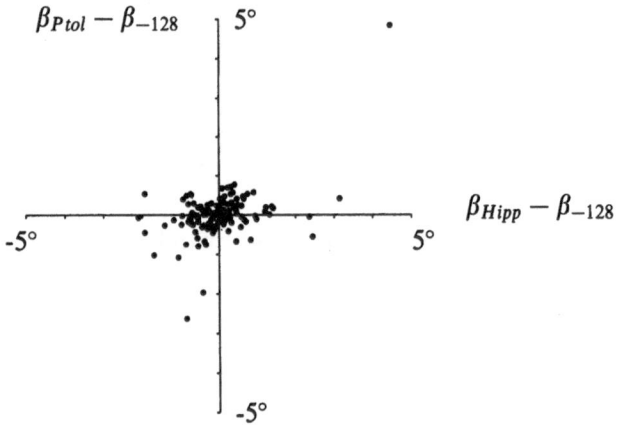

Figure 4.11: Errors in latitude, Almagest versus Vogt's reconstruction.

the diagonal, although there are also points representing formidable errors in the Hipparchan latitude with minimal Ptolemaic differences.

As a supplement, Pearson's Phi-Test should show whether this distribution contains significant correlations.[45] The null hypothesis is tested, that the differences in latitude are asymptotically normally distributed around the mean value $\mu_\varphi=0$ with a standard deviation of $\sigma_\varphi = \sqrt{1/N}$, as is to be assumed in the case of independent coordinate errors.

With the χ^2-test the probability of the rejection of the null hypothesis is decided for

$$\chi^2 = \varphi^2 N \tag{4.4}$$

with

$$\varphi = \frac{(ad - bc)}{\sqrt{((a + b)(c + d)(a + c)(b + d))}}. \tag{4.5}$$

whereas a, b, c, and d are the number of errors Δ in the coordinates for the four possible combinations

[45]Lienert, G. A. (1973), *Verteilungsfreie Methoden in der Biostatistik*, 2 vols. and tables, 2. ed., Meisenheim, vol. I, p. 528.

count	Δ Hipp.	Δ Ptol.
a	$\Delta \geq 0$	$\Delta \geq 0$
b	$\Delta \geq 0$	$\Delta < 0$
c	$\Delta < 0$	$\Delta \geq 0$
d	$\Delta < 0$	$\Delta < 0$

Vogt's reconstructed latitudes are distributed as:

$$a=42 \; ; \; b=23 \; ; \; c=19 \; ; \; d=38 \; ; \; N=122$$

The resulting $\chi^2 = 11.9$ with one degree of freedom indicates that the null hypothesis has to be rejected with a probability of more than 99.8%. If the alternative to the null hypothesis states that Ptolemy had taken over all the stars by a simple mathematical transformation then this is justified with a very high degree of significance. There remains the possibility, which will be examined later, that a common systematic error in the measurements of both astronomers affected the resulting coordinates and hence led to a correlation of errors.

Nevertheless, one must consider whether Ptolemy had taken over a large portion of the coordinates and, in so doing, whether he corrected a subtotal of the Hipparchan values, especially the ones with the larger errors, or whether he took these over from other sources. When there are taken out from all the values those that Vogt had already clearly identified as of Hipparchan origin, the test shows with a=39, b=23, c=19, d=34 and N=115 that $\chi^2 = 8.87$, and there is 99% probability of a correlation of the errors in latitude. Even with this stricter test, Vogt's results are refuted.

There is no significant difference from a normal distribution and one has to assume that a coherent and large group of stars is measured or calculated in a way that the errors of the reconstructed Hipparchan coordinates and those of Ptolemy correlate significantly.

1/4 and 1/6 Degree Stars

The two groups of stars with 1/4 and 1/6 degree accuracy in the coordinates could reveal a different correlation with the reconstructed Hipparchan positions. Stars with an recognizable accuracy of 1/6 degree in the Almagest are catalogued with the fraction of the degree 10′, 20′, 40′, and 50′, and those with a degree fraction of 15′ and 45′ as 1/4 degree stars. The number of 1/4 degree stars with a reconstructed Hipparchan coordinate is so small that no safe statistical statement can be expected. The correlations in latitude as well as in longitude are not large at all for these stars. The error distribution in latitude is plotted in fig. 4.12.

The deviations from the accurate latitude show, as in the longitude, no significant correlation. The hypothesis stating that the 1/4 degree stars of the Almagest correlate with the reconstructed Hipparchan coordinates cannot be justified by a statistical interpretation of the positional errors.

Not so the 1/6 degree stars. Here the correlation is even more pronounced than for the reconstructions as a whole.

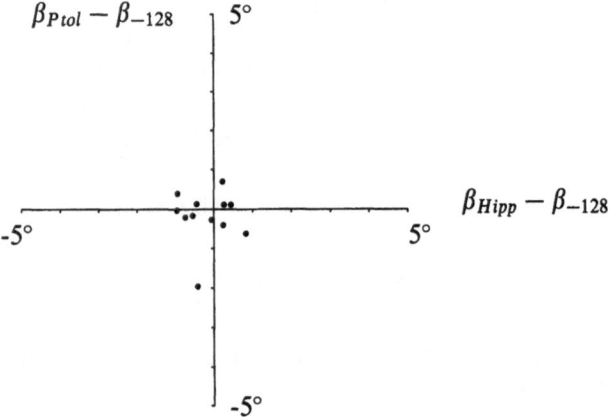

Figure 4.12: Errors in latitude, for 1/4 degree stars.

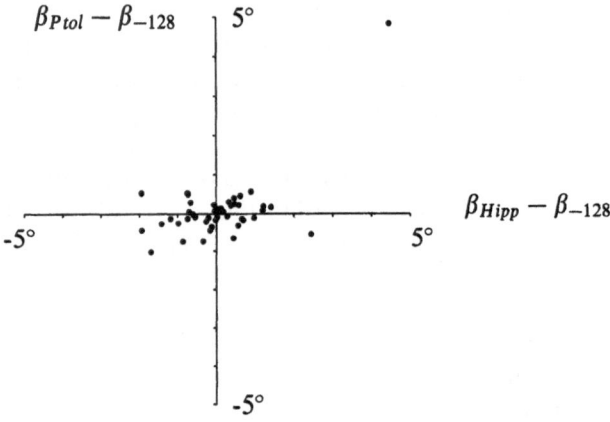

Figure 4.13: Errors in latitude, for 1/6 degree stars.

Figure 4.13 shows the deviations from the actual latitude for the recognizable 1/6 degree stars. The same strong correlation is found for the deviations from the accurate longitudes (fig. 4.14).

The only circumstances under which similar errors of the Hipparchan and Ptolemaic coordinates would not speak for a genetic identity would be if a general systematic error had been introduced in both observations. Nor can a periodic error of the solar theory, which could lead to systematic errors especially in the longitudes, explain the similarly large differences.

The highly significant correlation of the 1/6 degree stars of the Almagest with the reconstructed Hipparchan coordinates provides strong evidence that they do indeed stem from the same source. Both the examination of the differences in longitude as well as the differences in latitude show correlations which transform Vogt's result into its opposite. Not only does the sole argument on which the independence of

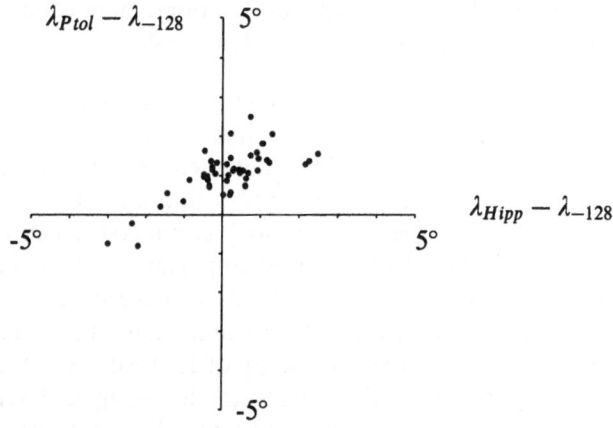

Figure 4.14: Errors in longitude, for 1/6 degree stars.

the Ptolemaic coordinates had been in large measure founded collapse: even more, Vogt's reconstruction supplies the material by which the dependence of the Ptolemaic values on those of Hipparchus can be demonstrated.

Newton argued against Vogt's proof of independence saying that it could actually tell us nothing about the relationship between the Ptolemaic and the Hipparchan registers, for his initial material stems from an early manuscript of Hipparchus and because Hipparchus could have repeated the measuring of the coordinates at a later date. The correlations illustrate in a convincing way that this suggestion is false. A major part of the reconstructions show such strong similarities with the Ptolemaic material that a complete and new observation of the Hipparchan fixed star register is ruled out.

4.2.6 Dating

For the dating of the reconstructed data, Vogt uses the variability of the coordinates as a consequence of the precession motion. Just like the ecliptical longitudes, so are the equatorial coordinates – the right ascensions and declinations – so strongly dependent on precession that it is possible to use them for dating. Vogt dates four groups of Hipparchan coordinates.[46]

(i) For "77 zodiacal and equatorial stars of the actual Aratus Commentary, excluding the star clock attached, the year -150 with a mean error of ±3.77 years is yielded as the mean time of observation of the right ascension".

(ii) "In contrast, 18 right ascensions of the star clock which are amenable to this strategy produce as mean observation time the year −130 ± 7.45".

(iii) "18 of the declinations taken from the actual Commentary yield the mean observation year −156 ± 10".

---

[46]Vogt, H. (1925), cols. 31f. It is difficult to reproduce Vogt's calculations, because he excludes values with "large errors" without further specification.

(iv) "In contrast, 16 Hipparchan declinations, handed down from Ptolemy and
Strabo, point to the mean year −130 ± 6.46".

It still remains unclear which distinct sections of the Aratus Commentary Vogt
relied on in the dating in (i). Vogt had maintained that the Hipparchan text
consisted of the "actual" part of the Commentary and two appendices, from which
the first is an "extract, related to Rhodes (latitude 36°), of the treatise on the 42
constellations".[47] When the "actual part" is understood in this manner, then Vogt
did not carry out a dating of the most important part of the Aratus Commentary,
namely, the second major part with the data concerning Rhodes.

Manitius divides the Aratus Commentary into three main parts: the first part
containing the critique of the astronomy of Eudoxus and Aratus (Book I and Book
II Chs. 1–3); the second major part with the rising and setting conditions of the
constellations (Book II Ch. 4 to end and Book III Chs. 1–4); and finally the appendix
with the star clock (Book III Ch. 5). This division is consistent with the passage
of Vogt discussed just above. Nevertheless, there is evidence that Vogt deviates
substantially from the customary terminology in his dating of different parts of
Hipparchus' work.

(i) A reading of the first major section of the Aratus Commentary in no way
turns up 77 right ascensions of the stars in the zodiacal and equatorial regions.
Vogt himself mentioned previously only " ... 22 real right ascensions of stars
lying on one or another parallel circle which are found dispersed from page
44 to page 150".[48]

Should it really prove to be the case that Vogt had only included the first
section of the work in his initial dating procedure, then for some inscrutable
reason he must have neglected to date the important second section, even
though it contains the most exact data upon which his reconstruction of the
Hipparchan coordinates is fundamentally built.

(ii) In his description of the section to be dated (i), Vogt explicitly excludes only
the star clock, not the second major section.

Therefore, it must be assumed that the first and second parts were dated together
in (i). It remains unclear why Vogt did not date the second part separately, since
it differs considerably from the first part as far as the geographical latitude of the
observation sites and the astronomical terminology. The material for the data of
the second section is more comprehensive than all of the others put together. Only
parenthetically does Vogt make mention of a dating of the right ascensions taken
from the second part of the Aratus Commentary.[49]

"In the second part of his Commentary, Hipparchus gives about 650
degree entries which are equivalent to right ascensions: 400 = 8/13 of
them in full degrees, 250 = 5/13 in half degrees. When one interprets
the full degrees as standing for the beginning of the degrees, as Manitius

---

[47] Vogt, H. (1925), col. 17.
[48] Vogt, H. (1925), col. 28.
[49] Vogt, H. (1925), col. 28.

did, one obtains, taking into consideration the true precession of the equinoxes, the mean time of those observations as -150, that is to say, in agreement with the entire character of the Commentary, before the discovery of the precession."

Independently of the direct dating of the numerical material taken from the Aratus Commentary, Vogt's reconstructed longitudes offer a means of dating.

Although the reconstruction combines heterogeneous sources, the figures from the second part of the Commentary are numerically more strongly represented. The longitudes, therefore, reflect the epoch belonging to this section.

Vogt computes the mean value of the reconstructed errors of longitude relative to the year -127, as $-0°.16 \pm 0°.08$.[50] These would have no systematic error for the time 11 years before -127, namely for -138,[51] after the errors in longitude exceeding a value of $\pm 2°.5$ are excluded from the sample.

Independent of Vogt's calculation, 167 simultaneous culminations can be dated from the comprehensive numerical material of the second part of the Aratus Commentary. From the simultaneous culminations M, one can calculate with the help of equation (4.6) the right ascensions α.

$$\tan \alpha = \tan M \cos \epsilon \qquad (4.6)$$

Their epoch is then given as the time t for which the calculated right ascensions best approximate the accurate positions. The distribution of values shown in histogram (fig. 4.15) has the mean value of the observation time of -131 and the standard deviation of 75 years for the individual values and 6 years for the standard deviation of the mean. While the mean value allows us to identify the second part of the Aratus Commentary as an early text, the large standard deviation points up to the uncertainty involved in the dating procedure. The scattering of these data does not allow an exact dating.[52]

Maeyama repeats the dating of ancient declination measurements and obtains results similar to Vogt's.[53] According to his findings the coordinates from the first part of the Aratus Commentary were compiled during the time from -140 to -156 and can be therefore dated before the Hipparchan texts which try to come to terms with the precession motion (presumably about -130). The Hipparchan ecliptical longitudes reconstructed by Vogt belong to the period before these writings (with an epoch of -138). With that, the existence of a comprehensive Hipparchan fixed star register prior to the discovery of the motion of precession is strongly supported. The dating of various Hipparchan data is summarized in table 4.5.

The examination of the Hipparchan stars reconstructed by Vogt reveals that a considerable number of them are based on the same measurements as the coordinates of the Almagest.

---

[50]The difference is calculated as $\Delta = \lambda_{Hipp} - \lambda_{-127}$.

[51]Vogt, H. (1925), col. 25.

[52]Vogt calculates from 18 declinations in the Commentary an epoch of -156 ± 10, after the exclusion of 7 numbers with exceptionally large errors.

[53]Maeyama, Y. (1984), Ancient Stellar Observations Timocharis, Aristyllos, Hipparchus, Ptolemy –; the Dates and Accuracies, *Centaurus* 27, p. 292.

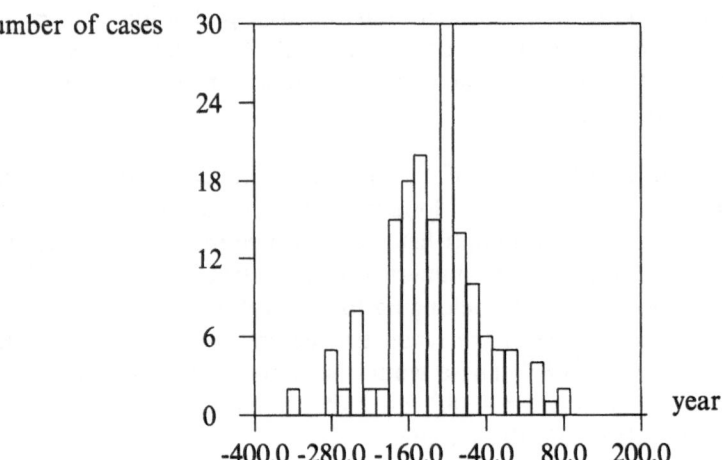

Figure 4.15: Epoch of coordinates in Aratus Commentary.

Data	Reference	Epoch	Author
Timocharis			
declinations	Almagest	-290±10	Maeyama
Aristyllos			
declinations	Almagest	-260±5	”
Hipparchus			
declinations	Aratus Commentary	−145 ~ −150	”
	Part I	±10	
declinations	”	-156±10	Vogt
right ascensions	Aratus Commentary	-156±10	”
right ascensions	Aratus Commentary	-130±7.45	”
	Appendix II		
declinations	Almagest	-130±6.46	”
declinations	Almagest	-130±5	Maeyama
ecl. longitudes	mainly Aratus Com.,	-138±5	Vogt
	Appendix I		
Ptolemy			
declinations	Almagest	+130±10	Maeyama

Table 4.5: Dating of the star data.

It remains unclear whether a systematic register of all stars with ecliptical coordinates like those in the Almagest existed. Notwithstanding that it is certain that Hipparchan coordinates were indeed taken over by Ptolemy.

Gundel's list of stars with ecliptical longitudes goes beyond Vogt's reconstructions in showing a correlation with the errors of the Ptolemaic coordinates. Besides the genetic identity of the Hipparchan coordinates with their counterparts in the Almagest, it also demonstrates that the Hipparchan coordinates had already been noted down and possibly, did not have to be converted in the first place by Ptolemy from other Hipparchan values.

4.3   Gundel's Stars

Just like the reconstructed coordinates of Vogt, the ecliptical longitudes of the stars in Gundel's list, which date to the time of Hipparchus, allow us to compare their errors with the Ptolemaic values and to test them for genetic identity.

Gundel proposed to classify the differences in longitude of the Hermes text from those of the Almagest into two groups. One group contains stars with longitudes with a difference of about 2°40′, the other contains differences with more than 3° compared to the longitudes of the Almagest. Besides the characteristics in the terminology of the star names, the few cases of "corrupted" coordinates, which Gundel could identify as simultaneous phenomena on the zodiac, provides the evidence that the coordinates represent authentic Hipparchan data. However, the two groups of longitude differences seduce Gundel into suspecting even older coordinates besides the Hipparchan longitudes.[54]

> "When one takes a glance at the longitudes given in our star catalogue and compares them with those that Ptolemy offers, one finds in every case smaller values; in fact, no single position is even as much as one degree more than the corresponding Ptolemaic value. We conclude from this that this catalogue is older than that of Ptolemy. When we distinguish amongst the given figures, we find 22 positions that are 2°20′ − 2°50′ smaller, they go back, therefore, to the time of Hipparchus or his student ... For 31 stars, this difference is even bigger, about 3° to 3°40′ less than the longitudes of the Ptolemaic star catalogue. This information stems from someone, then, who lived 335–370 years before Ptolemy ... They are traceable to the measurements of the astronomer Timocharis and Aristyllos and their school."

The Hipparchan origin of at least a part of the coordinates is proven beyond doubt. Gundel's method of dating the two classes of differences has been, quite rightly, criticized by Neugebauer on two grounds:[55]

(i) the longitudes in "Hermes" are only accurate to the full degree, while the Ptolemaic counterparts are catalogued with an accuracy of 1/6 degree. The two longitudes, then, cannot be compared neglecting the additional error resulting from the rounding of numbers. Neugebauer accordingly considers two possibilities: either the numbers were rounded to the next full degree or they were simply truncated to the full degree by leaving out the fraction of the degree. The last procedure is common to both Babylonian and Greek mathematics, so that Neugebauer favors the second method. This implies that all longitudes of "Hermes" should be smaller than the values from with they are derived.

(ii) The dating of the stars cannot use the longitudinal difference from the Almagest; rather, the coordinates have to be compared with the accurate positions. Neugebauer compares 59 values with the recalculated longitudes of

---

[54]Gundel, W. (1936), p. 131.
[55]Neugebauer, O. (1957), p. 68.

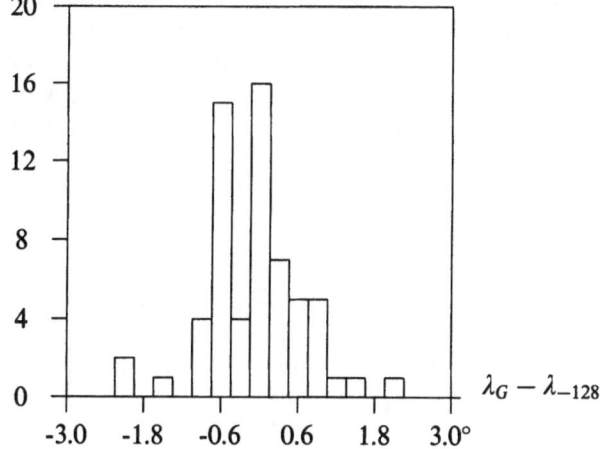

Figure 4.16: Histogram of $(\lambda_{Gundel} - \lambda_{-128})$.

Peters-Knobel and finds that 96.5% of them refer to the time between -130 and -60.[56] "Thus they were taken either from Hipparchus's catalogue itself or from the catalogue of an astronomer of the next generations. Gundel's hypothesis, however, of a star catalogue which preceded Hipparchus and which gave the positions in ecliptical coordinates is disproved."[57]

The longitudes of "Hermes" could date from one of Hipparchus's successors. It would explain the large number of coordinates whose longitudes are somewhat larger than is to be expected for the epoch of -130. Neugebauer finds support for this possibility through the finding of three successive stars in Gundel's list whose longitudes appear to be in another manuscript called "Hermes, De XV stellis".

We repeated Neugebauer's calculation using the accurate positions of the stars from Appendix B. The distribution of the differences in longitudes, relative to the epoch -128, is shown in fig. 4.16.

The distribution shows two significant maxima, one before and the other slightly after the Hipparchan epoch. The mean value of the differences is about $-0\overset{\circ}{.}28$, just below the mean value obtained from the reconstructed longitudes of Vogt. This congruence confirms Neugebauer's statement that "Hermes" probably contains genetically homogeneous material.

In proceeding with the question of whether the coordinates of the Almagest are identical with those from which Gundel's list was produced, it should be settled whether "Hermes" data stem from the supposed large Hipparchan register. In summary, there are two pieces of supporting evidence:

(i) The list of the stars of Hermes does not systematically use one coordinate system. In particular, four stars are contained that are not noted with ecliptical

---

[56] Hence, Neugebauer reckons two cases with larger differences.
[57] Neugebauer, O. (1957), p. 69.

coordinates. This mistake points to the carelessness with which the list is assembled. It proves that the copies of the text were transcribed without a later check for plausibility. Gundel proves that these numbers are the degrees on the ecliptic which simultaneously rise or culminate with the star. These data are identical with the figures Hipparchus gives in his Aratus Commentary.

(ii) Neugebauer suspects that three stars from Gundel's list are identical to stars from another Hermes text.[58]

Star	λ Hermes	λ Hermes
ϕ Per	15°	15°27'
β Per	27°	27°20'
α Tau	39°	39°28'

Table 4.6: Longitudes of Hermes "De XV Stellis".

The longitudes of the second Hermes also do indeed point to the time of Hipparchus. Their degree fractions are, however, presented with an accuracy which could not possibly have been observed. They must be read as conversions from other coordinates.

Many procedures can be imagined by which the Hipparchan coordinates of the stars were transformed into the ecliptical longitudes of "Hermes". Apart from a direct loan of ecliptical coordinates and their correction for the precession, it is conceivable that data of the simultaneous phenomena of the ecliptic, as they are contained in the Hipparchan Aratus Commentary, were converted for example by a globe to the ecliptical longitudes. It might explain the mixing of ecliptical longitudes with phenomena data. It would also explain why the longitudes of certain stars of "Hermes" deviate by more than one degree from those of the Aratus Commentary as well as from their corresponding Ptolemaic values.

The dating of the "Hermes" longitudes is dependent on which rounding procedure one supposes for generating the full degrees. The mean value of the longitudinal differences is -0°.28, relative to the epoch -128. Neglecting the effect of rounding, it would relate to a time earlier than that of the Commentary. The explicit and unambiguously late-Hipparchan terminology of the star names firmly excludes such an early period. Neugebauer's proposal, that the Hipparchan longitudes were first converted to a different epoch by a correction for precession, and the longitudes then truncated to the next smaller full degree, is to be pursued. Since a truncation diminishes the mean difference of longitudes by about half a degree, the data of Hermes would have to be on average half a degree larger before the truncation took place. Furthermore, an autonomous observation of coordinates after the time of Hipparchus is implausible. The carelessness with which the longitudes were compiled, which produced mixing of earth-related values among the ecliptical longitudes, as well as the complete lack of historical reports of astronomical observations in the first century B.C.E., clearly suggest Hipparchan authorship for the coordinates.

---

[58] Neugebauer, O. (1957), p. 69.

Seen in this perspective, the Hermes Trismegistos text appears as a list of ecliptical longitudes from the time of Hipparchus which, although they stem from the lost Hipparchan star register, do not completely coincide either with the data in the Aratus Commentary or with the reduced longitudes of the Almagest. This result is significant for the comparison of Hermes with the Almagest for, in the event of identical origin for the two registers, the longitudes of Hermes, provided one follows the suggestions put forth by Neugebauer, are exactly the Hipparchan ones that were truncated and to which Ptolemy added 2°40′. A Hipparchan star with a longitude of for example 97°40′ ought to be catalogued in the Almagest with 100°20′ and in Hermes with 97°. Even a superficial perusal of the list, though, refutes this possibility.

Suppose that the longitudes of Hermes were not obtained directly from a star catalogue composed of ecliptical coordinates, but via some intermediate procedure such as, for example, reading off of a globe. Then the longitudes of "Hermes" would not be numbers truncated from the Hipparchan coordinates, but would differ randomly from the original entries of a Hipparchan star register.

In the following simulation the truncation procedure is represented by the function $int(x)$ which just cuts off any fraction of a full degree, a normally distributed random variable $Err3$ with a standard variation σ_3 and the mean value $\mu = 0$. The longitudes of Hermes λ_{Her} and the reduced longitudes of the Almagest λ_{Ptol} are related by to the equation (4.7). To the original Hipparchan λ_{Hipp} the precessional adjustment of $0°35$ is added.[59]

$$\lambda_{Her} = int(\lambda_{Hipp} + 0°35 + Err3)$$
$$\lambda_{Ptol} = \lambda_{Hipp} + 2°67 \tag{4.7}$$

As we have seen in the case of Vogt's reconstructed Hipparchan coordinates, it is possible here too to compare both longitudes with each other and to test them for genetic identity. In the following diagram the difference in longitude of 56 stars from Gundel's list is correlated to the actual positions for the epoch -128 with the differences of the Ptolemaic longitudes for the same epoch. The deviations of Gundel's longitudes from the actual position $(\lambda_{Her} - \lambda_{-128})$ is on one coordinate, and the difference of the longitudes in the Almagest from the accurate positions of the epoch -128, $(\lambda_{Ptol} - \lambda_{-128})$ on the other (fig. 4.17).[60]

As in Vogt's case too, the differences between the longitudes correlate (here with r=0.51). The right side of the distribution of points clearly shows a greater average deviation from the accurate position of the year -128 than the side with the strongly negative difference of the Hermes longitudes. As with Vogt's reconstructed longitudes before, there is evidence that the investigated coordinates of the Almagest were not measured independently, but stem from Hipparchan sources.

The internal structure of the differences is apparent. The values are not homogeneously distributed over all quadrants in the diagram; they exhibit clearly discrete lines parallel to the diagonal. Such a distribution is generated when the values of

---

[59]The added constant for precession does not change the result of a regression test.
[60]In total 61 data could be used.

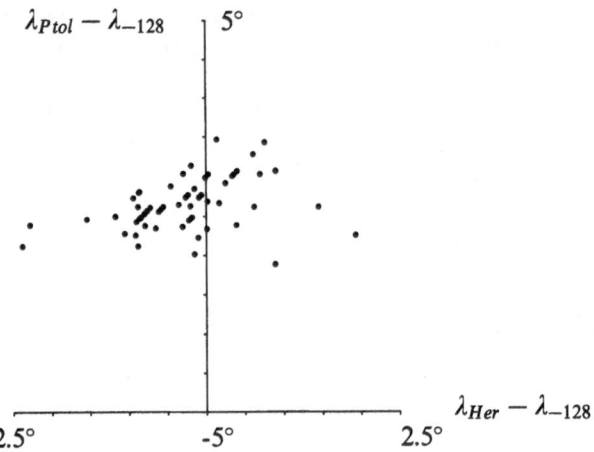

Figure 4.17: Distribution $(\lambda_{Gundel} - \lambda_{-128})/(\lambda_{Gundel} - \lambda_{-128})$.

the longitudes do not cover all real numbers but discrete values. So the differences between them also can combine to few discrete values as well. Since Vogt's reconstructed longitudes are given accurate to a hundredth of a degree, their errors correlate with the ones of the Almagest without visible discrete structure. In contrast to them the Hermes longitudes can differ from the errors of the Almagest by only those fractions found in the Almagest: one sixth and (less frequent) a fourth of a degree. For this reason, the lines lie parallel to each other, separated by one sixth of a degree. Such a structure can occur with dependent as well as with non-correlating data.

To illustrate the possible dependence of the Hermes longitudes from the Ptolemaic longitudes, the results of a simulation of dependent and independent longitudes are depicted in the next two diagrams.

The lost Hipparchan longitudes λ_{Hipp}, for which it is assumed that they, like the stars of the Almagest, were given with an accuracy of 1/6 degree, can be calculated from the accurate longitudes λ_{-128} for the epoch of the Aratus Commentary with a normally distributed error of the standard deviation σ around the mean value $\mu = 0°$ according to (4.8). From the accurate longitudes the value of the precession $(0°.14)$ is subtracted to adjust to the earlier epoch of the Aratus Commentary.

$$\lambda_{Hipp} = \frac{\text{int}((\lambda_{-128} - 0°.14 + Err1) * 6)}{6} \tag{4.8}$$

Under the assumption that Ptolemy increased the longitude by the $2°.67$, the difference $\lambda_{Ptol} - \lambda_{-128}$ can be computed according to (4.9):[61]

$$\lambda_{Ptol} - \lambda_{-128} = \lambda_{Hipp} + 2°.67 - \lambda_{-128}$$

---

[61]The constant $2°.53$ is the mean difference of the Ptolemaic longitudes for the epoch -128 and it takes into account the possible addition of 2°40′ by Ptolemy and the 0°.14 by which the older Hipparchan longitudes are smaller than the standard epoch of -128.

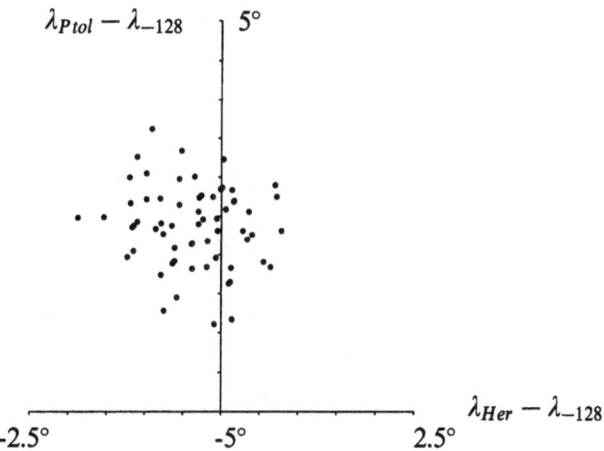

Figure 4.18: Simulation of independent errors.

$$= \frac{\text{int}((\lambda_{-128} + Err1) * 6)}{6} - \lambda_{-128} + 2\overset{\circ}{.}53 \qquad (4.9)$$

The difference of the Hermes longitudes from the calculated position using equation (4.7) and (4.8) amounts to:

$$\lambda_{Her} - \lambda_{-128} = \text{int}(\text{int}(\frac{\lambda_{-128} - 0\overset{\circ}{.}14 + Err1) * 6}{6}) + 0\overset{\circ}{.}35 + Err3) - \lambda_{-128} \qquad (4.10)$$

In the case that the Ptolemaic positions were measured independently of the co-ordinates in the Hipparchan register, both of the random errors of the measurements in (4.8) and (4.9) are independent of each other. Here, the Ptolemaic measurements are not afflicted by the same Hipparchan error $Err1$, but rather with a different error $Err2$, and the longitudinal difference of the independent Ptolemaic errors is derived according to (4.11):

$$\lambda_{Ptol} - \lambda_{-128} = \frac{\text{int}((\lambda_{-128} + Err2) * 6)}{6} - \lambda_{-128} + 2\overset{\circ}{.}53 \qquad (4.11)$$

In the first figure, the differences in longitude generated by random variations according to (4.10) and (4.11) are independent of one another. The simulated Hermes values are thereby scattered around their mean value to a larger degree than the more exact Ptolemaic positions (fig. 4.18).[62]

The errors are distributed like those in (fig. 4.18) if the Ptolemaic longitudes were independently observed from Hipparchus, on which the Hermes longitudes are based, and if there is no common systematic error in both measurement procedures. On either side of the y-axis the points are distributed around the same mean value for the differences of the Ptolemaic coordinates from the accurate positions. In other words, the slope of the regression line is close to zero. The distribution is different if the errors of the Ptolemaic coordinates are dependent on the Hipparchan

---

[62]The standard deviation of the random variables Err1 and Err2 is $0\overset{\circ}{.}5$, that of Err2 is $0\overset{\circ}{.}3$.

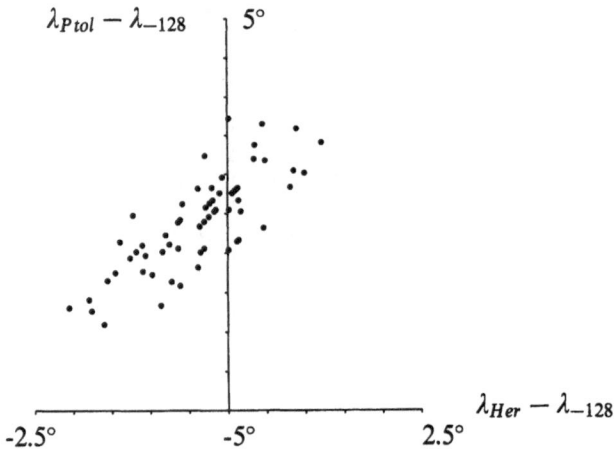

Figure 4.19: Simulation of dependent errors.

ones, as is to be expected if the positions of the Almagest are only mathematical transformations from a Hipparchan register. A simulated example with errors in longitude according to equations (4.9) and (4.10) is shown in (fig. 4.19).

In this case the positive errors of the Hipparchan longitudes correspond significantly to larger Ptolemaic errors than the negative ones. This different visual distribution in the diagram can be confirmed by a regression analysis. As seen before, the high correlation coefficient of r=0.51 contradicts the thesis of independent stellar coordinates. Assuming the absence of common systematic errors in the measurement procedures, further evidence for the genetic identity of the Ptolemaic coordinates with those leading back to Hipparchus has been obtained.

Our results can be summarized as follows:

(i) The star list in the Hermes Trismegistos text contains ecliptical longitudes from the time of Hipparchus.

(ii) The ecliptical longitudes are not reported results of direct measurements. The comparison with the accurate values reveals that they were truncated from more exact values.

(iii) The errors in the longitudes correlate significantly to those of the Almagest. This can mean either that both are measured with a similar systematic error, or that they are genetically identical.

5. Structures in Ptolemy's Star Catalogue

5.1 Star Maps

In the last two chapters the relationship of the Ptolemaic star catalogue to its assumed precursors was examined and discussed. In the following sections the analysis of the internal structures of the stellar data will reveal parts of their history.

No observational program with the rigor and comprehensiveness of the one which led to the star catalogue of the Almagest is devoid of systematic errors which, retrospectively, might characterize the measuring and evaluation procedures involved. Even a catalogue completely free of errors can reveal a recognizable structure such as varying accuracies of the coordinates or preferences for certain arrangements of the data. The causes of these peculiarities in the scientific documents should be explained by their historical reconstruction.

Causal effects – and this is confirmed by the history of the interpretation of the Ptolemaic star catalogue – can be rarely proven directly. Often enough causal explanations provoke alternative hypotheses which are equally able to explain a phenomenon with similar internal rigor. Tycho Brahe explained the errors in longitude by Ptolemy's mistakes in the transformation procedures of Hipparchan coordinates, but shortly after him Laplace could explain the deviations of the stellar positions by referring to errors in Ptolemy's solar theory with the same methodological correctness. In the time that followed the task was to find further evidence differentiating between those two fundamental interpretations. The work of Boll, Björnbo, Dreyer, and Vogt can be seen as belonging to this period. As a result of this development, more aspects of the catalogue have been included in the over-all evaluation which in turn deepened the knowledge of underlying ancient astronomical methodology.

The exact reconstruction of the Ptolemaic star catalogue and its evaluation by computer lead to the discovery of new details which up to now had been inaccessible due to the plethora of the data.

On the next four pages all of the stars of the Ptolemaic catalogue along with their deviations from the accurate positions with respect to the epoch +137 are plotted both in the ecliptical and the equatorial coordinate system. The catalogued position of the star is marked by a dot, and a dash branches off representing the positional error, i. e. the other end of the dash corresponds with the accurate position of the star. The first map shows the stars in the ecliptical coordinate system. To facilitate the identification, the catalogue numbers of the stars are plotted in a second map. The second pair of maps shows the stars in the equatorial coordinate system.

The dashes on the maps represent the positional deviations on a constant scale. Therefore, a deviation in longitude of one degree in a very northern declination is depicted with the same length of dash as on the equator. By contrast, for an observer of the sky the apparent angle of a coordinate difference of one degree close to the pole is shortened and very much different from the same coordinate difference at the equator. This effect of perspective need not be considered in the following error maps. Both Delambre's thesis of a simple mathematical conversion of Hipparchan coordinates as well as the explanation of the errors by the faults of the solar theory assume initially unbiased observations. Both hypotheses account for the systematic error either through an incorrect precession constant or else through an error in the position of the mean sun; consequently all longitudes of the stars are on the average affected by the same systematic error of about one degree, independent of the declination of the respective star.

The systematic error in the ecliptical longitude is evident. In general the corresponding accurate longitudes of the stars are larger by about one degree. The dashes therefore stretch out from the star markings to the right. Several characteristics are obvious:

(i) The band of the ecliptic:

   The density and homogeneity of the deviations are greatest for stars around the ecliptic. Their distinctive band not only reflects the higher density of stars in that particular region of the sky, but is more the result of a long tradition of observations which customarily included each star of the ecliptic.

(ii) Star groupings:

   The impression of an ecliptical band arises not only because of the higher density of stars. The positional errors of the stars are very homogeneous, compared with other areas. The similarities of the positional errors within zones or groups is striking and is in agreement with the method of observation described in the Almagest. According to Ptolemy the positions of the individual reference stars were determined with the aid of the moon and the position of the sun. Then the positions of the other stars were measured relative to the places of the reference stars. Such a procedure implies that a positional error of the reference star is transferred to all the stars whose position are measured relative to it. Conforming to this description, we find individual groups with similar positional errors (for example $\lambda = 280°, \beta = -18°; \lambda = 150°, \beta = 80°; \lambda = 190°, \beta = -36°$). At the same time there are areas with nearly vanishing errors, for example the constellation Lepus. The positional measurement in groupings of stars is obvious from (fig. 5.5) in which the error of latitude is plotted as a function of the catalogue number in the Almagest.

Sections of the catalogue with consistently positive or negative errors in latitude are likewise easy to identify. The second constellation of the zodiac, Taurus, with stars starting at No. 380, marks the beginning of a section of the catalogue with predominantly negative errors in latitude which continues

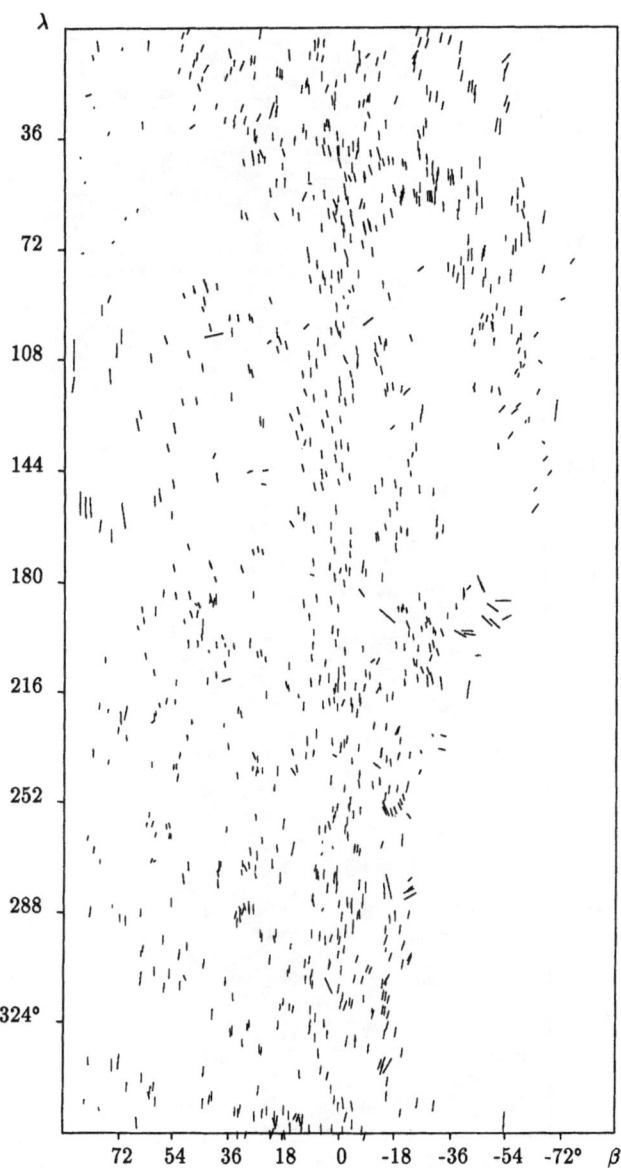

Figure 5.1: Ecliptical coordinates.

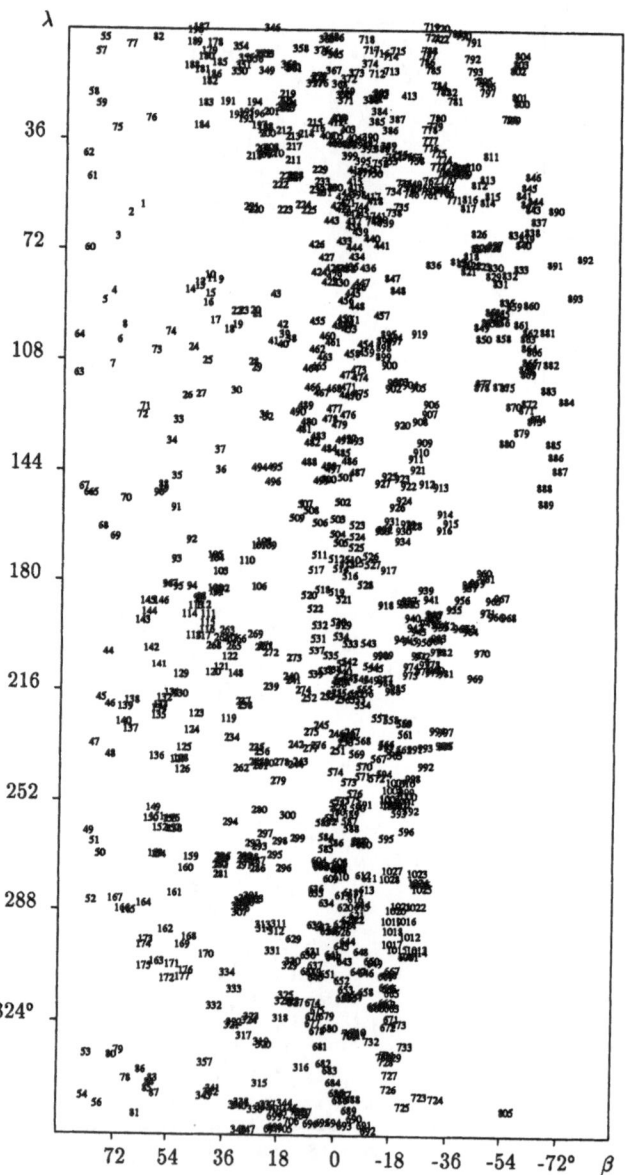

Figure 5.2: Ecliptical coordinates.

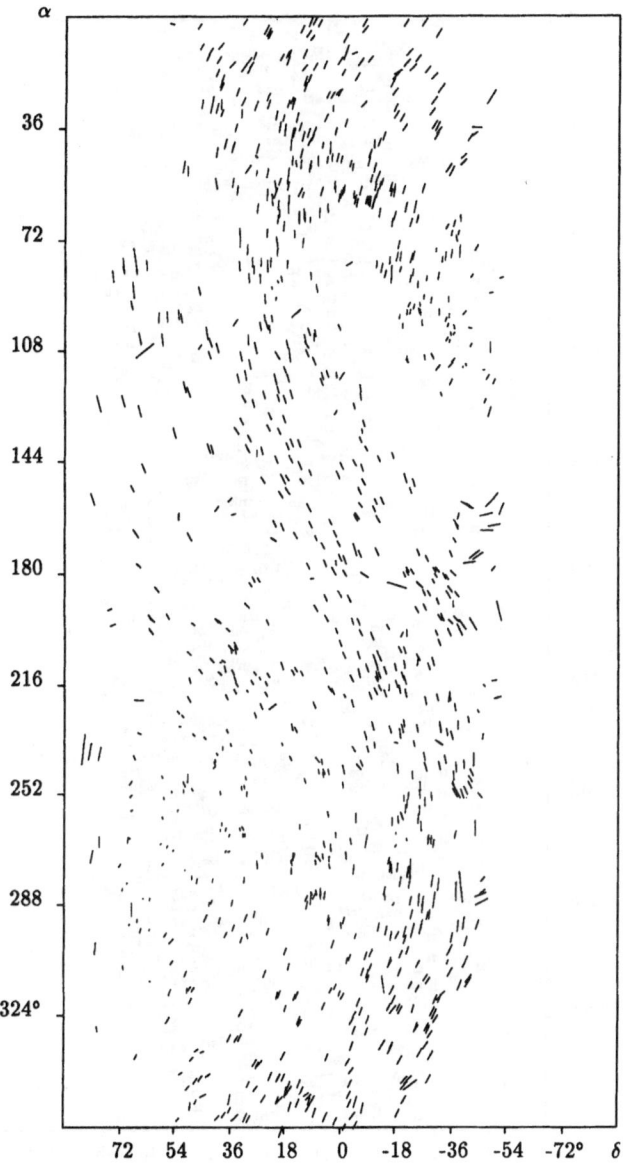

Figure 5.3: Equatorial coordinates.

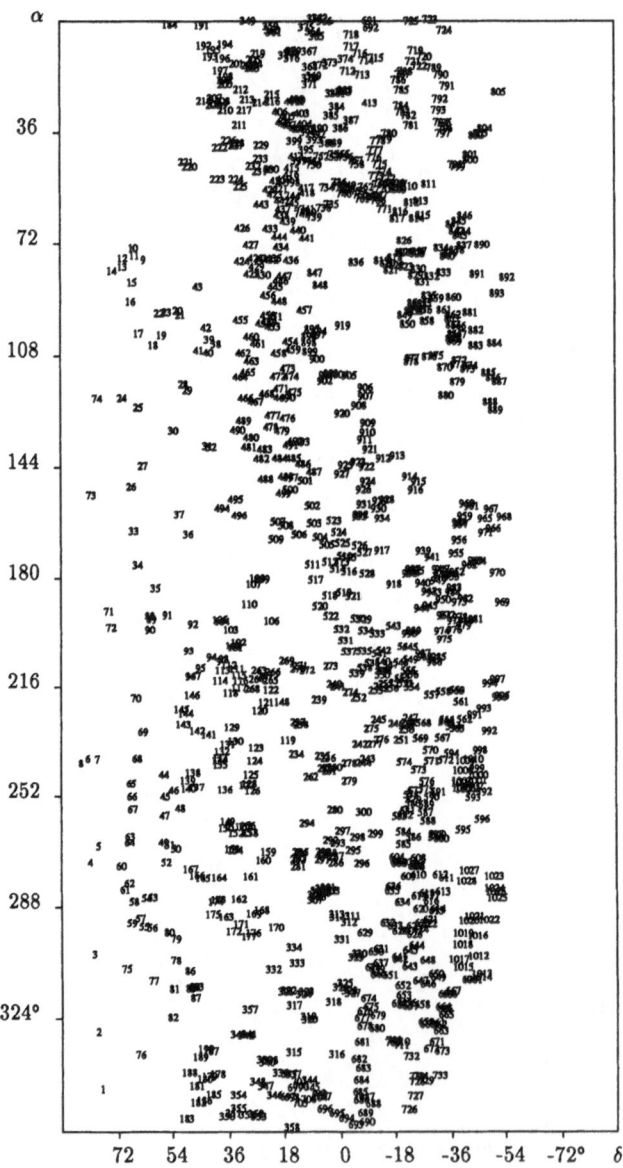

Figure 5.4: Equatorial coordinates.

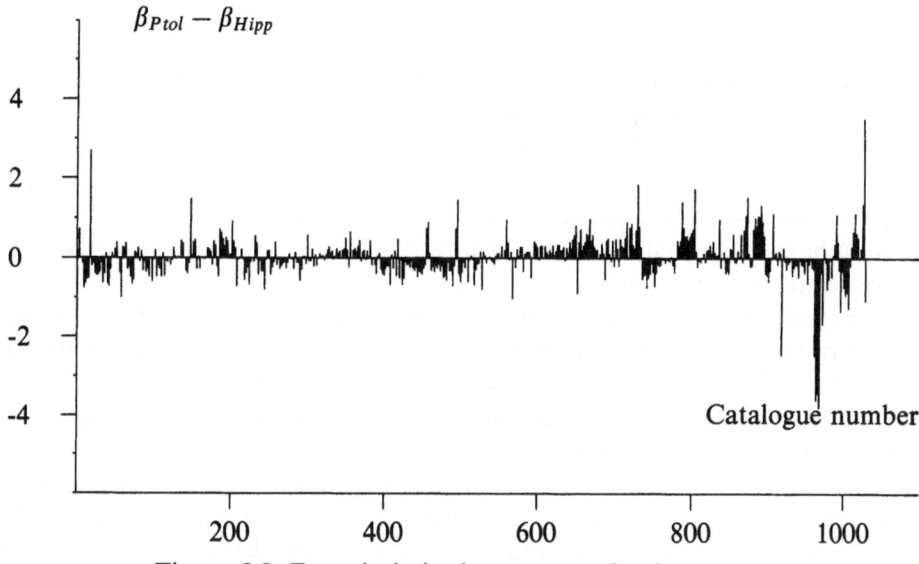

Figure 5.5: Error in latitude, average of pairs.

approximately up to the constellation Virgo. This constellation is at the same time the last zodiacal sign of the northern hemisphere. With the first southern constellation of the zodiac, Libra (beginning with No. 529) a series of small positive errors in latitude begins.

Another conspicuous grouping starts with the constellation Orion (beginning with No. 735)[1] and ends in the middle of the constellation Eridanus. The stars of Cetus and the subsequently catalogued stars of Eridanus display in the majority of cases positive errors in latitude which are clearly distinguishable from the negative differences of Orion.

The southern limit for the group of circumpolar stars is not detectable by any specific change of the systematic errors. The longitudes of the northern stars are indeed much harder to determine precisely than those in other areas of the sky. Such inaccuracies prevent an exact fixing of the limiting declination for circumpolar stars and with it an unambiguous criterion for latitude of the observational site.

Ptolemy did not add stars to the catalogue which were visible to him but not to Hipparchus from the more northerly Rhodes. If, though, he actually added his own measurements to a smaller Hipparchan register, as the results of Vogt's analysis seem to imply, then he only supplemented the traditional constellations.

The next sections discuss more detailed aspects of the structures in the data of the catalogue.

5.2 Multiple Sources

After Boll's article numerous authors considered the possibility of a multiple origin of the catalogue in the Almagest. Especially Björnbo, Dreyer and Vogt have suspected

---

[1]The exception is star λ Ori. In the catalogue it is described as nebular and one has to assume with Toomer that the observations pointed to the nebular around λ Ori and ϕ^1 Ori, viz. Ptolemy, C. (1984), p. 382.

that the star catalogue is based both on Hipparchan and Ptolemaic measurements. Björnbo even considered the possibility of a further source.

The extreme northern and southern stars have large errors in latitude. In these cases one has to assume complex systematic errors. In most of the other areas in the sky the errors in latitude are minor. Therefore the large systematic error in longitude cannot be caused by effects which imply a large error in both coordinates, e.g. rising times, extinction or refraction of the star.

The two alternative explanatory hypotheses about the origin both account adequately for the average longitudinal error. Whether the coordinates were biased by Ptolemy in the transformation of Hipparchan coordinates or whether Ptolemy corrupted his positional measurements systematically by the faulty solar theory, an average error in longitude of about one degree results in either case. Therefore an analysis of the differences in longitude cannot differentiate between the two interpretations.[2] Moreover, these two causes are by no means the only ones imaginable. Although it is improbable, Ptolemy could have carried out positional measurements independent of the solar theory and in doing so he might have obtained correct coordinates. Also, he could have examined and evaluated all of the star registers available to him in order to include as many stars as possible in the catalogue being compiled for the Almagest, and converted them to the same epoch of reference according to his precession formula.

If these measurements lead to correct coordinates and the later methods of evaluation are free of error, the differences of the catalogued longitudes from the calculated ones, $\lambda_{Ptol} - \lambda_{137}$, are proportional to the time of their observation. The absolute differences in longitudes become larger the older the original source is on which Ptolemy based the transformations.

In what follows special subsets of the catalogue will be examined to test for significantly different errors in longitude.

5.2.1   Dreyer's Paradigm

When Dreyer attempted to explain the different levels of accuracy of the stellar coordinates in his "On the Origin of Ptolemy's Catalogue of Stars", he presupposed that the catalogue is composed from different sources: the stars with an accuracy of 1/4 degree in ecliptical latitude were appended by Ptolemy to the roughly 850 stars of the lost Hipparchan star register, and all positions were reduced to the same epoch of +137.[3]

Vogt's critique demonstrates that Dreyer had underestimated the total number of 1/4 degree stars in the Almagest. Although his calculations exclude the possibility that the 1/4 degree stars exactly fill the gaps between the number of stars in the Almagest and Boll's estimation of the Hipparchan stars, this does not disprove the thesis that the coordinates of the 1/4 degree stars were obtained by another measurement procedure at another epoch. If the errors in longitude arose essentially due to Ptolemy's use of the precession constant, then the stellar longitudes originally

---

[2]The differentiation fails in particular when only the mean error in longitude is considered. Discriminating small variations in the longitudinal errors will be discussed later.

[3]Dreyer, J. L. E. (1917), pp. 528–539.

coming from Hipparchus should have an average error of one degree and the older coordinates a respectively larger error. In general the relation holds:

$$\Delta\lambda = \Delta p * T \qquad \text{with T in years, and } \Delta p = 0.0039°/y \qquad (5.1)$$

This relation assumes that the directly measured coordinates are free of a large systematic error, in other words, they are also independent of the mean error of the solar theory. Thus an assumption about the use of the solar theory is only unnecessary when the data are taken from Hipparchan measurements, because for that time the error of the mean sun is small and the average difference of the stellar longitudes from the accurate positions of Ptolemy's time should be 2°40′.

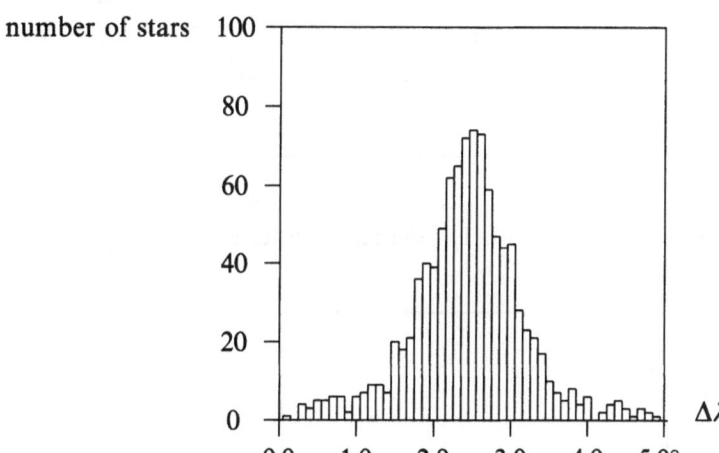

Figure 5.6: Distribution of errors in longitude, all stars.

The next histogram (fig. 5.6) shows the distribution of errors in the longitudes for all stars of the Almagest, while the map (fig. 5.7) includes only those stars whose ecliptical latitudes are recorded with an accuracy of 1/4 degree, i.e. those catalogued with a 15′ and 45′ fraction.

These stars are well distributed over the entire visible sky. Also, the distribution of their errors in longitude does not deviate significantly from the distribution of the differences of all stars, as can be seen from fig. 5.8.

Dreyer suggested subdividing the catalogue into the stars originally coming from Hipparchus and the genuine Ptolemaic data recognizable by their accuracy of 1/4 degree. If the Ptolemaic measurements were unbiased[4], they should be distinguished significantly from the originally Hipparchan coordinates by the distribution of errors in longitude.

As is apparent from (fig. 5.8), the error distributions of the longitudes of the 1/4 degree stars do not differ significantly from the error distribution of all stars.[5] It implies that the 1/4 degree stars were either observed at approximately the same time as the others – and with that, the thesis that they had been observed by Ptolemy

---

[4]The measurements are unbiased when they are independent of the solar theory.
[5]The distribution of errors for stars with a fraction of 10′, 20′, 40′ in latitude.

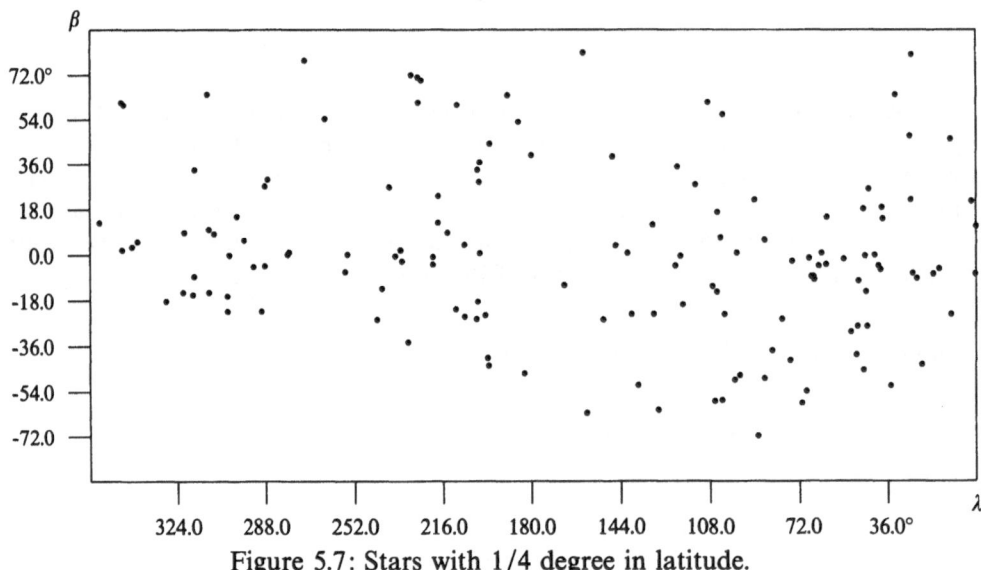

Figure 5.7: Stars with 1/4 degree in latitude.

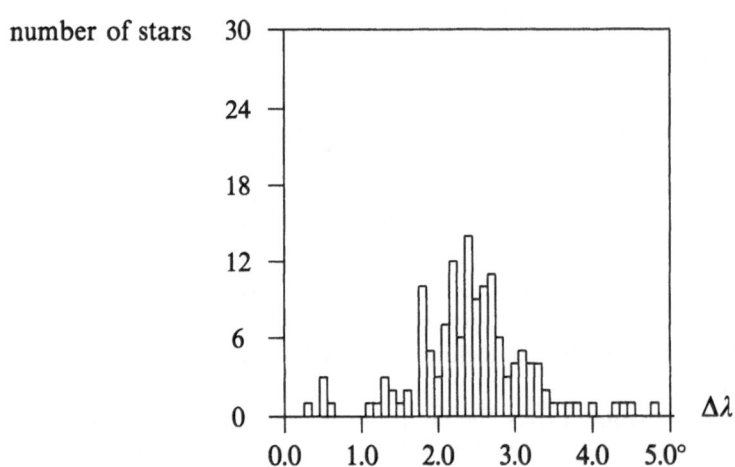

Figure 5.8: Distribution of errors in longitude, 1/4 degree stars.

would be weakened – or else the Ptolemaic measurement of the stellar positions includes a systematic error which distorts the ecliptical longitudes of his observations by exactly the amount of the practically error-free Hipparchan coordinates.[6] The 1/4 degree stars are not distinct from the others with respect to the mean error in

---

[6]This is the proposition of Dreyer, J. L. E. (1917), pp. 538f: "To sum up, the following conclusions may reasonably be drawn: [...] 2. That if he did borrow the relative positions of these 850 stars (of which there is no proof), 175 stars, probably identical with the stars observed with an instrument, of which the latitude circle was only divided to 1/4° , must at any rate have been observed by Ptolemy. [...] 4. That the method adopted by Ptolemy for determining the longitudes of standard stars of necessity introduced large systematic errors in the results."

longitude or their distribution in the sky.

The only apparent distinction of the 1/4 degree stars is their brightness: On average they are fainter. Among them there is no star of the first magnitude and only one star of the second magnitude (Pollux, β Gem). The distribution of the magnitudes for all stars is shown in the histogram (fig. 5.9).

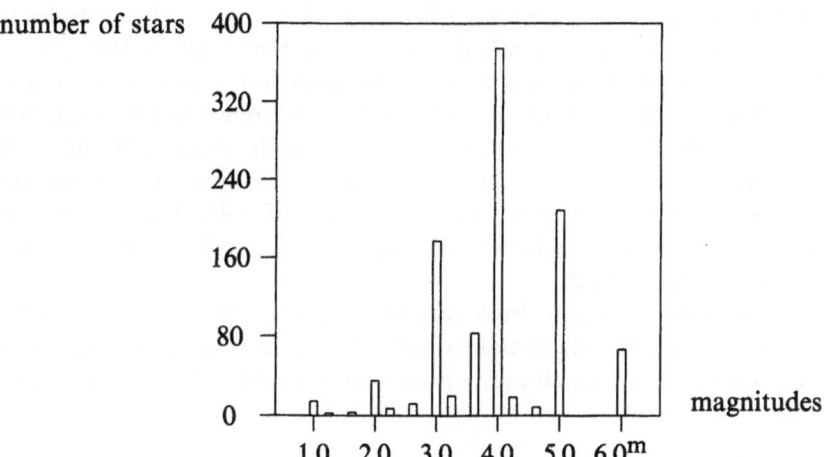

Figure 5.9: Distribution of magnitudes, all stars.

The distribution of magnitudes for the 1/4 degree stars is significantly different (fig. 5.10). It could be argued that fainter stars are more difficult to observe and therefore catalogued with less accuracy. On the other hand one cannot assume that the 1/4 degree stars cover those faint stars which were overlooked in the first survey and later added to a large Hipparchan register, for the bright star Pollux of Gemini had to be included in any star catalogue from the very beginning.

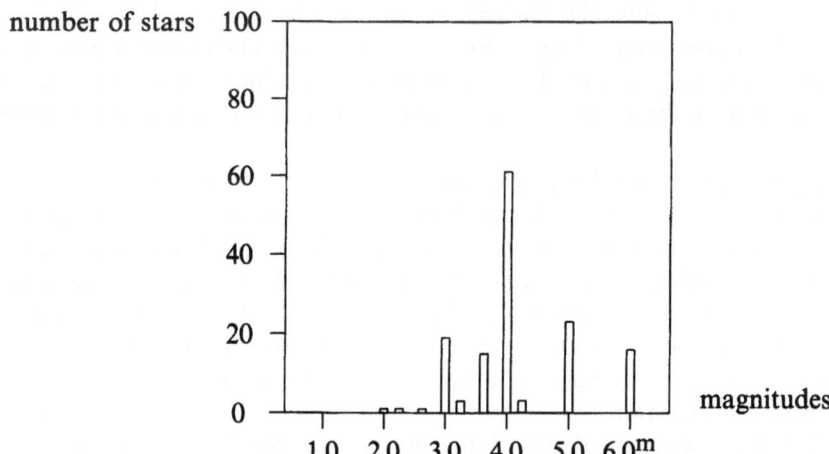

Figure 5.10: Distribution of magnitudes, 1/4 degree stars.

Dreyer's view that the stellar positions in the Almagest are of heterogeneous origin cannot be confirmed with respect to the 1/4 degree stars on the basis of coordinate errors. The general idea can be still followed, namely to search for subsets of stars in the catalogue with significantly different longitude errors. On the premise that most of the stars have no large systematic error in longitude besides the error of either the mean sun or the precession transformation, one can use the difference of the catalogued longitude from the position at the standard epoch to determine the observation time. The average deviation of the ecliptical longitudes for the stars of the Almagest from the recalculated ecliptical longitudes at the epoch of +137 must be small if a group of stars without systematic error had been observed by Ptolemy. Relative to the epoch of -128 the average difference in longitude must correspond to the value of the precession motion between the years -128 and +137, i.e. 3°40′. If in turn Hipparchus had originally observed a group of stars in the year -128 and Ptolemy added the wrong precession constant of 2°40′, then the average longitude is one degree too small relative to the epoch 137, and 2°40′ larger relative to the positions of the epoch -128.

If the systematic error is due to Ptolemy's precession calculation or, according to a different argument due to the mistaken solar theory, then the probable time of observation t_o for a subset of longitudes with a difference $\Delta\lambda$ relative to the epoch -128 is given by

$$t_o = \frac{\Delta\lambda}{\Delta p} + t_e \qquad \text{with } \Delta p = 0.0039°/y \text{ and } t_e = -128. \tag{5.2}$$

In the following a statistical test determines whether there is a group of stars distinguished from the others by a significantly different error in longitude.

5.3 Method of Selective Error Distribution

The coordinates of the stars around the northern pole are extremely difficult to determine. Depending on the geographical latitude of the observer, the stars in those regions do not set below the horizon. A star exactly on the pole would have no apparent daily motion and all stars close to it move on much smaller arcs than the stars at the equatorial equator. For this reason the positional measurement, in either the equatorial or the ecliptical coordinate system, is less accurate for those stars.

As Ptolemy reports in the Almagest, ancient astronomy used either lunar and planetary conjunctions with stars for position measurements, or the rising and setting times combined with the meridian observation. These simple methods cannot be used in the case of circumpolar stars. In addition to all the difficulties involved in observations with the astrolabe one had no other accurate alternative method available. Therefore the positions of the circumpolar stars must be much less accurate than the others. For the geographical latitude of Rhodes ($\varphi = 36°$), all stars are circumpolar with a declination of $\delta > 54°$.

Stars with a large southern declination rise just over the horizon in the south and they are not visible at all from the geographical latitude of Rhodes if their declination is more southern than about $-54°$.[7] The following histogram shows the

---

[7]The visibility conditions in the south are influenced by two factors: The refraction of the light in

distribution of the differences $\Delta\lambda = \lambda_{Ptol} - \lambda_{-128}$ of the longitudes in the Almagest compared to the accurate data of the epoch -128 only for those stars which were visible not higher than 15° above the horizon of Rhodes (fig. 5.11).

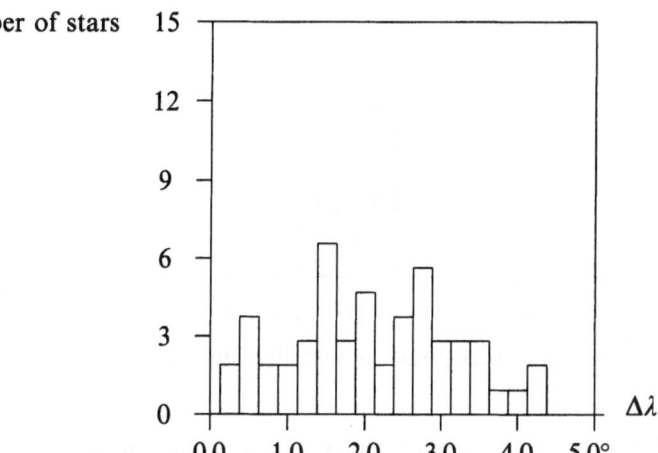

Figure 5.11: Differences in longitude for southern stars.

The differences are widely distributed, reflecting the inaccuracy of the measurements. Conspicuous is a group of stars with a longitudinal difference of much less than 2°40', pointing to much older measurements, if other sources of error in the measurement are excluded.

One can safely assume that the measurements were not organized according to coordinate fields, but rather, the observations were made according to constellations or parts of them. The constellation Lepus and parts of Eridanus are discriminated from the rest by a very small error in longitude for the epoch of Ptolemy. In the histogram with the reference epoch of -128, the longitudes of those constellations cluster around the mean differences of 3°40' – the accurate precession between the times of Hipparchus and Ptolemy (fig. 5.12).

The stars in the ecliptic show a uniquely coherent distribution of errors. These stars had already been observed in the earliest history of astronomy together with the motion of the sun, moon and the planets. They naturally could serve as a standard for the positions of the other stars in the sky. In the Aratus Commentary Hipparchus describes the stellar positions in terms of their simultaneous rising and setting with the ecliptical degree intersecting the horizon. The later discovery of the precession motion in the plane of the ecliptic had the effect that the simultaneous phenomena must be adjusted for different periods with complicated transformation schemes. The ease with which stellar coordinates in the ecliptical coordinate system can be converted to arbitrary epochs is the major rationale for a star catalogue

---

the atmosphere makes the celestial object appear to be higher above the horizon than its geometrical position. Close to the horizon this amounts to half a degree. All objects therefore rise a little bit earlier and set later than calculated on the basis of the geometrical construction alone. With an opposite effect the extinction in the atmosphere absorbs so much light that stars become visible only at a certain altitude above the horizon, depending on their brightness.

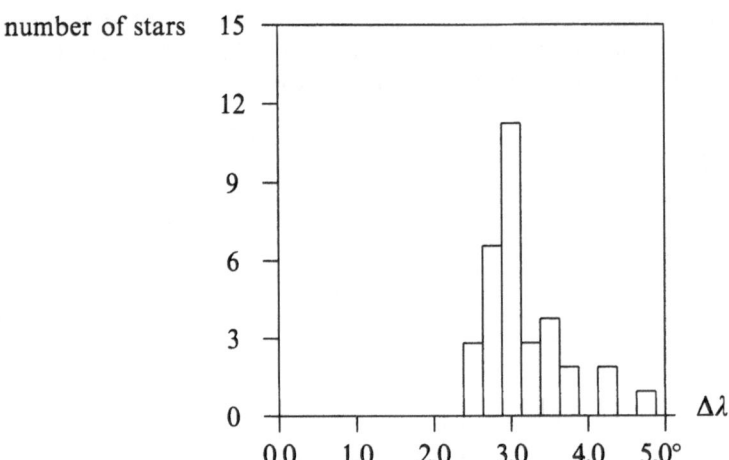

Figure 5.12: Differences in longitude, constellations Lepus, Eridanus.

in such a system. The particular significance of the stars in the ecliptic suggests that they had been observed by one astronomical school during a limited period of time. This is supported by the homogeneity of the distribution of the differences in longitude (fig. 5.13).

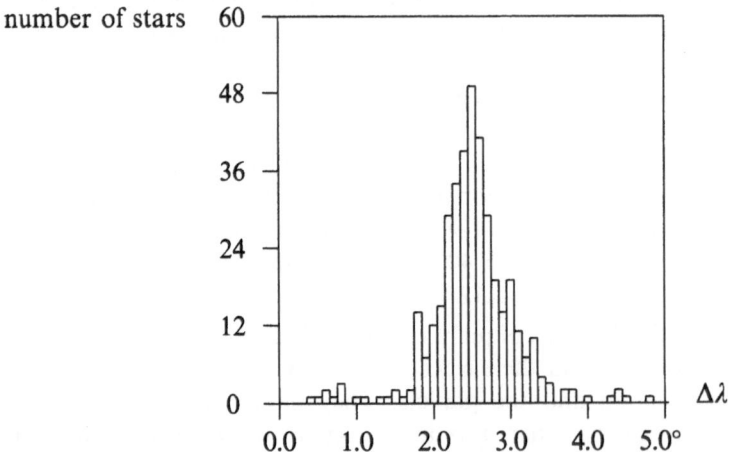

Figure 5.13: Differences in longitude, stars of the zodiac.

5.3.1 Cluster-Analysis

At the first sight only the stars with extreme northern and southern declinations deviate significantly in their average error in longitude from the rest. For those stars, though, substantial errors are easily explicable by the difficult measurement conditions. Consequently the peculiar errors do not necessarily point to the existence of another observer of a different epoch.

In the following we will analyze whether the data of the star catalogue contain extended subsets with significantly different distributions of errors in the ecliptical longitude. Differences might point to different times of origin if they are due to Ptolemy's mistaken precession transformation. However, it should be emphasized that there are alternative explanations for systematic coordinate errors of subsets of stars, such as positional errors of reference stars being transferred to all stars whose coordinates are determined with it.

In the first step the analysis concerns whether there is any subset of stars with a significantly different error in longitude. In general a given distribution of longitudinal differences can be tested as to whether it consists of a mixture or superposition of more than one normal distributions. Suppose there are n normal distribution in the whole set of data and some plausible assumptions about their shape and mean value can be made. This initial guess can be used in an iteration process to improve the parameters of the individual normal distributions, until a "best fit" to the given overall distribution is obtained. This local maximum of the measure for the fit is interpreted as the most probable superposition of n normally distributed sources of error. We vary the hypothetical number of distributions from 1 to 4 and estimate the probability of the distributions with the χ^2-test.[8]

On the assumption that the differences in longitude are normally distributed only around one mean value, the most probable distribution takes the form seen in (fig. 5.14), after the iteration process for the best fit is carried out.

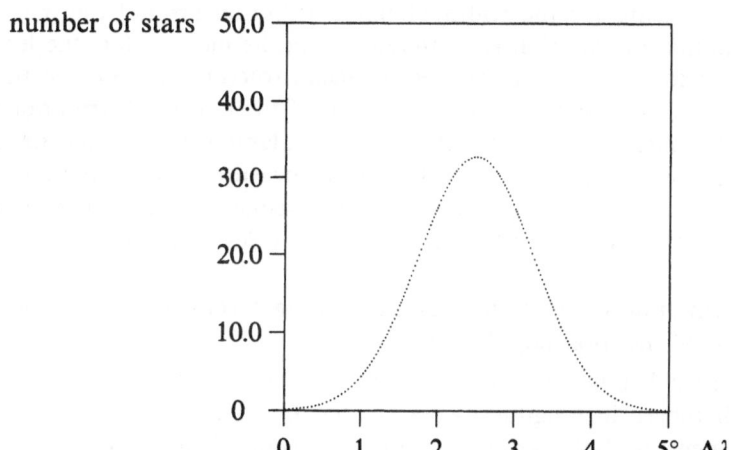

Figure 5.14: Distribution of differences in longitude for one source.

The parameters of the normal distribution hypothetically assumed serve as a testing distribution against which the probability of the hypothesis of two overlapping distributions is estimated. Two normal distributions superpose to the distribution shown in (fig. 5.15).

---

[8]The "best fit" is defined as the maximum of the likelihood function. For the iteration process we applied the computer subroutine NORMIX of the program "Clustan", Computing Laboratory, University of St. Andrews, North Haugh, St. Andrews, release 2.1. Statistical outliers with a difference in longitude $\Delta\lambda < 0°$ and $\Delta\lambda > 6°$ are excluded from the data set.

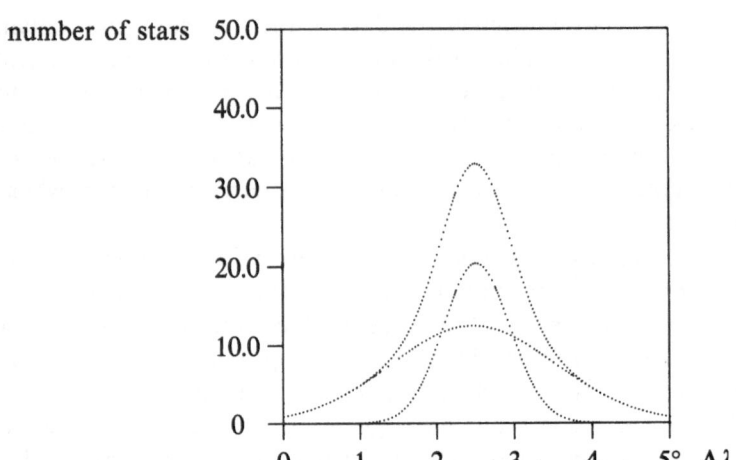

Figure 5.15: Distribution of longitudinal differences for two sources.

The first normal distribution iterates to the parameters $\mu = 2°.46, \sigma = 1°.07$ for 38% of the total population, the second normal distribution has a mean value of $\mu = 2°.51$ with a standard deviation of $\sigma = 0°.44$ and covers 62% of all stars. The mean values of both normal distributions are close together, while their variance differs radically.

The variance is a measure for the width of the distribution around the mean value. The wider a normal distribution is scattered around its mean value, the less precisely the differences in longitude are defined. In reconstructing the cause of the errors one is seeking for a well defined and sharp peak of the normal distributions indicating a real measurement effect. A distribution with a large standard deviation is more likely to be caused by random effects during the measurement. Such a large variance has been obtained for the first distribution. Consequently, the assumption of a superposition of two normal distributions hardly provides good ground for a sound interpretation.

Three normal distributions iterate to the best fit as in (fig. 5.16) with a proportion of 11%, 82%, and 7% of the total population.

This superposition yields a much better approximation with a surprising numerical result; if the differences in longitude were created by transformations of the Hipparchan measurements with the wrong precession constant, then the coordinates in the Almagest would have to have a longitude 2°40′ larger than the accurate values calculated for the epoch -128. Now, as we have seen, the iteration process leads to a best fit for the major distribution with a mean value of $\mu = 2°.53$, which is about $0°.14$ less then expected. On the other hand, the Hipparchan coordinates reconstructed by Vogt are dated to an epoch 10 years before -128. For them the mean difference of the Hipparchan longitudes from the accurate values of the epoch -128 amounts to

$$\overline{\lambda_{Hipp} - \lambda_{-128}} = 0°.14 \qquad (5.3)$$

If Ptolemy obtained the longitudes by a procedure equivalent to an addition

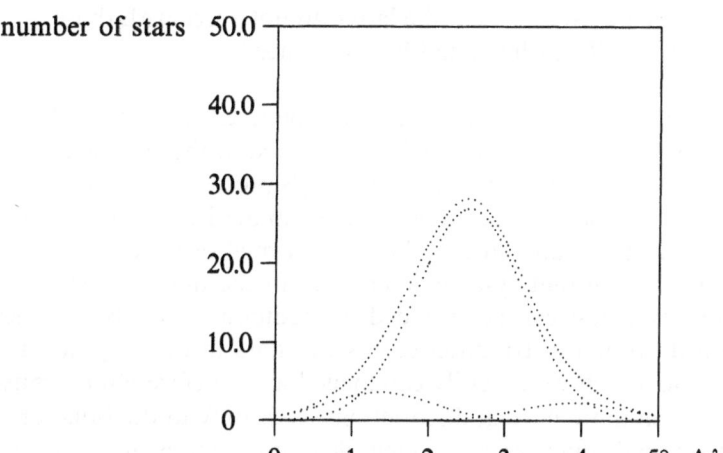

Figure 5.16: Best fit with three normal distributions.

of 2°40′ to Hipparchan longitudes, one should obtain a difference of the longitude from the epoch -128 as

$$\overline{\lambda_{Ptol} - \lambda_{-128}} = \overline{\lambda_{Hipp} - \lambda_{-128} + 2\overset{o}{.}67} = 2\overset{o}{.}53 \qquad (5.4)$$

The mean difference in longitude of the major distribution coincides exactly with the reconstructed coordinates of the Aratus Commentary. This is a strong and independent confirmation of the error correlation of the reconstructed Hipparchan stars and their counterparts in the Almagest.

It is still an open question, whether the two other clusters confirm a group of coordinates from a different observational period, or whether they can be interpreted as results of possible mistakes in the measurement procedures.

Had Ptolemy indeed observed himself without any systematic error, it would follow that the longitude of such stars would be about 3°40′ larger than the recalculated ones relative to the epoch -128. Such a difference comes close to the mean value of the third normal distribution, but it covers only 7% of the total population. It is therefore much to small to argue in favour of unique and error-free Ptolemaic observations. As mentioned before, the stars with a small error for the time of Ptolemy are clustered in the constellations Lepus and Eridanus. Whether this can be attributed to Ptolemaic observational activities or to other causes remains undecided.

Even more uncertain is the interpretation of the first normal distribution in the set of three. Their small difference from the epoch -128 could point to very old coordinates, e.g. from the school of Timocharis. The Almagest mentions declination measurements from this time:[9]

> "But when we consider the distances [of the stars] from the equator, as measured along great circles through the poles of the equator, we find [1] that those observed by us do not agree with those recorded in the

---

[9]Ptolemy, C. (1984), p. 330.

same way by Hipparchus, and [2] that the latter do not agree with those recorded even earlier by Timocharis and his associates"

The Almagest mentions observations of Timocharis between the years -293 and -271.[10] Coordinates from that time should have been entered in the catalogue, after a coordinate transformation and the addition of Ptolemy's precession constant, with a longitude that would be $1°9$ larger than those of the standard epoch -128.[11] This is significantly larger than the mean value of the third normal distribution, so that the existence of coordinates without systematic error from the time of Timocharis in the catalogue of the Almagest can be excluded. In addition to purely statistical considerations, the third normal distribution covers the stars in the very northern and southern declinations which were hardly catalogued at all before Hipparchus.

The assumption of even more normal distributions leads only to the isolation of smaller groups of stars with distinct errors in longitude that have to be interpreted as contingent groupings caused by the observations using reference stars. Furthermore, the χ^2-test discards the hypothesis of four normal distributions vis-a-vis the null hypothesis of three distributions, as is summarized in the following table.[12]

	k	μ_k	σ_k	t_k	p
n=1	1	$2°49$	$0°74$	100%	
n=2	1	$2°46$	$1°07$	38%	100%
	2	$2°51$	$0°43$	62%	
n=3	1	$1°37$	$0°69$	11%	82%
	2	$2°53$	$0°47$	82%	
	3	$3°89$	$0°70$	7%	
n=4	1	$1°35$	$0°66$	11%	32%
	2	$2°51$	$0°45$	80%	
	3	$3°41$	$0°35$	6%	
	4	$4°49$	$0°43$	3%	

Table 5.1: Probabilities of normal distributions.

The cluster analysis succeeds in isolating a homogeneous group of longitudinal differences. It works as a filter sorting out groups of stars with distinctive observational errors. The mean difference in longitude of $2°53$ supports the result of the correlations of Vogt's coordinates with those of the Almagest: a large proportion of the coordinates in the Almagest was originally measured by Hipparchus. The epoch of observation is about 10 years earlier than the year -128 on the condition that

---

[10]Ptolemy, C. (1984), p. 335 and p. 477.

[11]Calculated with 410 years between Timocharis and Ptolemy. Therefore the precession of about $4°1$ degrees had to be added to the old coordinates, while the accurate precession between Timocharis and Hipparchus is about $2°2$.

[12]n: number of normal distribution to account for the total population; μ_k and σ_k : mean value and standard deviation of the kth normal distribution; t_k: proportion of stars covered by the kth distribution; p: probability of n normal distributions against the null hypothesis of (n-1) normal distributions.

larger systematic errors in the Hipparchan longitudes (or equivalent coordinates) can be excluded. Even if there was originally a systematic reduction of the longitudes, e.g. by truncation of the coordinates, there remains the striking coincidence of the epoch of the stellar coordinates from the Almagest and the Hipparchan Aratus Commentary.

There is no evidence supporting the view that other observers contributed accurate data to the catalogue of the Almagest. Still, the method of cluster analysis cannot distinguish between two measurement procedures with systematic errors, both finally leading to the same error in the catalogue. Therefore it remains an open question whether the stars of the Almagest are of Hipparchan origin, or a large section of several hundred stars was in fact observed by other astronomers.

5.4   Errors of the Solar Theory

The previous interpretation of the cluster analysis correlates groups of stars with a significantly different longitude to different epochs of observation. Hereby it is presupposed that the longitudes of the stars could be obtained by an evaluation or measurement procedure free of non-constant systematic errors. If during one observational period the process of the measurement had introduced different errors for some groups of stars, it would be impossible to interpret the differences of the longitude from the accurate values as a linear function of time caused by the error of the mean sun or the wrong transformation due to the too small precession constant. The cluster analysis infers from a non-normal error distribution that it has to be composed of a multitude of normal distributions, i.e. from different observational sources. With varying systematic errors one cannot attribute a group of stars to a different observational epoch, hence there are limits to the achievements of cluster analysis.

Every single measurement of a stellar position is an evaluation process far more complex than just reading off numbers from an instrument: The coordinate system has to be defined, the instrument must be properly adjusted, the mathematical transformations must be suitable, and the auxiliary astronomical hypotheses should be correct. Every one of these steps in the construction of the final catalogue data can introduce a deviation from the accurate value that, retrospectively, is summed up into the total error. So far, the analysis of the stellar longitudes considered only total errors which are normally distributed around a constant mean value for all stars measured at one epoch.

The cluster analysis shows that the errors in longitude cannot be represented by a single normal distribution. Such a distribution implies in particular that the error is not a function of the longitude, i.e. that one cannot distinguish statistically between the average error in longitude for different parts of the zodiac. To test this assumption we divide the zodiac into 20 parts and calculate the mean error in longitude. As stars with a large northern and southern declination exhibit large inaccuracies in the observations, only stars with a declination of up to $\pm 20°$ are included in the testing set.[13]

The results are plotted in (fig. 5.17). The error bars represent the uncertainty of the mean value, calculated according to[14]

$$\sigma_m = \frac{2\sigma_i}{\sqrt{n}} \tag{5.5}$$

The strong dependence of the errors on the ecliptical longitudes is obvious. It

---

[13] The average is calculated for a zone around the equator and not the zodiac, because otherwise for the southern zodiacal constellations too many southern stars would be included. Control calculations with other divisions of zones have not revised the general conclusions of this section. In the calculation of the means we excluded extreme errors with more than four times the standard deviation from the mean value.

[14] The standard deviation of the mean value for n independent data with a standard deviation of σ is $\sigma_m = \sigma/\sqrt{n}$. In this case the standard deviation is enlarged, because the positions of the stars are measured in relation to a reference star. Hence their coordinate errors are not independent of each other. The factor 2, by which the standard deviation has been enlarged, is an estimate dependent on the size of the groups of stars with a joint measurement.

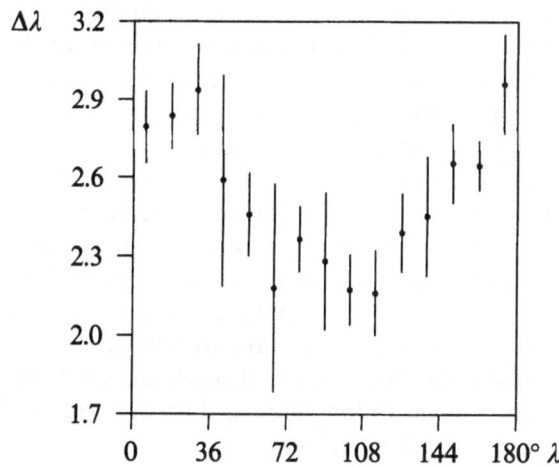

Figure 5.17: Error in longitude as a function of the longitude.

undermines the interpretation of the cluster analysis, that the non-normal error distribution could be due either to particular grouping effects of the observed star fields, or that it originates in data coming from different periods of observation.

Looking at the empirical error function in (fig. 5.17), one cannot see why stars with a longitude around 180° with their small difference from the accurate value of the epoch -128 should originally have come from a different observational source than the constellations close to the spring equinox. The clear periodicity of the errors has to be caused by a variable systematic error of the measurement method.

Ptolemy recounts how the stellar positions were measured relative to the vernal point: at the time of the observation the position of the sun was "determined". Then the positions of the reference stars were measured with the astrolabe either using the calculated position of the moon or one of the other reference stars.[15] The position of the sun had to be calculated from tables whose entries were based upon the Hipparchan/Ptolemaic solar theory. Although the orbit of the sun had been carefully investigated since the rise of the Babylonian astronomy, the inaccuracies of its theoretical description could well match the periodic error in the stellar longitudes.

The first error of the solar theory to be recognized, the error of the mean position of the sun, is due to inaccuracies in determining the solar year. The solar theory which Ptolemy took over from Hipparchus estimates the length of the tropical year[16] as 365 1/4 − 1/300 days, which is longer than the accurate 365 1/4 − 1/128 days.[17] The theoretical motion of the mean sun is therefore too slow. The error of the mean sun increased from zero at the time of Hipparchus up to one degree at the time of Ptolemy. This is the error which, according to one possible historical interpretation, caused the systematic error of the Ptolemaic star coordinates.

---

[15]Ptolemy, C. (1984), p. 219 and p. 339.

[16]The tropical year is the time in which the sun returns to the vernal point. The difference between this and the sidereal year, the time in which the sun moves through a full circle in relation to the stars, is equivalent to the precession motion of about 50″ per year, or 20 minutes of time.

[17]Neugebauer, O. (1975), p. 1083.

Though the mean sun was nearly free of errors for Hipparchus, still the errors of the eccentric motion of the sun, which describes the deviation of the sun from the uniform motion, could be carried over to the measurements of the stellar longitudes.

A detailed examination of the Hipparchan solar theory was carried out by Britton in 1967.[18] There he calculated not only the error of the mean sun but also the actual deviation of the theoretical position of the sun from the accurate place.

The apparent orbit of the sun around the earth is represented in the Hipparchan/Ptolemaic theory by an epicyclic motion of the sun around the earth or, alternatively, by the mathematically equivalent circular orbit of the sun around the eccentric earth.[19] The epicyclic or eccentric motion is a good approximation to the actual path of the sun and it accounts for the apparently variable speed of the sun on its orbit. These periodic corrections let the theoretical sun oscillate around the longitude of the mean sun. The mean longitude $\bar{\lambda}$, then, is corrected by the value g to derive the longitude λ of the theoretical sun:

$$\lambda = \bar{\lambda} + g \tag{5.6}$$

The error of the solar theory can then be estimated by the sum of the errors of the mean sun and the correction function g:

$$\Delta\lambda = \Delta\bar{\lambda} + \Delta g \tag{5.7}$$

The error of the mean sun $\Delta\lambda$ has been calculated by a number of authors.[20] The error of the mean sun for a given Julian century T is according to Britton

$$\Delta\bar{\lambda} = +0; 30.7° + 0; 25.15° \ T + 0; 0, 2.1° \ T^2 \tag{5.8}$$

For the time of Ptolemy the error of the mean sun is therefore about $1\overset{o}{.}08$ and it disappears about the year -123.[21] With the apogee A as the point of the solar orbit farthest from the earth, the mean anomaly $\bar{a}$ of the sun is defined by

$$\bar{a} = \bar{\lambda} - A, \quad \text{with} \quad A = 65\overset{o}{.}5 \tag{5.9}$$

Britton estimates the error of the correction function g as dependant on the mean anomaly $\bar{a}$ as[22]

$$\Delta g = c_1 \sin \bar{a} + c_2 \sin 2\bar{a} + c_3 \cos \bar{a} + c_4 \cos 2\bar{a} \tag{5.10}$$

The coefficients c_i are time-dependent. Their values for the time of Hipparchus and Ptolemy are given in table (5.2).

---

[18] Britton, J. P. (1967), On the Quality of Solar and Lunar observations and Parameters in Ptolemy's Almagest, PhD diss., Yale.

[19] Ptolemy, C. (1984), p. 149.

[20] E.g. Petersen, V. M., Schmidt, O. (1968), The Determination of the Longitude of the Apogee of the Orbit of the Sun according to Hipparchus and Ptolemy, *Centaurus* 12, pp. 73–96; Fotheringham, J. K. (1918), The Secular Acceleration of the Sun as Determined from Hipparchus' Equinox Observations; with a Note on Ptolemy's false Equinox, *Monthly Notices of the Royal Astronomical Society* 78, pp. 406–423. Britton, J. P. (1967), p. 51.

[21] Britton, J. P. (1967), p. 51.

[22] Britton, J. P. (1967), pp. 53f.

	c_1	c_2	c_3	c_4
Ptol	+0;23.4°	-0;1.3°	+0;9.3°	-0;0.2°
Hipp	0;23.0°	-0;1.2°	+0;1.1°	0;0°

Table 5.2: Time coefficients of the errors in the solar theory.

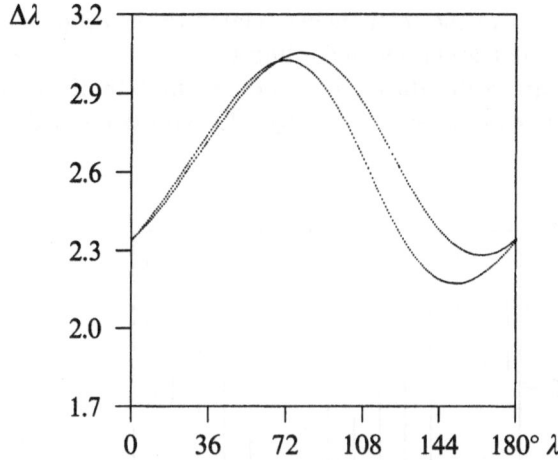

Figure 5.18: Error of the solar theory with the reference epoch = -128; a) Hipparchus +2;40°; b) Ptolemy.

Britton obtains a periodic error of the correction function g, which, he suspects, would have to enter the measurements of the stellar coordinates in addition to the error of the mean sun.[23] Britton had to leave his conjecture unproven, because there was no exact statistical analysis of the stellar positions available to him. In the next step we calculate the error of the solar theory as depending on the ecliptical longitude and juxtapose it to the mean errors of the catalogue stars. First, the longitudinal difference $\Delta\lambda$ is computed for the time of Hipparchus and Ptolemy as a function of the ecliptical longitude according to equations (5.7)–(5.10). The differences in longitude of the star coordinates refer to the standard epoch -128. Assuming that Ptolemy increased the longitudes by 2°40′ to adjust for the precession motion, one has to add this constant to the sum of the errors in the Hipparchan solar theory. The resulting total error function is compared with the erroneous solar theory for the time of Ptolemy (+137). In (fig. 5.18) both error functions are combined, where

$$\Delta\lambda_{Hipp} = \Delta\bar{\lambda}_{Hipp} + \Delta g_{Hipp} + 2°40' \tag{5.11}$$
$$\Delta\lambda_{Ptol} = \Delta\bar{\lambda}_{Ptol} + \Delta g_{Ptol}$$
$$\text{with } \Delta\bar{\lambda}_{Hipp} = 0°,$$
$$\Delta\bar{\lambda}_{Ptol} = -1°.08 + 3°.67$$

---

[23] Britton, J. P. (1967), p. 55.

Both functions have a periodic error similar to that in the stellar longitudes, only it is shifted by about half of a phase. The phase shift could be explained by the measurement procedures: the stellar longitudes have to correspond to those solar longitudes which were used at the moment of the positional measurement. Obviously the longitudes cannot correspond one to one, because then the stars would be too close to the sun and hardly visible to the observer. If those star positions were determined which were exactly in opposition to the sun – the stars which rise approximately when the sun sets and which culminate about midnight – then the longitudinal error of the solar theory related to the ecliptical longitude λ_{sun} is carried over to the stellar longitude according to $\lambda_{star} = \lambda_{sun} + 180°$. Only in this configuration are the stars visible during the whole night.[24] In diagram (fig. 5.19) the errors of the solar theory are phase-shifted by 180° and combined with the errors of the stellar longitudes.

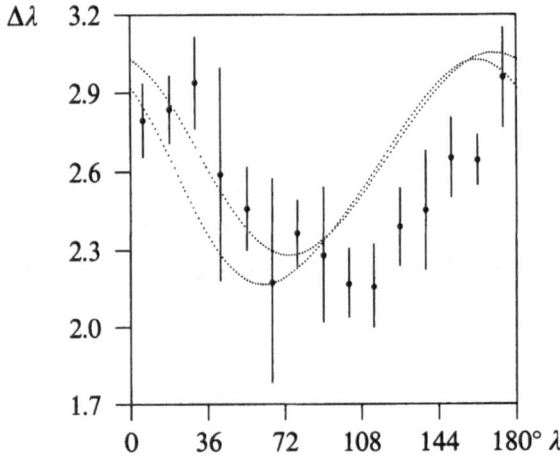

Figure 5.19: Error of the solar theory (-128), with phase shift 180° and errors of the star catalogue.

The strong similarity between the periodic errors of the solar theory and the errors in the star catalogue proves a causal relationship. We cannot see any sound alternative explanation of the periodic errors in the stellar longitudes. The use of the solar theory in the measurement of the star coordinates is assured.

For a better comparison of the errors of the solar theory with the errors in the stellar longitudes, one can approximate the error bars in (fig. 5.19) by the simple periodic function

$$l(\lambda) = a\sin(\lambda + b) + c \tag{5.12}$$

In (fig. 5.20) the interpolation function $l(\lambda)$ is superposed on the previous figure. It reveals important details about the origin of the star catalogue.

(i) It is plausible that the stellar positions were systematically measured with a limited variety of techniques. For if the positions of the reference stars, which

---

[24]This is of course an approximation, since e.g. the circumpolar stars are visible on every clear night.

define the reference system for the stars in their neighborhood, been measured unsystematically in various positions relative to the sun, the dependence of the errors in longitude would be less obvious.

(ii) The interpolation function has a phase shift of about 250° relative to the Hipparchan solar theory. Those stars whose positions were measured directly with the help of the solar theory rise about four hours after sunset. Thus, Hipparchus had not determined the positions of the reference stars at the time when the stars culminated, shortly after sunset. He used an observational arrangement which implies a longitudinal difference between sun and star of about 250°, e.g. he might have measured the positions of the rising stars in the evening.

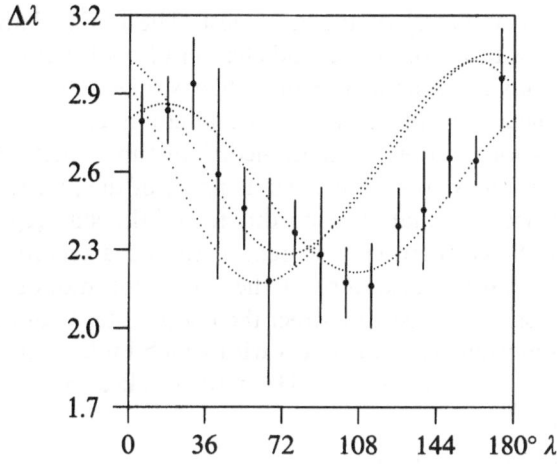

Figure 5.20: Error and interpolation functions.

(iii) The amplitude of the errors in longitude of the stars is slightly smaller than those of the solar theory. This can be explained by the limited number of reference stars. The amplitude of the catalogue errors decreases significantly, if the position of the reference stars is not measured at exactly the time of the maximal error of the solar theory. In addition the grouping of the stars around the reference stars has be arranged according to the shape of the constellation, and not to their ecliptical coordinates. Hereby the amplitude of the interpolation function is further decreased. However, the number of reference stars cannot be too small, because otherwise the longitudes would not reflect the periodic error of the solar theory.

(iv) The values of the sine function oscillate around the mean value of $2°.55$. With Ptolemy adding $2°.67$ to the Hipparchan longitudes, one derives a mean error of the stellar longitudes which is $0°.12$ smaller than would be expected for the epoch -128. Taking the precession as $50''/y$, the epoch of an error-free measurement must be set back 9 years, to the year -137. This agrees very well with the dating of Vogt's reconstructed Hipparchan longitudes, the analysis of the Hipparchan longitudes in the Hermes text, and the results of the cluster analysis.

Vogt calculated a mean difference of the reconstructed longitudes from those for the standard epoch -127 of $\Delta\lambda = -0°.16 \pm 0°.08$. Assuming that all coordinates were measured at the same period, one obtains a probable observation time of -139 ± 6 years. The chronological agreement of the epoch with the one previously derived for Ptolemy's star catalogue confirms the statistical analysis which indicated the genetic identity of Vogt's reconstructions with the coordinates from the Almagest. Also, the Hipparchan coordinates of "Hermes" point to the same period of observation. Unless one assumes a systematic decrease of all longitudes by a sixth of a degree, all three sources point to an epoch of observation 10 years earlier than Ptolemy mentions in the Almagest. These observations, then, are the common source for the second part of the Aratus Commentary, the Hipparchan longitudes in the Hermes text and for (at least major parts) of the Almagest. The initial dating of the lost Hipparchan star register is based on the second chapter of book VII of the Almagest. Ptolemy quotes here from the now lost Hipparchan works "On the length of the year" and "On the displacement of the solsticial and equinoctial points", where for the first time the precession motion is mentioned.[25] In this context, a Hipparchan observation of Regulus in the year -128 is cited as proof of the precession. Nowhere, however, can we find evidence that the coordinates of the star register had been assembled at that time. Since there was no reason to add a record of time to the observed coordinates before the discovery of the precession motion, Ptolemy may have conjectured the epoch -128 just to correct the coordinates with the appropriate value of the precession motion. The vagueness with which Ptolemy himself describes the time difference between himself and the Hipparchan observations is transparent in the following passage:[26]

> "Therefore the star on the heart of Leo has moved $2\frac{2}{3}°$ towards the rear along the ecliptic in the 265 or so years from the observation of Hipparchus to the beginning [of the reign] of Antoninus [137/38], which was when we made the majority of our observations of the positions of the fixed stars."

There is no evidence that Hipparchus actually did carry out his measurements in the year -128. The lost fixed-star register, different from the one behind the Aratus Commentary, with coordinates observed and assembled after the discovery of the precession and the appearance of a new star, is a myth. All indications point to the Hipparchan coordinates being measured about ten years earlier. The coordinates could only be attributed to the year -128 if there is a systematic truncation of Hipparchan longitudes by about a sixth of a degree. Since it is even not known in which coordinate system Hipparchus had observed or registered the stellar positions, there is hardly evidence for such a truncation mechanism. However, it cannot be excluded that Hipparchus measured the stellar positions at the later period; but then these observations were also the ones behind the rising and setting data in his Aratus Commentary.

---

[25]Ptolemy, C. (1984), p. 329.
[26]Ptolemy, C. (1984), p. 328.

The critics of Delambre's interpretation referred to the solar theory in order to explain the deviations of the star catalogue. The error of the mean sun did not lead to stellar longitudes significantly different from Delambre's interpretation. Therefore there was no criterion to distinguish between the two hypotheses. The accurate value for the precession between Hipparchus and Ptolemy is $3°.68$.[27] The error of the mean sun is $1°.13$,[28] which results in a mean difference in longitude of $2°.55$ with respect to the standard epoch of the year -128 and fits well with the findings in the star catalogue. The recognition of the additional periodic error of the epicyclic correction to the solar orbit led to different error functions, as can be seen in fig. 5.18. However, the differences between the amplitude of the periodical error and the phase shift are too small to allow one to discriminate between different types of stellar coordinates, unless the exact method of observation is known.

The very good correspondence of the error in the solar theory with the interpolation function of the catalogue errors makes it clear that the question of the origin of the star catalogue cannot be decided on the basis of the mean errors. It also shows that Ptolemy could very well have compared the adjusted position of the Hipparchan star register with actual observations made by him without obtaining significant differences. In the end, the false positions of the stars must have been confirmed by the observations.

The genetic identity of stellar coordinates from the Almagest with Hipparchan ones has to be proven on the basis of errors common to the Almagest and independent Hipparchan sources. The distribution of the fractions of the degrees in the catalogue fits coherently these findings, though it cannot decide between the two interpretations of the catalogue.

---

[27]Extrapolated from Lieske, J. H., Lederle, T., Fricke, W., Morando, B. (1977), Expressions for the Precession Quantities Based upon the IAU (1976) System of Astronomical Constants, *Astronomy and Astrophysics* **58**, p. 14.

[28]Britton, J. P. (1967), Figure II-5. Petersen, V. M., Schmidt, O. (1968), p. 9.

5.5 Fractions of Degree

The frequency of the degree fractions in the star catalogue depends on the measurement procedures and the mathematical conversions from the input data to the coordinates finally documented in the catalogue.

The reconstruction of the distribution of degree fractions in a catalogue cannot be neglected in any historical explanation of its origin. A correct deduction of the frequency distribution is a powerful criterion to decide between suggested historical interpretations, because it narrows the wide range of historically possible observational techniques.

In defending the originality of the Ptolemaic observations, Vogt's explanation of the frequency of fractions contradicts the claims of R. R. Newton. Before proceeding to a critical assessment of both hypotheses we revise the numerical data which the arguments of both authors rely on.

The fractions of the degrees in the ecliptical latitudes are distributed according to the histogram (fig. 5.21).

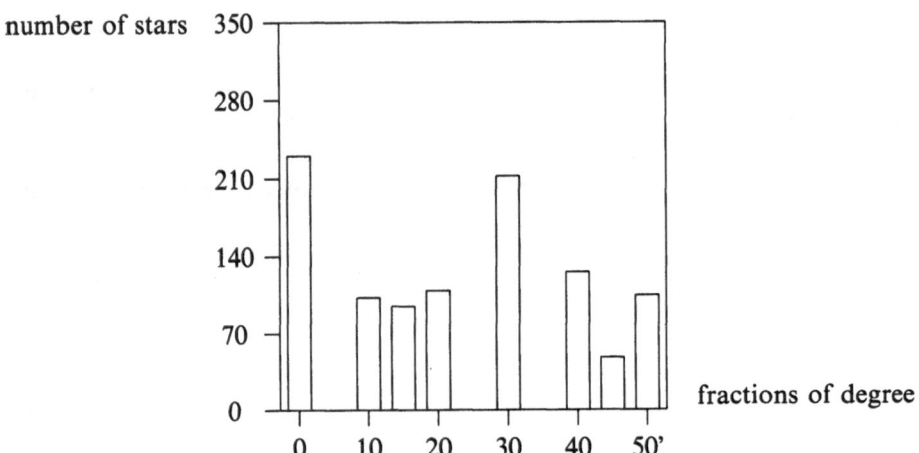

Figure 5.21: Fractions of the degree in latitude.

The frequency of the degree fractions in particular:[30]

0′	10′	15′	20′	30′	40′	45′	50′	Total
228	101	95	110	212	126	49	104	1025

Table 5.3: Distribution of fractions of the degree in latitude.

The histogram (fig. 5.22) shows the distribution of degree fractions in longitude as tabulated in table (5.4).

A number of obvious anomalies in the distributions of fractions require an explanation:

---

[30]The three stars which are catalogued a second time are not counted (No. 96-147, 230-400, 670-1011).

0′	10′	15′	20′	30′	40′	45′	50′	Total
222	180	4	179	98	241	0	101	1025

Table 5.4: Distribution of fractions of the degree in longitude.

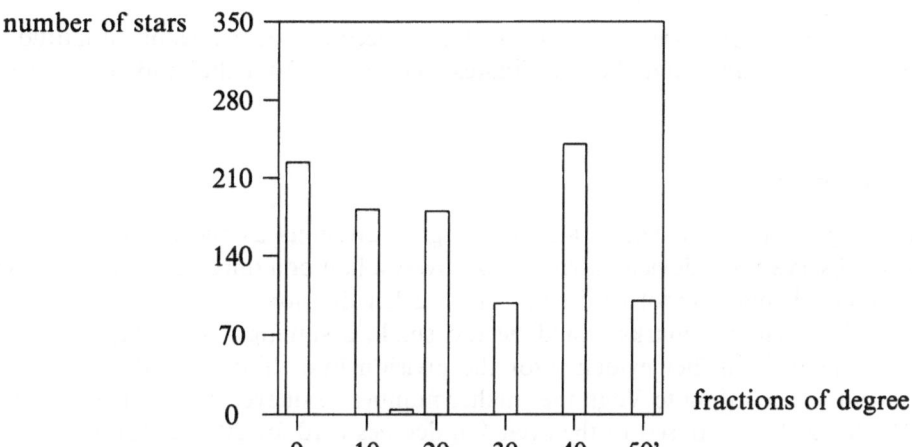

Figure 5.22: Fractions of the degree in longitude.

(i) There are four stars, three of them in the constellation Virgo, with a fraction of 1/4 degree in longitude.

No.	HR	Name	λ
502	4689	η Vir	158°15′
513	5100		177°15′
525	4955		172°15′
709	9087		332°15′

Table 5.5: Stars with 1/4 degree in longitude.

All of them are catalogued with a 15′ fraction which should not appear if the longitudes had been transformed by a simple addition of the precession constant. Nonetheless, the four exceptions cannot be omitted as transmission errors without any evidence from the manuscripts.

(ii) The proportion of 40′ values is much higher than any other. If the catalogued longitudes represent readings from an observational instrument, one would expect the full and half degrees to be more frequent than the others.

(iii) While the full degrees are almost as frequent as the 40′ values, the number of half degrees amounts to less the half of them.

(iv) Without knowing the specific graduation of the observational instrument, one would expect a similar distribution of the degree fractions for the ecliptical latitude and longitude, if the catalogue coordinates are direct readings of a graduated scale. This obviously is not the case.

5.5.1 Fractions of Degrees in Latitude

Vogt and Newton propose a different reading procedure for the ecliptical latitudes, although both assume that the coordinates were taken from the graduation of an astrolabe.

Two Instruments

Vogt interprets the 1/4 degree and 1/6 degree divisions as the outcome of two different observational devices having respectively a half or a third degree graduation. The graduated rings cannot be divided into further divisions, because otherwise the intervals between two marks would be too small, assuming a reasonable size for the instrument.[31] Further evidence for the division into thirds of a degree of one instrument is, according to Vogt, the smaller number of uneven 1/6 degrees (101 for 10', 104 for 50') as opposed to the even 1/6 degrees (110 for 20', 126 for 40').

The directly recognizable 1/4 degrees are problematic for Vogt's account. Making up a total of 95 stars, the 15' values are twice as frequent as the 45' values with 49 stars. This unequal distribution could not possibly arise had the latitudes been measured with an astrolabe graduated into 1/4 degrees. Vogt suggests interpreting this notable imbalance as the result of a graduation of half degrees.[32] Vogt does not carry out a more detailed analysis of the frequency distributions.

The number of full and half degrees is the sum of the respective frequencies of the 1/4 and 1/6 degree graduation, if the fractions were read off from two different instruments. The star catalogue in the Almagest contains 228 stars with latitudes of full degrees and 212 stars with half-degree latitudes. Since the recognizable 1/6 degrees, i.e. the 10', 20', 40', and 50' values, are equally frequent,[33] one should expect that the number of full and half degrees of this graduation is about the same.

In the following the distributions of fractions is modelled according to various observational arrangements. A measurement procedure is historically possible if the resulting frequency of fractions is compatible with the number in the star catalogue. Assuming that the number of full and half degrees in the 1/6 degree graduation is 110 stars in each case,[34] then the complementary amount of 1/4 degree stars would be 119 and 101 stars. The rare occurrence of the 45' (3/4) fractions contradicts such a straightforward model. If one wants to preserve the model in general, one has

---

[31]Vogt, H. (1925), col. 41. Vogt quotes Pappus, who attributes a size of one cubit to the astrolabe. Assuming that one Egyptian cubit is equivalent to 535mm, the size of the astrolabe should not exceed 1m.

[32]Vogt, H. (1925), col. 42: "The supposition of consistent reading-off of a graduation of quarter degrees is impossible in the light of these peculiarities. Possibly we may recognize them as the result of graduation in half degrees, which left the quarter degrees to be estimated and thus gave free play to subjective judgments of the observer and his mistakes."

[33]The χ^2-test shows, with $\chi^2 = 3.01$, that there is no significant difference.

[34]The mean frequency of the other 1/6 degree fractions is 110.5.

fraction	0′	10′	15′	20′	30′	40′	45′	50′
Frequency of fractions in latitude, two graduations.								
Rounding 1: 3/4 degrees rounded to full and half degrees.								
Rounding 2: 3/4 degrees rounded to full degrees.								
1/6° stars	110	110		110	110	110		110
1/4°, without rounding	91		91		91		R	
1/4°, rounding 1	28				10		53	
1/4°, rounding 2	38						53	
sum (rounding 1)	229	110	91	110	211	110	53	110
sum (rounding 2)	293	110	91	110	201	110	53	110
Almagest	228	101	95	110	212	126	49	104

Table 5.6: Theoretical models for the distributions of degree fractions.

to find an additional mechanism that decreases the number of 45′ fractions, e.g. a rounding step to the next full and half degrees. In this case it is necessary to add the complementary proportion for the full and half degrees to the average frequency of 110.

Since the 1/6 degree stars add up to about 660 stars, there remain 365 stars for the 1/4 degree graduation, which is about 91 stars for each fraction. If the 3/4 degrees are rounded either to the half or full degrees in order to obtain the best fit to the distribution in the catalogue,[35] then about 28 stars with a 45′ fraction should be rounded to the full degree and 10 to the half degree. So far nothing is known about the reason why this rounding should have occurred.

First we will analyze whether it is at all possible that such a rounding procedure was used. The following table compares the theoretical distribution of fractions for the rounding model 1 – which distributes the mean frequency of 91 stars to 53 for the 45′ fraction, 10 stars to the half degree, and 28 stars to the full degree – with the rounding model 2, which assigns 38 stars out of the 91 with a 3/4 degree fraction to the full degrees.[36]

Thus it appears to be possible to confirm the assumption of two differently graduated astrolabe readings with the documented data, if one assumes an unknown rounding mechanism for the 3/4 of a degree coordinates. Furthermore it is clear that a partial rounding to the full and the half degrees leads to an excellent agreement with the distribution of fractions as found in the Almagest. It is noticeable that rounding model 2 still provides a satisfactory description of the empirical distribution. The frequency counts alone cannot differentiate between the varieties of rounding models.

The rounding of numbers has other effects besides the change of frequency of the fractions. If for example the full degrees cover measurements which were consistently rounded from a particular fraction, then the mean coordinate error for this fraction should differ from those fractions which do not contain rounded coordinates. If for

---

[35]The best fit is the minimum of the sum of differences between the predicted frequency and the documented one of the Almagest.

[36]"R" stands for the fractions to be distributed to the other fractions according to the two rounding models.

example a larger proportion of the full degrees had been truncated, then the error for the full degrees should be smaller or more negative than for the other fractions.

In general, the mean μ_0 of the ecliptical latitudes β_{1i} and β_{2i}, coming from two different sources, is obtained by

$$\mu_0 = \frac{\sum_{i=1}^{n} \beta_{1i} + \sum_{k=1}^{m} \beta_{2k}}{n+m} \tag{5.1}$$

If the latitudes of one source are systematically rounded to the next larger value, this is equivalent to adding to each latitude a rounding value of R_i. It affects the mean value μ_β as

$$\mu_\beta = \frac{\sum_{i=1}^{n} (\beta_{1i} + R_i) + \sum_{k=1}^{m} \beta_{2k}}{n+m} \tag{5.2}$$

The difference between the two mean values is obtained with the mean rounding effect $\overline{R} = \sum_{i=1}^{n} R_i/n$ as

$$\mu_\beta - \mu_o = \frac{\overline{R}n}{n+m} . \tag{5.3}$$

If e.g. the proportion $n/m = 1/2$ of the 3/4 degrees had been rounded to the next larger full degree, then, with $\overline{R} = 1/4°$, one expects a change of the mean values by $\mu_\beta - \mu_0 = +0°.08$. With truncated fractions the rounding effect enlarges to $\overline{R} = 3/8°$ and with it the error to $\mu_\beta - \mu_0 = -0°.13$.

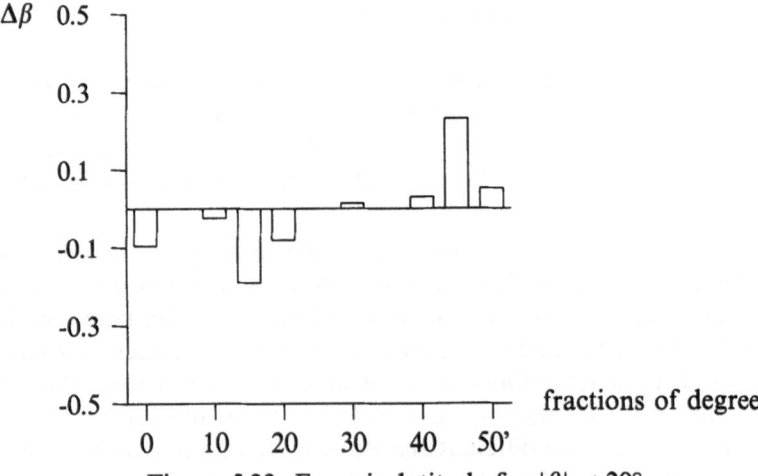

Figure 5.23: Error in latitude for $|\beta| < 20°$.

The histogram in (fig. 5.23) shows the average of the differences in latitude $\beta_{Ptol} - \beta_{calc}$ as a function of the fraction of the degree. It is calculated for the stars with a maximal distance of 20° from the ecliptic.[37]

---

[37]Statistical outliers with an error of more than four times the standard deviation are excluded from the sample.

The various errors in latitude for the individual fractions are clearly recognizable. The uncertainty of the mean values with a standard deviation of about 0.08 is nevertheless so large that there is a considerable chance for random variations without any significance for the particular rounding procedure.

A special mathematical property of the catalogued latitudes leads to a further indication of rounding effects. The ecliptical latitudes are catalogued as positive and negative values, depending on the hemisphere of the ecliptical coordinate system. The Almagest notes the absolute numerical value of the latitude followed by the sign. From many other mathematical examples it seems reasonable to interpret rounding procedures as defined only for the absolute value independent of the sign. For example, while a positive latitude is rounded down, e.g. from $+14;45°$ to $14;00°$, an analogous negative latitude, for instance, $-14;45°$ is not rounded to $-15;00°$ but rather up to $-14;00°$. The rounding of the absolute value must therefore have a different effect on either side of the ecliptic.

The diagram (fig. 5.24) relates only to the differences for the positive latitudes up to $\lambda = 20°$. There is a noticeable negative error in the case of the full degrees and a large positive error with regard to the 3/4 degrees. The error in latitudes for the full degrees changes dramatically when the mean is calculated for negative latitudes (fig. 5.24).

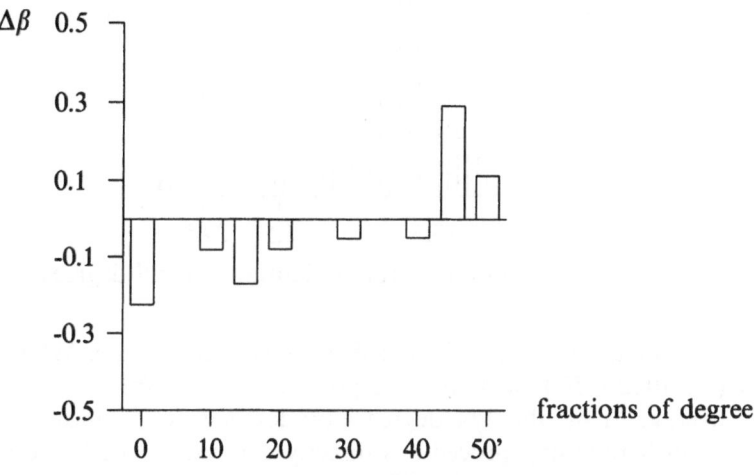

Figure 5.24: Error in latitude for $0° > \lambda > 20°$.

A change in the difference for the positive and negative latitudes of more than the double amount of the standard deviation is recognizable only in the case of the full degrees. For them, the difference Δd between the mean error of the northern and the southern stars is largest (table 5.7).

The table supports the view that the full degrees are mixed with rounded coordinates and in that way their number is increased. The average rounding difference of -0.12 could be interpreted as a rounding down of the 3/4 degree values, as discussed in the sample model. In a similar fashion the difference could be enhanced by star coordinates whose degree fractions were "lost" during the transmission process. Only the difference Δd for the fraction $40'$ comes close to that

	0′	10′	15′	20′	30′	40′	45′	50′
$\beta_+ - \beta_-$	$-0°24$	$0°04$	$0°06$	$-0°07$	$-0°05$	$-0°21$	$0°12$	$0°10$

Table 5.7: Differences Δd for the fractions of the degree.

for the full degree.

The distribution of errors should not be normal if there is a proportion of rounded values among the full degrees. Instead, it should have a second maximum for the different error of the rounded coordinates. Exactly this phenomenon can be seen in the distribution of errors for the full degrees (fig. 5.25).

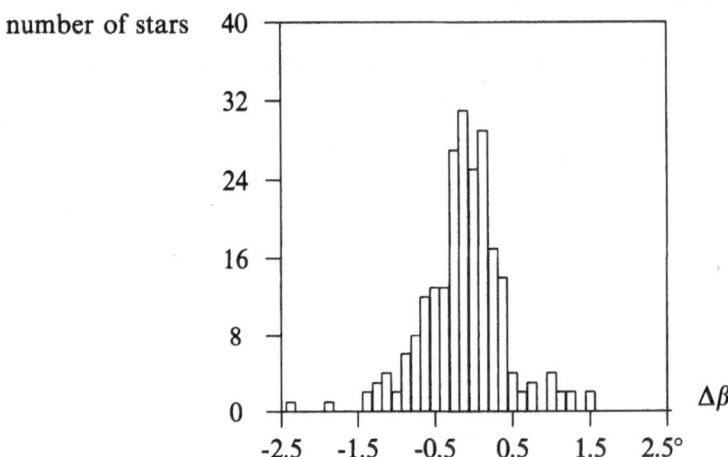

Figure 5.25: Distribution of errors in latitude for full degrees.

The frequency of the recognizable 1/4 degrees is peculiar, too. While the 1/6 degree stars (i.e. $10′, 20′, 50′$), with the exception of the $40′$ fractions, display no considerable average error, the 1/4 degree latitudes deviate significantly in their average errors. Such rounding procedures directly contradict Vogt's thesis of two different measuring procedures for the different accuracies.

One Instrument

The results also contradict Newton's hypothesis of a single observation procedure, in which the coordinates were noted down both in 1/4 and in 1/6 of a degree. According to Newton's suggestion, the degree fractions result from the estimations of intermediate values by the observer with a graduation ring divided into half degrees. Values just in between the marks had then to be noted as 1/4 degree and coordinates slightly closer to the marks to be catalogued in 1/6 of a degree.

Newton's only evidence for such a complicated procedure is taken from a fairly good agreement of the theoretical frequency distribution with the one found in the catalogue.[38] However, there are several arguments against his model:

(i) It is extremely difficult to estimate three different intermediate values. Instead of marking the graduation lines far enough apart to estimate three intermediate values, it makes more sense to mark another line on the graduation ring. There is no comparable historical case with places such a high demand on estimation skills.

(ii) The errors in latitude of the 15′ and 45′ values are different from the errors of the recognizable 1/6 degree latitudes. According to the procedure Newton proposes, these values lie in the middle of two graduation marks and should for that reason be estimated with a better accuracy than is in fact found.

(iii) There is only one star with a magnitude brighter than 3^m (Pollux) catalogued with a recognizable fraction of 1/4 degree. All other stars of that type are fainter. One would expect a similar brightness distribution (cf. fig. 5.26 and fig. 5.10) for the two classes of accuracies, if the 1/4 degree stars were measured in the same way as the stars with a 1/6 degree fraction in latitude.

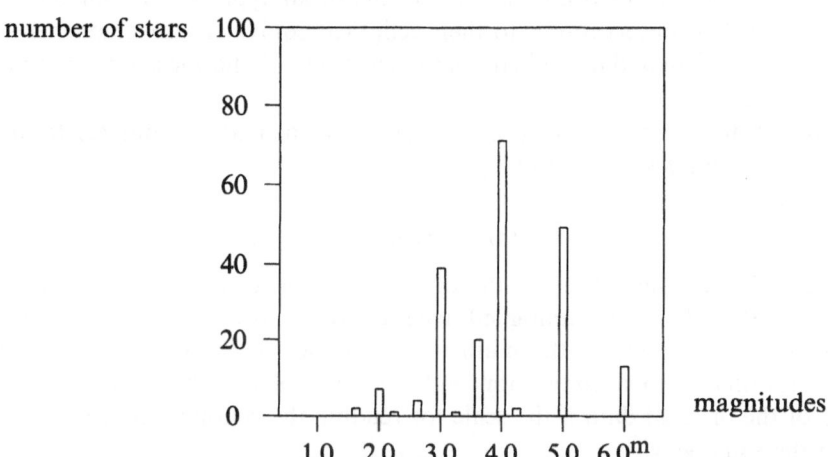

Figure 5.26: Histogram of magnitudes, 1/6 degree stars.

These arguments suffice to reject the observational model proposed by Newton. Up to now all interpretations assume a measurement with the astrolabe. Vogt was committed to this assumption, because he intended to prove the authenticity of Ptolemy's statements in the Almagest. There the astrolabe is the only instrument mentioned. Nevertheless, it is possible that the stellar coordinates were measured with an entirely different setup.

Measurements with the astrolabe have the disadvantage that the setting of the instrument must be permanently adjusted to the daily rotation of the sky. Before

---

[38] Testing with a χ^2-test one finds a significant difference. The theoretical distribution of the fractions is according to Newton's model: 171 stars with 0′, 128–10′, 86–15′, 128–20′, 171–30′, 128–40′, 86–45′, 128–50′. The difference from the empirical distribution is significant, with $\chi^2 = 59$. Only the additional assumption of "suitable" rounding effects can eliminate the severe differences.

measuring the position of a star, one has to turn the astrolabe until the ecliptical longitude of the outer ring coincides with the visual line of the reference star or the moon. Only then can the inner ring of the astrolabe be adjusted to the object of observation and its coordinates determined by the position of the rings. The whole procedure has to be carried out quickly, since the longitudes can increase within four minutes of time by up to one degree. Only a trained observer could obtain reliable results with such a complicated mechanism.

It would be much easier to determine the stellar positions by meridian measurements. The declinations of stars provided by the school of Timocharis, as quoted by Ptolemy to demonstrate the precession effect, provide evidence for meridian observations. Declinations δ are easy to derive from the measured altitude h of the star during culmination.

The ancient astronomer could obtain a complete set of coordinates, e.g. the ecliptical coordinates, by mathematical transformations, if in addition to the altitude the culmination time is recorded.[39] The accuracy of the culmination time of one minute of time is equivalent to an accuracy of 1/4 degree of arc. If this accuracy of measurement had been carried over to the catalogued coordinates, one could imagine a possible origin for the coordinates with a 1/4 degree accuracy in Ptolemy's star catalogue. However, the transformation of an equatorial type of coordinates, as obtained from meridian observations, to their ecliptical equivalent does not merely add constant terms, so that the original degree fractions of the measurements are not easily maintained.

In general, a transformation function t maps the measured coordinates (c_1, c_2) to the ecliptical coordinates of the catalogue.

$$t(c_1, c_2) \mapsto (\lambda, \beta) \tag{5.4}$$

If t is a non-linear function, as in the case of the conversion of equatorial to ecliptical coordinates, then the catalogued accuracy depends not on the measured accuracy but on the accuracy of the mathematical tables or the final steps in the numerical operations. Even a conversion with the help of a globe only preserves the accuracy of the original data if the scale for reading the ecliptical coordinates is graduated in the same way.

Thus the accuracies in the star catalogue do not necessarily reflect the accuracies of the measurement procedure unless the catalogued coordinates are records of the readings from the instrument, for example the readings from an astrolabe. Otherwise the variety of the transformation functions t hides the originally recorded precision of the data from the reconstructions of the historian.

5.5.2 Fractions of Degree in Longitude

Very different from the latitudes is the distribution of the fractions of the degree in longitude. The explanation of these has been considered a decisive criterion for the origin of the star catalogue. Since no explanation considers the ecliptical longitudes

---

[39]The methods are described by Ptolemy in the first two books of the Almagest. However, Hipparchus could not use spherical astronomy with the Menelaus theorems. He had to employ either graphical methods or approximations.

of Ptolemy's star catalogue as direct recordings of the measurements, the historian has to construct a model for the transformation function t. The number of premisses required by these models prevents one from drawing any immediate conclusion from the distribution of degree fractions to the original measurement procedure.

Methodologically, one can only generate plausible models and test the consistency of their predicted frequency distribution with the documented one. The initial simplicity of the fundamental models, intended to explain the main feature of the documented data, is typical for such a model design. Parts of the models are varied, generalized or restricted, leading to a more and more complex picture of the historical background for the scientific activities, about which the historian has only partial evidence in his documents.

In the following the explanation of the distribution of degree fractions in the longitudes is presented as a series of models, numbered according to their systematic order. An integer number is assigned to fundamental models (e.g. "3"), whose hypotheses are included in the more complex submodels. Submodels, with additional features and auxiliary hypotheses, are characterized by numbers with another digit appended to the number of the related more general model (e.g. "3.2").

Fundamental Model 1: The coordinates of the star catalogue are based on Ptolemy's astrolabe readings.

Vogt, who supports this general model, has to base his argumentation on his reconstructed Hipparchan coordinates. The independence of the coordinate errors in the two star registers is a crucial element in the defense of the authenticity of Ptolemy's observations. As was shown before, stars like π Hydrae already display a highly significant common error in Vogt's reconstructions and the Almagest, so that Vogt has to assume a common observational root for at least 10% of the stars. Furthermore, the destruction of Vogt's so-called proof for independent observations leaves model 1 without evidence, except Ptolemy's own statements. Only the implausible hypothesis that about 100 stars in Hipparchus' Aratus Commentary are not representative for the stars of the Almagest, and that therefore their properties cannot be generalized to the rest of the Ptolemaic stars, could save the core of model 1.

Model 1.1: A considerable proportion of the coordinates in the Almagest was measured by Ptolemy.

There are no stars in the Almagest with longitudes of a 3/4 degree fraction, and only four are catalogued with 1/4 of a degree. The number of longitudes with $0'$ and $40'$ fractions sums up to about half of the catalogue, and surprisingly the most frequent fraction is 2/3 of a degree.

Vogt states, neglecting the 10% of stars with Hipparchan origin proven by himself:[40]

> "I can only explain the different frequencies of the even and odd sixths of the degree by the assumption that Ptolemy's astrolabe was graduated in even sixths, i.e. in thirds of a degree, so that only those

---

[40]Vogt, H. (1925), col. 42.

values could be read directly and the uneven 1/6 of a degree had to be estimated. It is the only plausible way to explain why the half degrees count 73 stars less and the 5/6 degrees 69 less than the average of 170 stars, while the 4/6 = 2/3 exceed it by 70."

Vogt's interpretation seems to avoid the difficulty of the too small number of half degrees. Besides the existence of a subset of star coordinates for which a common origin has been proven, there is another internal difficulty in the distributions of fractions contradicting Vogt's model.

One would expect the full degrees to be the most frequently recorded coordinates using a astrolabe graduated in 1/3 of a degree. Even if we leave out of consideration those stars for which the fractions of the degree were omitted in the long tradition of copying the star catalogue and whose longitudes were therefore rounded to the full degree, any astrolabe must have had the full degrees marked more prominently than the fraction graduations. Already from this visual difference the full degree values should have attracted statistically more recordings by the observer.

Nonetheless, the number of 2/3 degrees is about 10% larger than those of the full degree.[41] Also, according to Vogt's model the number of odd 1/6 degree estimations should be about the same, and they should be smaller than any other of the even 1/6 degree coordinates. This is refuted by the number of 10′ longitudes, with double the number of stars compared to the 30′ and 50′ longitudes and 1 star more than the stars with a 1/3 degree longitude. Such uneven distributions cannot be explained by an instrument graduated as proposed by Vogt. The arguments against Vogt's model contradict every interpretation of the longitudes of the star catalogue as direct recordings of an astrolabe reading. In any case there has to be an intermediate step of converting the longitudes to the finally documented data.

Model 1.11: A considerable proportion of the coordinates in the Almagest was measured by Ptolemy. The longitudes were not directly read off the graduation ring but converted to their final form.

Filling the details of this model is tantamount to constructing a conversion procedure t which operates on a given pair of stellar coordinates and derives the documented ecliptical coordinates with the proper distribution of fractions. The most prominent models of this type assume a final addition of a constant term c to the longitude, i.e.

$$\lambda = c + t'(\lambda', \beta') \tag{5.5}$$

From all the peculiarities of the distribution of fractions as discussed before one can conclude that the constant c should be an integer number plus 2/3 of a degree. As possible realizations one could imagine

(i) Ptolemaic measurements with the astrolabe using reference stars with longitudes of the standard epoch of Hipparchus. In a subsequent step all longitudes are converted to the later period. Ptolemy could have used standard longitudes

---

[41]The exact ratio of the number of 2/3 degree longitudes to the full degree is 1:1.081. It is nearly identical to the ratio of the full degree to the half degree in latitude with 1:1.085.

for the reference stars, adjusted the astrolabe with these (false) coordinates and measured the position of the other stars in relation to these fundamental stars. After the measurements Ptolemy then had only to add the precession value of 2°40′. This procedure introduces a small periodical error into the final coordinates, because the ring system of the astrolabe is misplaced by 2°40′ in longitude. This error is so small for both coordinates that it cannot be significantly tested.[42]

(ii) measurements of only the difference in longitude between the reference stars and the stars being measured. With a freely rotating graduation ring one could determine the longitudinal difference to the reference star and later add the precession value to the Hipparchan standard coordinate.

The arguments in favor of these models suffer from the fact that at least in minor parts Ptolemy's description of the use of the astrolabe has to be changed and, more seriously, that they cannot explain why a considerable number of coordinate measurements go back to Hipparchan sources, as was proven in the analysis of the Aratus Commentary, while the remaining coordinates were determined later in a fairly indirect fashion.

However, the existence of such models demonstrates that the distribution of fractions cannot decide the history of Ptolemy's star catalogue.

Fundamental Model 2: The star coordinates in the Almagest were originally measured by Hipparchus.

Vogt's attempted proof of the independence of the reconstructed Hipparchan coordinates from those in the Almagest is quoted as the standard refutation of the fundamental model 2:[43]

> "The worst excess of the modern notion of the strict dependence of the Almagest on Hipparchus is the belief that we can obtain Hipparchus' catalogue simply by taking Ptolemy's catalogue in Almagest VII and VIII and lopping 2 2/3° (to account for precession) off the longitudes. This was conclusively refuted by Vogt, who showed by a careful analysis of the coordinates of 122 stars derived from the commentary on Aratus that in almost every case there was a significant difference between Hipparchus' and Ptolemy's data."

In fact Vogt's argumentation is much less than reliable. Newton, who accepted Vogt's propositions without control, had to infer that Hipparchus compiled a new star catalogue after he wrote his Aratus Commentary. Only those later coordinates, of which is no direct historical trace, could then be used by Ptolemy and converted to his epoch by adding 2°40′ to the longitudes.

---

[42] We cannot follow contrary claims by Rawlings. A detailed discussion of this variant can be followed in: Gingerich, O. (1980), Was Ptolemy a Fraud?, *Quarterly Journal of the Royal Astronomical Society* **21**, pp. 253–266; Gingerich, O. (1981), Reply to Newton, *Quarterly Journal of the Royal Astronomical Society* **22**, pp. 40–44; Newton, R. R. (1979), On the Fractions of Degrees in an Ancient Star Catalogue, *Quarterly Journal of the Royal Astronomical Society* **20**, pp. 383–394; Newton, R. R. (1980), Comments on "Was Ptolemy a Fraud?" by Owen Gingerich, *Quarterly Journal of the Royal Astronomical Society* **21**, pp. 388–399.

[43] Toomer, G. J. (1975), p. 217.

Model 2.1: The star coordinates were originally measured by Hipparchus. Ptolemy converted the Hipparchan data by adding 2°40′ to the longitudes and by rounding the original 15′ fractions to 0′ (i.e. 15′ + 40′) and the 45′ fractions to 20′ (of 45′ + 40′).

The method of rounding the 1/4 degrees suggested by Newton does not grow out of an historical examination of Greek mathematical and astronomical practice, but was proposed with the sole intention of constructing the best approximation to the frequency distribution of the fractions in longitude. This strategy can prove neither the correctness of the initial hypotheses nor the origin of the star catalogue. It only shows that one set of hypothetical assumptions can derive the documented properties of the data.

The number of implicit assumptions in every model is enormous and usually they cannot be confirmed directly by historical evidence. Just how concealed the implicit presuppositions can be is evinced very clearly by Newton's proposed rounding practice.

The original distribution of degree fractions in longitude, before Ptolemy converted them according to Newton's rounding procedure, is set equal to the distribution of fractions in latitude. In this assumption there seems to be no commitment to a particular method of observation. At this point Newton merely assumes that the measurement techniques were the same for the latitude and the longitude.

From the model which applies the assumed rounding procedure for the 1/4 degree stars one derives a distribution of fractions in longitude strongly resembling the documented distribution.[44]

fractions of degree	β	number of instances for:		λ
		+40′, no 1/4°	+40′, incl. 1/4°	
0′	229	108	203	222
10′	103	211	211	181
15′	95	0	0	4
20′	108	126	174	180
30′	211	105	105	98
40′	126	229	229	222
45′	48	0	0	0
50′	105	103	103	100

Table 5.8: Newton's theoretical distribution of fractions in longitude.

Newton's rounding procedures require that one increase the number of the 0′ and 20′ fractions by the respective amount of the rounded 55′ = 15′ + 40′ and 25′ = 45′ + 40′. There is no other way to define the rounding practice if one wants to

---

[44]Meaning of columns: "fractions of degree": The fractions of the longitude and latitude in minutes of arc; "β": number of instances in longitude; " +40′, no 1/4°": theoretical number of fractions in longitude after addition of the precession constant without consideration of the 1/4°; "+40′, incl. 1/4 deg.": theoretical number of fractions in longitude after addition of the precession constant including the 1/4 degrees; "λ": documented distribution of fractions in longitude.

approximate the documented distribution.

We have already refuted Newton's proposal that all observations were done with one graduation circle.[45] It is not yet apparent how important the rejected auxiliary hypothesis turns out to be for the consistency of the over-all model.

All evidence supports the thesis that the 1/4 degree coordinates were determined by another instrument or evaluation practice than the stars with a 1/6 degree coordinate. If this is the case, one should also assume that both coordinates were measured either with the instrument of a 1/4 degree accuracy or by an instrument with a 1/6 degree graduation. A different accuracy in the coordinates is unlikely, because otherwise one would obtain no 1/4 of a degree longitudes at all – or one has to assume the use of one instrument with latitude graduations of 1/4 degree and longitude graduations of 1/6 degree, and the existence of a second instrument with exactly the opposite arrangement of graduations.

The latter arrangement of instruments is implausible because whatever reason led to a different graduation for one instrument should also apply to the second one. Therefore, if one interprets the coordinates of the star catalogue in the Almagest as direct readings from the astrolabe and assumes that there were originally longitudes with 1/4 degree fractions, one should conclude that there were two instruments with an identical graduation in both coordinates.

This leads directly to inconsistencies with Newton's rounding procedures. Since the 1/4 degrees in longitude are rounded, after the addition of the precession constant, to the catalogued fractions $0', 10', 20'$, and $40'$, these are the only fractions in longitude with which one can possibly combine 1/4 degree fractions in latitude. In particular the fractions $15'$ and $45'$ in latitude cannot be combined with $30'$ and $50'$ in longitude. In fact, there are 30 stars in the Almagest with $15'$ or $45'$ in latitude and half a degree in longitude, and 10 stars with a 1/4 degree in latitude and $50'$ in longitude. This is a clear contradiction to the observational practice as described by Newton.

The internal structures of the data link the methods of observation closely with the distributions of coordinate fractions. Either the division of coordinates coincides with the existence of two sets of instruments (or evaluation methods) with different accuracies for both coordinates – in which case Newton's conception of the rounding procedures has to be discarded – or the different accuracies cannot be attributed to the same set of stars, which is an implausible solution.

Model 2.2: A large proportion of the star coordinates was originally observed by Hipparchus who also used them for writing the second part of his Commentary on Aratus. Ptolemy transformed the Hipparchan data to ecliptical coordinates referred to the epoch $+137$, using the precession constant $2°40'$.

A consistent picture emerges from the examination that has been conducted so far.

(i) Especially for the stars of the Almagest whose latitudes are divided into 1/6 degrees, the Hipparchan coordinates reconstructed by Vogt and the direct correlation analysis reveals such a strong dependence of errors that their taking

---

[45]Cf. section 3.6.4.

over by Ptolemy is certain. The subset of 122 of reconstructed coordinates is not related to a special group of stars. Hence one cannot deny that they represent all stars in the Almagest. Therefore one has strong evidence for the hypothesis that all stars with an accuracy of a 1/6 degree in latitude were originally observed by Hipparchus. This is consistent with the fact that the star catalogue of the Almagest contains no star that is only visible from Alexandria but not from Rhodes.

(ii) The analysis of the Solar Theory.

The periodic errors of the solar theory are able to explain the variations in the longitudinal errors of the stars. The positions of the reference stars are dependent on the solar theory. Therefore their coordinates correlate to the error of the theoretical sun at that position where the sun was when the measurement was taken. This implies a phase shift between the errors in the stellar positions and the solar theory by exactly their difference in longitude at the time of observation.

With the assumption that the ecliptical equivalents of the Hipparchan coordinates were increased by Ptolemy by 2°40′ it was shown that the errors in longitude are not minimal for the epoch -128, but for a time about 9 years earlier. Vogt's independent dating of his reconstructed coordinates indicates the same observational period. These results suggest very strongly that there never was a second Hipparchan star catalogue composed after the compilation of the second part of the Aratus Commentary. It appears that Hipparchus had copious observational data at his disposal from which he derived the simultaneous phenomena of the stars for his Commentary either with the help of mathematical tables, or, and more probably, with the aid of a globe.

(iii) Further support for the Ptolemaic use of Hipparchan data is provided by the peculiar error in the magnitude of the star θ Eridani.

The six classes of magnitudes are used for the first time in the Almagest. The Hipparchan Commentary vaguely portrays the brightness of a star as "glaring", "bright" or "dark".

The scaling of the magnitude classes has been maintained until today. The constancy of the concept was guaranteed by the way it was introduced in the Almagest. Ptolemy does not mention how the magnitude classes were determined. Therefore it is not the physical or astronomical definition that has maintained the astronomical standard but the huge list of standard samples in the form of a list of well defined stars and their attributed magnitude number.

Though the modern definition of magnitude classes agrees well with Ptolemy's numbers, one must be careful in analyzing individual magnitudes. Since nothing is known about Ptolemy's measurement of magnitudes, it is fallacious to interpret deviations from the corresponding modern magnitude class with the standards of the modern physical theory of brightness determinations.

An example would be using a model of the extinction effect with the parameter of the geographical latitude to approximate the Ptolemaic stellar magnitudes in order to determine the latitude of their observation. This can be extremely misleading. The effect of the extinction should be noticeable only in very large southern declinations. Those Ptolemaic stars show, besides odd values of magnitudes, also large deviations in their position. Without knowing how the magnitudes were determined, one cannot show whether the extinction is responsible for an average decrease in the magnitude of the southern stars.

Another, much more plausible interpretation of the magnitude numbers sets them in relation to the arcus visionis of a star. If the magnitude of the stars is a simple function of the arcus visionis or a similar condition of the first visibility of a star, then one could explain the method of measurement and the reason for how and why a numerical concept was introduced by Ptolemy. For the first time one can explain why Ptolemy added the magnitude of a star to his catalogue at all. The effort in measuring and recording magnitudes would have been much too great, if their only use had consisted in producing an impressive representation of the sky on a globe. And since the only astronomical application of the stellar magnitudes is the calculation of the visibility conditions, one could conversely conjecture that at least some brighter magnitudes were determined by the recordings of those conditions. Only the magnitudes of the circumpolar and very faint stars then had to be determined by different methods, presumably by conventional estimations.

The Almagest catalogues the star θ Eridani (No. 805) as a star of the first magnitude. Furthermore this star has an exceptionally large error in longitude of more than $3°$ in the Almagest as well as in the reconstructed Hipparchan coordinates. Today this (non-variable) star has a magnitude of only 3.42^m. Such a difference cannot be explained by the usual errors of estimation. A star of the first magnitude catches the attention of any observer immediately and it would play an important rôle in the mythical complex of the constellation. The star θ Eridani was never mentioned in such a context. It would also have been recorded if such a bright star had suddenly disappeared from the sky. Furthermore, the star is also listed as a first magnitude star in Ptolemy's book on the "Phases of the Fixed Stars", and it is called a "brilliant" star in Hipparchus' Aratus Commentary. The frequent reference to this star in a variety of documents excludes a possible scribal error in the Almagest or later.

A sound interpretation must correlate the large error in magnitude with the error in longitude. Since both errors can be found in Hipparchus' Commentary as well as in the Almagest, there is no doubt that the Hipparchan observations of θ Eridani were taken by Ptolemy.

Model 2.21: Together with the properties of model 2.2 it is assumed that the 1/4 degree and 1/6 degree accuracies in the coordinates are caused by two different observation or evaluation methods.

The discussion of Newton's proposal for the rounding practice showed the importance of a detailed analysis of the combination of the degree fractions in both coordinates. The properties of the mean error in latitude for the individual fractions lead to the thesis that the two accuracies were caused by two different instruments or evaluation methods.

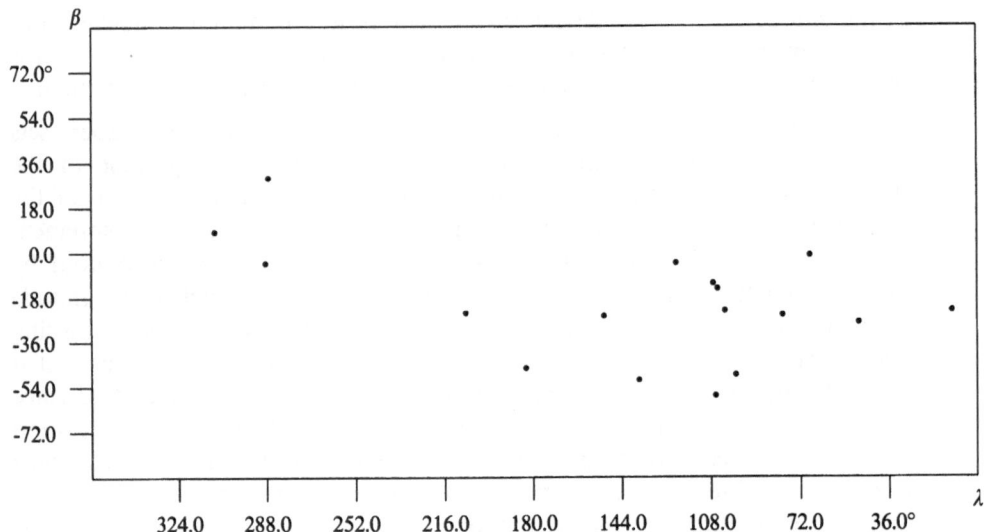

Figure 5.27: 1/4 degree latitude in combination with 1/2 degree in longitude.

Independently of the question of who observed the stars with 1/4 degree accuracy, with a consistent rounding procedure they should be found in combination with only 4 accuracies in longitude. The following table shows that in fact the 1/4 degree latitudes are found in (well-distributed) combination with every accuracy in longitude.[46]

fraction in λ	Z_v	Z_{v-}	Z_v/G	h
0′	34	19	0.238	0.217
10′	26	13	0.182	0.117
20′	29	11	0.203	0.176
30′	16	14	0.112	0.096
40′	28	13	0.196	0.234
50′	10	5	0.070	0.098

Table 5.9: Combinations of fraction in latitude and longitude.

It follows immediately that either the division into two observation procedures or the assumption of the rigorous rounding practice is inconsistent with the documented data.

The table reveals one irregularity which could point to a rather complex rounding procedure. For all longitude fractions with the exception of the half degree, the 1/4

---

[46]"Z_v"= number of 1/4 degree stars with a 15′ or 45′ fraction in latitude. "Z_{v-}"= number of 1/4 degree stars with southern latitude. "Z_v/G "= ratio of the number of 1/4 degree stars of the particular fraction in latitude to the total number of 1/4 degree stars (143). "h" = ratio of the number of the respective fraction in longitude to the total amount of stars (1025).

degrees are distributed equally in the northern and southern ecliptical hemispheres. Only the half degrees in longitude combine in 12 out of 14 cases with 1/4 degrees with negative latitudes. Such a distribution cannot be explained by a random distribution of the fractions among all stars, as indeed Newton's proposed measurement procedure implies.

Map 5.27 shows all stars with 15′ or 45′ in latitude combined with a half degree in longitude.

An entirely different distribution is shown in map (fig. 5.28), which covers all stars with 15′ and 45′ in latitude combined with 50′ in longitude.

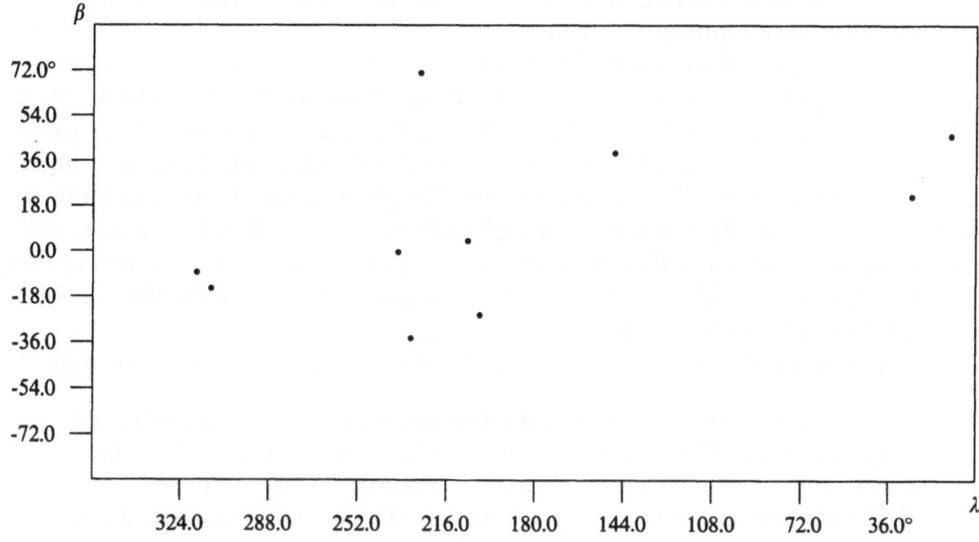

Figure 5.28: 1/4 degree latitude in combination with 50′ in longitude.

The irregularity in the half degrees could provide a clue to a special rounding procedure. On the one hand the free combination of the 1/4 degrees with the other fractions in latitude excludes observations with two instruments and a rigorous rounding procedure, on the other hand the complexity of the data demands more than one graduation ring as a measuring scale. The only possible solution maintains the division into two different observation or evaluation procedures, but rejects the interpretation that the 1/4 degree stars were directly measured by readings from the scale of an astrolabe.

5.6 Hipparchus' Commentary on Aratus

5.6.1 Sources

Hipparchus expressed the major purpose of his "Commentary on Aratus and Eudoxus" in the introduction addressed to Aischrion. There he restricts the scope of his work to the astronomical (i.e. mathematical) analysis of the celestial phenomena described by Aratus in his poetic "Phainomena". Aratus, who derived his astronomical knowledge from Eudoxus, pictures the constellations and their arrangement in the sky according to the apparent locations and brightness of their stars.

Book I of the Commentary contains a detailed criticism of Aratus' description of the constellations; Hipparchus points out inconsistencies or mere misrepresentations of facts. All of these investigations are based on the visibility conditions of the sky at the geographical latitude of Athens.

The second part of the Commentary gives, in contrast to that, an extensive systematic description of phenomena "as they appear in reality"[47] for all constellations relative to the latitude of Rhodes ($\varphi = 36°$). In this part of the work Hipparchus abandons the more narrative style of his criticism of Aratus and presents a highly formalized presentation of the constellations. The phenomena of a constellation are always described in the same pattern and with the same technical notation. Only the names of the stars and the corresponding numerical values vary. The description shows a strictly repetitive form – like a catalogue – it omits only the graphical representation of two-dimensional arrangements.

The first constellation described by Hipparchus is Bootes, and it begins with:[48]

> "Bootes rises simultaneously with the zodiac from the beginning of Virgo until the 27th degree of Virgo. While Bootes is rising, the zodiac culminates from the middle of the 27th degree of Taurus until to the 27th degree of Pegasus. The first star of Bootes to rise is the one in the head, and the last is the one in the right foot."

The passage continues with a list of the stars culminating at the moments of the start and finish of the rising of the constellation. Since Hipparchus has already given the degree of the zodiac culminating at those times, this is equivalent to stating the ecliptical longitude of the degree on the zodiac culminating simultaneously with those stars. Finally the text mentions the time taken by the constellation to rise completely above the horizon.

In these descriptions we find five different types of different phenomena (table 5.10).[49]

Without atmospherical effects like refraction and extinction the star would rise and set on the geometrical horizon. Simultaneously, the ecliptical degree Φ_1 on the zodiac intersects the geometrical horizon and its value is given by Hipparchus in the description of the rising stars. At the same moment the degree Φ_2 on the zodiac culminates and that value is provided by Hipparchus as a phenomenon of type two.

---

[47] Hipparchus (1894), p. 6.

[48] Hipparchus (1894), p. 186.

[49] The *degree of the ecliptic* is the ecliptical longitude of a point on the zodiacal circle and its value for the five types of phenomena will be abbreviated as Φ_i with an index i for the i-th.

Type	Symbol	Description
1	Φ_1	The degree of the ecliptic rising simultaneously with a star
2	Φ_2	The degree of the ecliptic culminating when a particular star rises
3	Φ_3	The degree of the ecliptic setting simultaneously with a star
4	Φ_4	The degree of the ecliptic culminating when a particular star sets
5	Φ_5	The degree of the ecliptic culminating together with a particular star

Table 5.10: Types of stellar phenomena.

If a star culminates simultaneously with a degree P on the ecliptic, whereby both P and the star cross the meridian at the same time, Hipparchus reports this as a phenomenon of type five with the value of Φ_5. The values for the setting stars are described in the same way as for the rising stars.

5.6.2 Numerical Values

Hipparchus follows what he calls the "old Mathematicians"[50] in placing the vernal point at the beginning of Aries. The zodiacal signs are then arranged according to the traditional order.

The method of counting of degrees within a zodiacal sign in the Commentary could cause some confusion. The beginning of a sign is called 'ἀρχή' and the following degrees are referred to by the greek numbering system using letter symbols. They start with the symbol 'β' for '2' and end with '30'. The symbol 'α' is missing in the text and Manitius correctly points out that the beginning of the sign ('ἀρχή') should be substituted for it. Because there is also a '30' in the text, it is obvious that Hipparchus expressed degrees of a sign as ordinal numbers.

Accordingly, in the first paragraph (quoted above) the 27th degree refers to a longitude of 26° in Virgo. With the additional information about the position of the vernal point we derive the equivalent expression $\lambda = 150° + 26° = 176°$ in longitude. The interpretation of the 'middle of the degree' follows the same principle of counting: '*The middle of the 27th*' has to be translated into '26°.5'.

This coherent way of counting degrees is not preserved in the translation by Manitius. The numerical values for integer degrees are simply taken over, e.g. '27°' for 'the 27th degree'. On the other, hand the half degrees are translated into their cardinal value, e.g.'26°.5' for 'the middle of the 27th degree'.

To avoid any confusion, the following analysis cites the Hipparchan values according to a consistent conversion into the cardinal counting system. Only column v of the catalogue in Appendix C shows the numerical data according to the translation of Manitius.

Book II contains in total 619 numerical entries for 292 stars. Usually the number of simultaneous culminations exceeds the number of values of the other phenomena (Table 5.11).

In the case of simultaneous culmination (phenomenon type five) we occasionally find positions slightly west or east of the meridian (table 5.12).

---

[50]Hipparchus (1894), p. 132.

Type of numerical entry : number	
For all phenomena	: 619
Phenomena of type 1	: 84
Phenomena of type 2	: 82
Phenomena of type 3	: 87
Phenomena of type 4	: 85
Phenomena of type 5	: 281

Table 5.11: Distribution of phenomena in the Commentary.

Type No.	positional description	: number
1	1 diameter of the moon west of the meridian :	26
2	nearly :	1
3	1 diameter of the moon east of the meridian :	18
4	2 diameters of the moon east of the meridian :	2
5	a little west of the meridian :	7
6	a little east of the meridian :	8
7	1 1/3 diameters of the moon east :	7
8	1 1/3 diameters of the moon west :	1

Table 5.12: Additional specifications in the phenomena.

Hipparchus did not provide any direct statement of the size of these units in relation to angular degrees. Appendix C therefore notes in column *a* the type number of the additional positional qualification.

In some instances the interpretation of the text is uncertain because:

(i) The identification of a star cannot be firmly established.

(ii) The original text of the manuscript cannot be reconstructed.

(iii) Two or more stars are jointly mentioned in reference to one numerical value although the stars cannot rise, set, or culminate at the same time.

In order to avoid the inclusion of less precise material the analysis considers only stars not affected by any of these uncertainties. In the tables they are referred to as *valid* star entries.

For the majority of the stars in the Commentary we have more than one reference so that in total 619 data entries refer to 292 different stars (table 5.13).

The longitudes are fairly well distributed over the whole range of values in a zodiacal sign, with significantly less frequent half degrees. There also appears to be a periodical fluctuation with a minimum around the degree 22, but it is not clear whether this is of any significance. The next diagram (fig. 5.29) shows the frequency distribution of longitudes in the zodiacal signs:

Appendix C lists all *different* entries of Book II of the Commentary for the five types of phenomena. The (few) cases of repetitive entries are omitted.

class of stars	: number
different stars in total :	292
valid stars :	271
valid stars with no addition :	245
valid stars for phenomenon type 1 :	69
valid stars for phenomenon type 2 :	67
valid stars for phenomenon type 3 :	74
valid stars for phenomenon type 4 :	75
valid stars for phenomenon type 5 :	194

Table 5.13: Number of stars associated with phenomena.

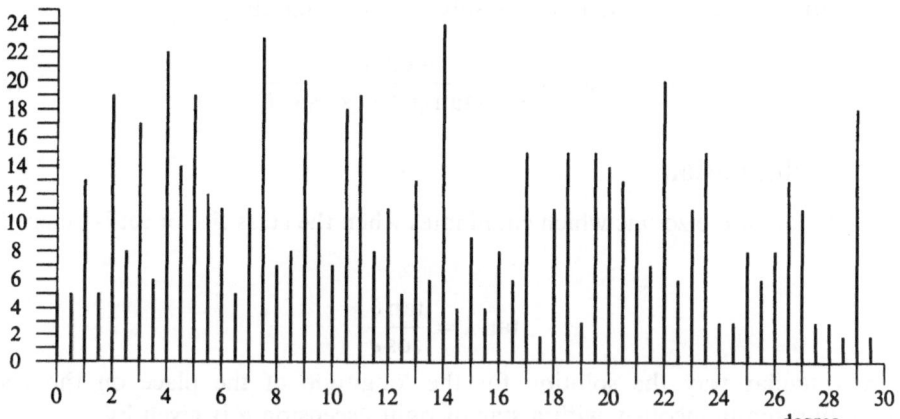

Figure 5.29: Distribution of the degrees of the phenomena.

5.7 Calculation of Phenomena

5.7.1 Local Sidereal Time

A convenient way to calculate simultaneous places of the zodiac by means of modern spherical astronomy first derives the local sidereal time of the phenomenon in question.

Given are the equatorial coordinates α, δ of a star. From the equations for the transformation from equatorial to horizontal coordinates one derives with

$$\sin h = \sin \delta \sin \varphi + \cos \delta \cos \varphi \cos H \tag{5.6}$$

and the altitude $h = 0°$ the hour angle H of the event by

$$\cos H = \tan \delta \tan \varphi \tag{5.7}$$

or the equivalent expression

$$\tan^2 \frac{1}{2}H = \frac{\cos(\varphi - \delta)}{\cos(\varphi + \delta)} \tag{5.8}$$

The local sidereal time θ is then easily obtained with

$$\theta = H + \alpha. \tag{5.9}$$

The hour angle H of a celestial object is defined as the rotation of the equatorial system since the moment of the object's culmination. It follows that the culmination of a star takes place at the local sidereal time

$$\theta = \alpha. \tag{5.10}$$

5.7.2 Simultaneous Rising and Setting

From the local sidereal time the longitude of the simultaneously rising and setting point of the zodiac can be derived by solving the equation

$$\tan \lambda = \frac{-\cos \theta}{\sin \epsilon \tan \varphi + \cos \epsilon \sin \theta} \tag{5.11}$$

5.7.3 Culmination

The longitude of the zodiac which culminates when the stars rise or set is determined by

$$\tan \lambda = \frac{\tan \theta}{\cos \epsilon} \tag{5.12}$$

In a similar way the solution for the longitude of the place on the zodiac culminating simultaneously with a star of right ascension α is given by:

$$\tan \lambda = \frac{\tan \alpha}{\cos \epsilon} \tag{5.13}$$

Hereby all types of phenomena of the Commentary have been derived from familiar astronomical concepts.

5.8 Deviations from Reality

5.8.1 Comparing two Catalogues

The question of the strongest historical interest centers on the relationship between a hypothetical "Hipparchan Star Catalogue" and the stellar positions as they are described by Ptolemy in the Almagest.

Vogt's approach to answering the question is straightforward: the test hypothesis claims that the positional information of both the Commentary and the Almagest refer to the same set of observational data; the alternative hypothesis asserts that the two sets of stellar coordinates have a different observational origin. The decision about these hypotheses can be made on the basis of an analysis of the errors: the test hypothesis is true if both sets of coordinates show similar errors, and must be rejected if the errors differ significantly.

Only if both star registers were perfectly precise and therefore both catalogues contained only true stellar positions, would there be no criterion to decide between the both hypotheses mentioned above. In that case the question of a common origin would have had to remain unanswered even if a complete Hipparchan star catalogue had been found.

In fact the situation is not so indecipherable. All measurements necessarily involve errors. The particular kinds of error depend on the observational methods. Errors in declination measurements made with a meridian instrument are usually smaller than observations of horizon phenomena; while the former typically involves errors in the measurement of height and timekeeping, the measurements with an astrolabe increase the number and size of possible errors.

In our analysis it is necessary to distinguish systematic from random errors. The difference can be defined in terms of repetitiveness: an error is systematic when it is repeated in the sequence of measurements in the same way; a random error cannot be reproduced in that way, and repeated measurements involving random errors float around some central values.

Usually a series of measurements involves both types of errors. Systematic errors distort results because for example the instrument is not perfectly adjusted or some auxiliary hypotheses necessary for the interpretation of data entail an error. Random errors occur accidentally during the measurements.

This very last type of error is a useful way of distinguishing between both possible interpretations of the origin of the star catalogues. Errors in the catalogues can be either systematic or random errors. Random errors are unlikely to be repeated during a second sequence of observations, hence both catalogues should have independent random errors in the case of independent observations and they should incorporate the same random errors otherwise.

There is a further complication. So far only the hypothesis of a common observational origin for a complete set of coordinates has been considered. One also could imagine partial adaptations, i.e. only parts of one catalogue are used for the other one. All proofs based on comparing random errors should therefore consider the less presumptive hypothesis related to *some* pairs of coordinates from the catalogues. For example, if it can be proven that pairs of coordinates do have the same observational origin and if there is no evidence to support the supposition that others of the catalogue do not have the same observational origin, then one is justified in claiming a common observational basis for the whole catalogue. Applying this technique, one obtains four proof situations:[51]

Some coordinates same origin	Some independent	Conclusion
+	+	Only partly same origin
+	−	Catalogues probably of same origin
−	+	Catalogues probably are independent
−	−	No conclusion possible

---

[51] "+" indicating positive evidence and "-" no evidence. In general it seems to be impossible to define exactly the conditions when positive or negative evidence is provided. This has to be decided from case to case.

A proper historical evaluation of the documents can only be achieved by using two different lines of evidence. First one analyzes whether some pairs of coordinates have a common origin:

> *The thesis of a different observational basis for some pairs of coordinates is justified, if the related data show independent random errors in the measurements. It is required that the errors under consideration be random errors involved in the measuring or evaluation process; differences in the random errors must be significant and for example not due to any method applied in the analysis.*

Evidence for a common observational origin of some pairs of coordinates takes the form

> *A common observational basis for some pairs of coordinates can be established if the random errors do not differ significantly. It is required that the errors under consideration be random errors involved in the measuring or evaluation process and furthermore that the similarity of the random errors be significant and for example not due to systematic errors or any method applied in the analysis. The probability of the similarity being generated accidentally has to be small.*

On the basis of these preliminaries Vogt's investigation can be reconsidered. Vogt calculates the ecliptical coordinates of a star if at least two of either the rising, setting or simultaneous culmination phenomena are given or there is some other information from other sources.

As a justification for the necessity of a "reconstruction" he claims that in doing so any auxiliary hypothesis would be avoided.[52]

As already pointed out in chapter 4.2, Vogt's statement is clearly mistaken. During the course of reconstruction Vogt had to assume that

(i) the geographical latitude relative to the phenomena is 36°.

(ii) the obliquity of the ecliptic to the equator is $23°.86$.

(iii) both assumptions are valid for Hipparchan data from all other sources.

(iv) all data refer to the same observational time.

(v) all data used for reconstructing one stellar position represent the position of the star equally well: they should contribute to the resulting coordinates in the same amount.

At first sight one might consider these assumptions admissable. Alternative assumptions might not differ from them enough to influence the discussion of the observational origin of the two catalogues. However, as outlined above, it must be shown that the errors under consideration are random errors in a measuring or

---

[52]Vogt, H. (1925), col. 3.

evaluation process and not either systematic errors or even errors resulting from the assumptions adopted during the course of analysis.

Vogt's "proof" of independence for the majority of coordinates is invalidated by any investigation into the nature of his errors. He calculates the differences of both the reconstructed coordinates and the coordinates of Ptolemy's Almagest from the true positions and calls these differences "errors". Considering only the stars for which he was able to reconstruct the Hipparchan coordinates, Vogt counts the number of instances for which the errors fall into a particular interval in the case of (a) the reconstructed Hipparchan coordinates, (b) the coordinates of the Almagest and (c) when they both fall into the same interval.

We again reproduce Vogt's table of errors in latitude:[53]

error interval	number of stars in the interval		
	Hipp.	Ptol.	shared errors
$[-0.10°, +0.10°]$	15	32	7
$[+0.10°, +0.33°]$	17	30	5
$[-0.10°, -0.33°]$	17	32	5
$[+0.33°, +0.67°]$	15	14	3
$[-0.33°, -0.67°]$	18	11	3
$[+0.67°, +1.50°]$	12	4	0
$[-0.67°, -1.50°]$	20	5	2
$> +1.50°$	4	1	1
$< -1.50°$	4	2	0

Table 5.14: Vogt's table of error classes in latitude.

According to this table 20 stars of Vogt's reconstruction have an error in their latitude larger than $-0°.67$, while only 5 stars of the Almagest show an error of that size and only 2 stars have an error both in the Almagest and in the reconstruction falling into the same interval.

According to the criteria developed above, Vogt should have provided evidence that the errors in the table are random errors of measurements (or of evaluation) and not differences introduced by any other circumstance. Vogt's conclusions based on his counting of coinciding errors are artificial and unfounded.

To repeat the argument developed in chapter 4.2: one could imagine such small error intervals that no error of the reconstructed Hipparchan coordinates would lie within the same interval as the corresponding one from the Almagest. Therefore the intervals have to be large enough to account for rounding errors and the errors resulting from the applied method of reconstruction. Even when the intervals are large enough, there always will be pairs of errors not sharing the same error intervals although they might originate from the same observational data.

Coincidences in error classes have to be interpreted statistically. After such an analysis the errors of Vogt's reconstructed coordinates show a significant correlation with the errors of the positions in the Almagest. This type of analysis will not be

---

[53]Vogt, H. (1925), col. 23.

followed here any more, but will instead be confirmed by a much broader and more significant investigation.

5.8.2   From Observation to Phenomena

We know hardly anything about ancient astronomical techniques of observation, especially in the case of stellar positions. Ancient astronomers somehow had to measure equivalents to stellar positions and evaluate them, for example by transforming them into a suitable coordinate system. These transformations could be done by calculation or by a graphical procedure such as a globe. The Commentary does not mention anything about the techniques used. Neither do we find any remark about the type of observation on which the Commentary is based.

A historical analysis, therefore, has to rely entirely on the description of the phenomena as the final result of a long line of evaluation steps, and it must seek for structures within these data to uncover a more detailed history.

Certainly the Hipparchan data have a real astronomical background and they are based ultimately on real observations. Any way of transforming observations to the numerical value of a phenomenon of type j in respect to a star i will be abbreviated by the map $\mathscr{T}$:

$$\mathscr{T}: \quad (data) \to \Phi_{ij} \tag{5.14}$$

It should be noted that the transformation map also accounts for all kinds of manipulations like truncating numbers, accidental misreading of numerals by the observer or the scribe in the scriptorium. It also accounts for all auxiliary tools like the application of globes.

A complete description of a stellar position has at least two independent coordinates. Usually (in modern notation) they are expressed in one of the coordinate systems, e.g. in the ecliptical coordinate system the position of a star i would be catalogued as λ_i, β_i.

As 'observations' the appropriate values of the graduation of an instrument are reported, for example the height of a star above the horizon and the culmination time. For a star catalogue these data are evaluated in a sequence of steps and from these sources finally the descriptions in the Commentary are derived.

For example, suppose that a fictitious astronomer observed all stellar positions with a meridian instrument and measured the culmination height and noted the time t of the event. Knowing the geographical latitude ϕ he could derive the declination δ of a star.

Knowing the culmination time θ_0 of the vernal point leads to the right ascension α of the star by

$$\alpha = (t - \theta_0)\frac{24^h}{23^h 56^m}. \tag{5.15}$$

The astronomer calculated the coordinates with very high precision and all transformations were done without any error. In the next step he plotted these coordinates on a globe and used it to read off the degrees of the phenomena by a simple rotation of the globe.

The process from the raw observations to the final numerical values in the text book could be ordered in steps. The generation of ideally correct data, indicated by the sign "*", can be formally described by

$$(h_i^*, t_i^*) \xrightarrow{\mathcal{T}_1} (\alpha_i^*, \delta_i^*) \xrightarrow{\mathcal{T}_2} globe \xrightarrow{\mathcal{T}_3} \Phi_{ij}^* \qquad (5.16)$$

When we leave this specific case and generalize the situation to any astronomical procedure of the derivation of phenomena, then in the last diagram the abbreviations for concrete astronomical data have to be replaced by variables standing for any record of data. This also includes mechanical or graphical methods. We call any chain of procedures a *generation tree* for the values of the phenomena, because they describe the path from the origin of the data (the observations) to their final value in the text.

The historical question of a common origin of two star catalogues can now be understood as the question of whether two generation trees have a common part, as in

$$(observations) \longrightarrow t_1 \longrightarrow t_2 \longrightarrow \quad t_3 \quad \longrightarrow \quad t_4$$
$$\Big\downarrow \qquad\qquad\qquad (5.17)$$
$$u_1 \quad \longrightarrow \quad u_2$$

Each of the transformation steps in real astronomy adds a systematic and random error to the values. Any ideally precise value $(coord_n^*)$ will therefore be slightly "distorted" by the errors:

$$(coord_i)^* \xrightarrow{\mathcal{T}_n} (coord_{n+1}^* + error_{sys} + error_{random}) \qquad (5.18)$$

The next transformation step adds further errors to the previous one and all errors are carried over to the final result. As a very good approximation one can assume that all random errors preserve their randomness during a generation path, and that the systematic errors might preserve their systematic character for a reasonably short length of the generation path. This allows us to compress a path with several steps into a single one, so that the whole process of generating a star catalogue from observational data may be represented in e.g. two major transformation steps. With such a simplification the historical question of a common observational origin can be replaced by the question whether a generation tree of the following form can produce the values in the astronomical manuscripts.

$$(Observer) \xrightarrow{\mathcal{T}_1} \quad (coord_1) \quad \xrightarrow{\mathcal{T}_2} \quad (\lambda_i, \beta_i)$$
$$\Big\downarrow \mathcal{u} \qquad\qquad\qquad (5.19)$$
$$(\Phi_{ij})$$

There is only one way to decide whether both catalogues share a generation path. At the knot of the path, where both paths separate, a significant amount of *random* errors has to exist. Only when these random errors are found at both ends of the generation paths, it is shown that they originate from the same observer.

Systematic errors could be shared by both catalogues, even though all positions were observed independently. If for example both authors of the star catalogues had wrongly assumed an obliquity of the ecliptic of 30°, it would change the catalogued positions in the same way. A common error does not in general imply that one of the authors did not observe.

Common systematic errors could be introduced by a false historical analysis as well. Supposed that the geographical latitude is relevant in calculating the error-free situations. If a historian assumes a wrong latitude in his interpretation, then it again might introduce common errors. The importance of proved common *random* errors cannot be underestimated.

5.8.3 New Ways of Comparison

Vogt's motive for his reconstruction of coordinates was founded in the desire to avoid unnecessary historical assumptions until the documents were compared. It was mentioned before that in fact several assumptions were tacitly introduced, but in addition to that there is a weakness to the method in general.

Any reconstruction requires a combination of at least two phenomena from either a rising, a setting or a simultaneous culmination. Hence, only stars with a description satisfying that condition could be used for the comparison. Restrictions of the amount of data leads to a serious decrease of significance in a test, especially since the horizon phenomena concern less frequently mentioned stars with far southern declinations, which were observed under bad conditions. These stars are likely to show large positional errors, but they could be not taken into account by any method which starts out by trying to make a reconstruction.

The best possible test would consider all available information from both catalogues for comparison. Furthermore the test must guarantee that no additional historical assumption invalidates the conclusion.

Two prejudices might make a Vogt type of analysis plausible. The first alleged advantage is more psychological: positional errors can easily be seen in the coordinates, especially when one compares ecliptical coordinates directly with the Almagest. It might seem as if the coordinates of the Almagest could represent the genuine Hipparchan catalogue as well as the reconstructed coordinates, and hence one has only to look at both catalogues and compare the juxtaposed entries. This is at once seen to be illusory when one set of coordinates only *approximates* Hipparchan data. In a proper analysis it makes no difference in principal whether the ecliptical coordinates of star catalogues or the calculated values of phenomena are compared. The later has the enormous advantage that it considers all data.

The second alleged advantage of a Vogt type of test concerns the assumption of an epoch at which the observations were done. For the reconstruction no epoch has to be assumed. Because of the precession motion the positions of the stars move nearly constantly around the pole of the ecliptic, and the phenomena change their value together with that motion. It looks as if only the reconstructed coordinates can eliminate any temporal assumption, whereas comparing errors in the phenomena cannot do without it. Ecliptical latitudes do not change over a reasonable historical period and the longitudes increase linearly with time, so that one can easily be compensate for this effect in a test for common origin.[54]

---

[54] Vogt did not consider the effect of precession motion in his comparison of longitudinal errors. In a proper correlation analysis a constant added to one of the errors can be neglected.

Precession increases the longitudes about 1.39 degrees per century at the time of Hipparchus. The calculation of phenomena is effected by it if the historical epoch for which the error-free phenomena are calculated differs significantly from the true time. One has to make sure that the reference time is not very wrong (e.g. more than thirty years). This condition can easily be satisfied. The estimated time of reference is determined by the minimal differences between the phenomena of Hipparchus' Commentary and the computed values of the phenomena for a given time t. The time t is not very much different from the underlying time of the Commentary, if the overall differences sum up to a minimal value. Instead of assuming any time of reference one could calculate it.

Thus the second reason for introducing reconstructions also loses its appeal.

The change of numerical values following the path of a generation tree can historically be tested at two places. The first is the stellar positions expressed in ecliptical coordinates. They can be juxtaposed to the coordinates in the Almagest. For comparing the errors of both historical documents one has to reconstruct the ecliptical coordinates of Hipparchan stars from the phenomena.

$$(Observer) \xrightarrow{\mathscr{T}_1} (coord_1) \xrightarrow{\mathscr{T}_2} (\lambda_i, \beta_i) \quad \succ Almagest$$
$$\downarrow \mathscr{U} \qquad\qquad\qquad\qquad\qquad\qquad (5.20)$$
$$(\Phi_{ij}) \qquad\quad \succ \quad Commentary$$

In the paths of the diagram one might start from the phenomena Φ to $(coord_1)$ and then derive the positions (λ, β). Alternatively one could go back from (λ, β) and derive Φ. Besides all the other advantages of this method which have already been mentioned, the tests will be much more significant than in a Vogt type of analysis, because the hypothetically assumed Hipparchan sources $(coord_1)$ can be constructed with higher accuracy. That is due to the fact that if there was a common Hipparchan source then the positions of the Almagest should reflect the original Hipparchan positions much better than the reconstructed places of the original Hipparchan text, because positions in the Almagest are catalogued with an accuracy of up to 1/6 of a degree, while the Commentary contains only half degrees at most. For that reason the test about a common knot at $(coord_1)$ should be more significant when the analysis starts from the more precise data.

The best method for comparing the two catalogues will analyze the errors of the phenomena.

The transformation map $\mathscr{U}^*$ determines the error-free value of a phenomenon Φ^* with given (true) positional coordinates $(coord_1^*)$.

$$\Phi^* = \mathscr{U}^*(coord_1^*) \qquad\qquad (5.21)$$

The differences between the error-free phenomenon and the documented Hipparchan values are called $\Delta\Phi_H$, and the differences between the phenomena calculated on the basis of the Almagest are called $\Delta\Phi_P$.

Phenomena calculated on the basis of the Ptolemaic coordinates should first go back as close as possible to the transformation from $(coord_1)$ the Hipparchan material and then follow the Hipparchan procedure as closely as possible to reproduce the values for the phenomena. There is no *a priori* knowledge about the exact form of the transformation procedures, but it is certainly a very good approximation to transform Hipparchan data to Ptolemy by simply adding a constant of 2°40′ to all longitudes. If Ptolemy did use Hipparchan data, then this constant is equal to the (wrong) precession constant he believed in, and if Ptolemy did not use Hipparchan data, then his error in the solar theory requires an adjustment of exactly the same size.

All errors already accumulated in ($coord_1$) form the error components ϵ_i between the equivalent position to ($coord_1$) expressed in (λ_i, β_i) and the error-free position (λ_i^*, β_i^*) of a star.

$$\epsilon = \begin{pmatrix} \lambda_i - \lambda_i^* \\ \beta_i - \beta_i^* \end{pmatrix} \tag{5.22}$$

In the following we use the abbreviation π for the vector of true coordinates (λ^*, β^*). Hipparchus' method of calculation adds some error δ to the derivations according to the strict formulas of spherical astronomy. If, furthermore, he based the evaluation on erroneous coordinates with the accumulated error ϵ, one derives the phenomena:

$$\Phi_{Hipp} = \mathscr{U}^*(\pi + \epsilon) + \delta \tag{5.23}$$

These two errors generate a corresponding error $\Delta\Phi$ in the phenomenon by

$$\Delta\Phi_{Hipp} = \Phi(\pi + \epsilon) + \delta - \Phi(\pi) \tag{5.24}$$

An approximation to the error is given by the linear term of the series

$$\Delta\Phi_{Hipp} = \frac{d\Phi(\pi)}{d\epsilon}\epsilon + \delta \tag{5.25}$$

In the case of the phenomena in Hipparchus' Commentary the error ϵ is the sum of the systematic and the random errors in his list of stellar positions.

In addition to the errors of the phenomena attached to the stellar positions of the Almagest one first has to admit a possible error ζ in the subtraction of 2°40′ from the longitudes in order to obtain Hipparchan coordinates. To these the rigorous formulas $\mathscr{U}^*$ are applied to derive the phenomena:

$$\Phi_{Ptol} = \mathscr{U}^*(\pi + \epsilon' + \zeta) \tag{5.26}$$

This implies an error in the phenomena

$$\Delta\Phi_{Ptol} = \frac{d\mathscr{U}^*(\pi)}{d(\epsilon' + \zeta)}(\epsilon' + \zeta). \tag{5.27}$$

There is a whole range of different types of errors which arise in the various parts of the derivation:

$$(Observer) \xrightarrow{\epsilon,\epsilon'} (coord_1) \xrightarrow{\zeta} (\lambda, \beta) \quad > Almagest$$

$$\downarrow \delta \tag{5.28}$$

$$(\Phi) \quad > Commentary$$

When the derivatives are abbreviated by κ and κ' respectively, the errors can be summarized as

$$\Delta\Phi_{Hipp} = \kappa * \epsilon + \delta \tag{5.29}$$
$$\Delta\Phi_{Ptol} = \kappa' * (\epsilon' + \zeta)$$

This long analysis of error types and the propagation of errors has now resulted in a proper criterion for deciding the question whether the Almagest positions are (at least) partly of the same observational origin as Hipparchus' phenomena:

Both stellar data originate from the same observations, if the total errors in the phenomena $\Delta\Phi$ are originated by the same error ϵ and if the differences between ϵ and ϵ' is due to random errors ζ.

The diagram of an error distribution for Hipparchus and Ptolemy can easily be visualized when the factors κ and κ' are given the value 1.[55]

We then obtain a set of two equations

$$\Delta\Phi_{Ptol} = \epsilon' + \zeta \tag{5.30}$$
$$\Delta\Phi_{Hipp} = \epsilon + \delta$$

In the case of independent observations the errors ϵ and ϵ' are independent, provided there is no common systematic error. Consequently, in a two-dimensional representation the errors of the phenomena would be distributed independently of each other around their axis of the coordinate system.

Only in the case of a shared generation path could the errors ϵ and ϵ' be identical. In that case one obtains a nice linear relationship

$$\Delta\Phi_{Ptol} = \Delta\Phi_{Hipp} + \delta + \zeta \tag{5.31}$$

If furthermore the error δ is 0, Hipparchus' sources would be identical with the Ptolemaic catalogue minus the precession constant, and if ζ could get the value 0, then the phenomena of Hipparchus should have the same value as those calculated on the basis of the Almagest in the case of a shared observational origin.[56]

In a two-dimensional representation of the errors all points should be on a diagonal line. If ζ and δ do not disappear then the error distribution may look

---

[55]It is almost the case that both are equal. The common factor can be summarized in the error ϵ, which then allows the approximation of both values being set to 1.

[56]This would require that Hipparchus' calculation of the phenomena was absolutely precise.

as in the following diagram. On the left side the case of independent observations is represented by five stars. Four of them show the outer limits of a whole group of errors inside the box. Since there is no common error ϵ, all error points are distributed equally over all quadrants.

On the right side is shown how the distribution differs when the errors ϵ and ϵ' are the same. The box with the majority of errors is tilted to the right and the one star with an exceptional error is close to the diagonal. The cases with a large random error should be an especially good test for a common origin.

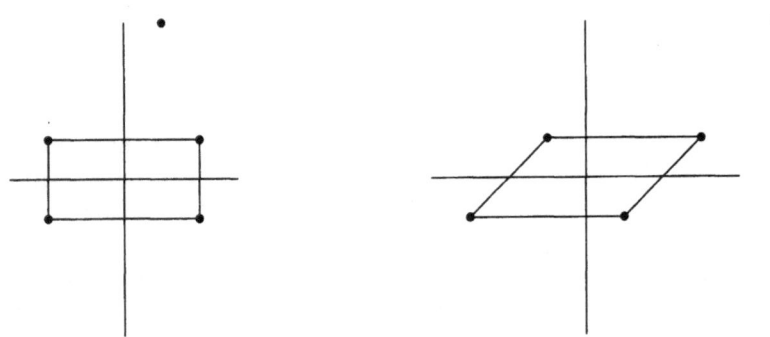

The time of reference is determined as the time t when the sum of errors for n instances of the Commentary is minimal. We define this quality as the *fit* of the data:

$$fit := \sum_{i=1}^{n} |\Phi_{Hipp} - \Phi_{calc(t)}| \qquad (5.32)$$

One may notice that here it is not the sum of error squares which is minimized, but the sum of absolute differences. This has the advantage that errors in the Commentary do not influence the result as much as if the square of errors were minimized. Such fit functions are also called *robust estimators*.

The minimum of the fit finds a reference time of about -143 with no sharp minimum, so that some ten years plus or minus cannot be excluded. This reference time is about 15 years before the time attributed to most of the Hipparchan stellar declinations in the Almagest. The difference in time is so small that we can take the usual reference time, the year -128, as standard for the further calculations.

In the following diagram the errors of the phenomena for all valid stars in book II of Hipparchus' Commentary are plotted. The time of reference is -128, the geographical latitude $\varphi = 36°$ and the Hipparchan obliquity of the ecliptic is $23°.86$. The scale of the axes is in degrees.

The close correlation between the differences in the phenomena is obvious without any statistical test. Several large common errors up to 10 degrees guarantee a high correlation coefficient. They also obviate the need to discuss possible systematic errors being the cause of the correlation. There is hardly any systematic error which could account for very large common errors larger than two degrees. The list of

Name	No
ζ Cas	178
β Sge	285
θ Gem	426
i Cnc	455
β Vir	498
ε Vir	509
β Sgr	592
χ Psc	707
ρ Cet	719
θ Eri	805
α Car	892
π Hya	918
α Cen	969

Table 5.15: Stars with large common error.

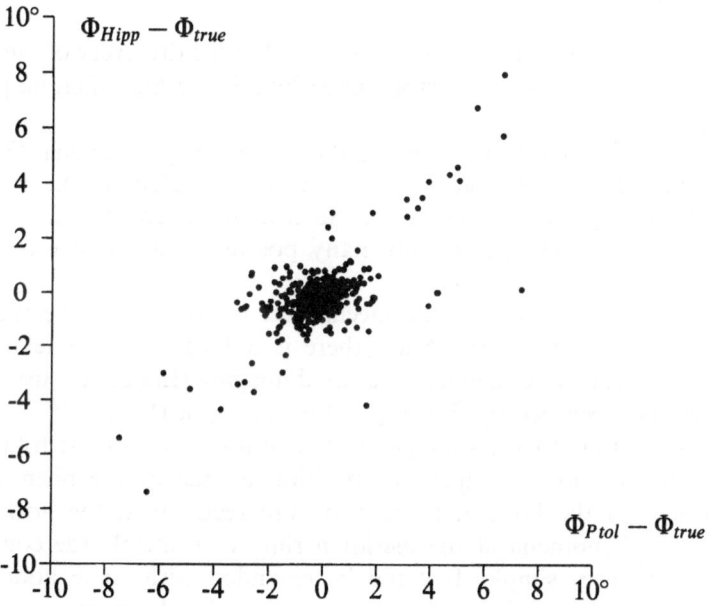

Figure 5.30: Correlation of errors for all phemonena.

stars with a common big error is lengthy. In the following table the names of the stars with a common error larger than one degree are listed (table 5.15).

The only exceptional error is related to the star θ Crt with a large Hipparchan error for the setting phenomena of about 10 and 7 degrees while the position in the Almagest is rather precise. This could be caused by a wrong identification of the star, later corruptions of the text, or simply a mistake in one of Hipparchus calculations. However, it seems implausible that such an instance provides evidence for some positions being independently observed.

A mistake in the identification would of course introduce common errors in the phenomena. But the list of stars contains so many well known stars that a correlation induced by the interpretation can be excluded. Stars like α Centauri or α Carinae (Canopus) are well known and it is impossible to misidentify them.

Furthermore, these stars do not share their extreme positional errors with any of the surrounding stars. That implies that they were not well known reference stars in the sky whose positions were established once and then served as reference points for related measurements. In that case their error would have had to be carried over to other positions as well.

For these reasons we can exclude the possibility that the large common ϵ error in the phenomena is due to a major systematic error. It can only originate from a random error on the path from observations to a prepared star catalogue or star list. What this catalogue looked like, and especially which coordinate system was used, can only be discussed after an analysis of the other types of errors. And even then there might be no good evidence for one or the other solution.

5.8.4 The Globe

An analysis of the other types of errors might lead to the discovery of more historical details. The δ errors characterize errors made by Hipparchus when he prepared his Commentary.

There are two possible ways to derive the values for phenomena. One way uses the mathematical tools described in the Almagest to calculate the corresponding phenomena from a given star catalogue. Such a procedure would require a long sequence of tedious calculations, with many possible sources of errors along the way.

The alternative, extremely simple mechanical method to derive the values for the phenomena requires a globe. Since there is not the slightest textual evidence reporting the construction of a globe, one could imagine Hipparchus using a complex globe like the one described by Ptolemy in the Almagest (VIII 3).[57]

Once all stars are plotted on the globe, it requires only a simple rotation around the axis of daily motion to adjust the position to one of the phenomena. Then either the values of the horizon phenomena are read off at the horizon ring, or the culmination phenomena at the meridian ring. In principle the construction of the globe can be very simple. It could be extended with some additional rings, or a system of coordinate lines, as reference points for the process of transferring positions to the globe. The globe would of course require a more sophisticated mechanism if it is also to account for the precession motion. In the latter case an additional axis through the pole of the ecliptic is needed, as described in the Almagest.

The numerical values reported for the phenomena provide good evidence for the supposition that Hipparchus in fact used a globe. The Commentary repeats culmination phenomena in several instances. For each constellation Hipparchus mentions the stars culminating during the rising and setting time. Because there

---

[57]Nadal, R., Brunet, J. P. (1983/84), Le "Commentaire" d'Hipparque I. La sphère mobile, *Archive for History of Exact Sciences* **29**, pp. 201–236.

is always more than one constellation rising and setting at a time, the list of culminations is repetitive.

If someone were to calculate the culminations of stars mathematically, then these culmination numbers should repeat themselves strictly. Errors could of course occur under certain circumstances, but they should not consist simply in a small variation of the value for a number of stars. Such errors would, however, be expected when a globe was used for the derivations. A globe is adjusted for a particular situation and the numbers are read off the rings, and the same culminating star is adjusted later on for a different constellation rising or setting. One can expect some variations in the results following this procedure. In that case one would get, for the same star, two different culmination degrees with a variation only within the dimensions of the graduation of the rings and the zodiac. Several of such errors would, by themselves, provide strong evidence for readings taken off a globe.

Furthermore, if different stars are mentioned as culminating with the same degree on the ecliptic, and if at a different place in the Commentary the same pair of stars is again mentioned as culminating together, but with a slightly different degree on the ecliptic, that seems to be explicable only by globe readings. Hardly any mathematical calculation would lead to such significant variations in the data. The Commentary contains several instances of that type. The following table contains pairs of stars culminating at the same time but with different phenomena at different contexts in the book.

star 1	star 2	page 1	Φ 1	page 2	Φ 2
β Cnc	ι UMa	187	94.5	267	96
κ Cas	π And	205	340.5	227	342
δ Cas	χ Psc	205	349.5	229	352
β And	η Cet	231	349.5	247	350

Table 5.16: Examples of varying phenomena for the same stars.

The δ errors therefore strongly support the view that Hipparchus used a globe for the final preparation of his Commentary. The size of the variations, about one degree, suggests that the resolution of the graduation is not smaller than the half a degree which also is the smallest fraction of a degree mentioned in the Commentary.

5.8.5 More Details

While the first figure displayed the errors of all five types of phenomena, the next figure contains only the errors in the simultaneous culminations.

It is apparent that the variations of errors are much smaller. This is partly due to the effect that randomly distributed positional errors usually generate larger differences in the phenomena on the horizon. There also might be some impact from the specific method of positional measurement which lies behind the data. By far the most influential factor is that for the culmination only selected stars along the meridian were mentioned in the Commentary, whereas on the horizon, for the rising and setting of a constellation, exactly the first and last stars had to be reported.

Therefore these stars also include cases of lower positional precision and much more difficult observational conditions.

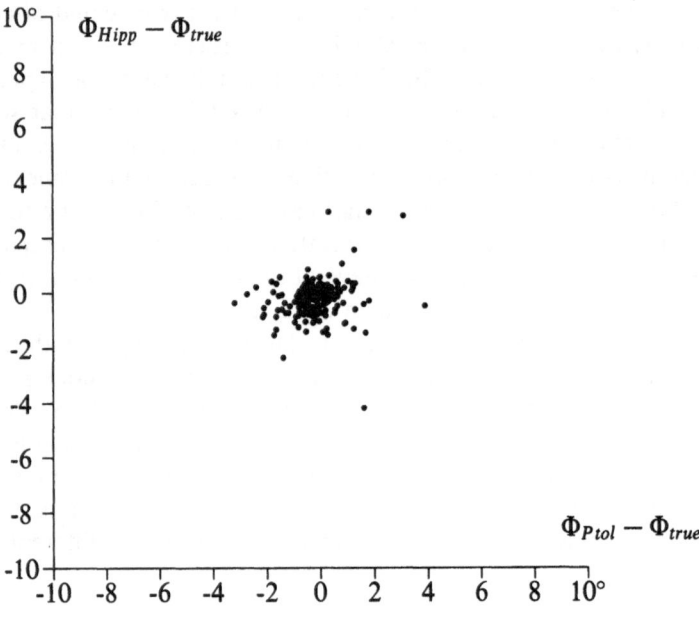

Figure 5.31: Correlation of errors for culminating stars.

The distribution shows a higher variance for the errors of horizon phenomena. The figure below shows the correlation of the phenomena without the simultaneous culmination.

The distribution does not change very much when only errors of the two rising phenomena are plotted. Although the absolute size of the error distribution is smaller, the variation of the errors is not changed very much because of the further reduction of the number of stars considered.

5.9   Reconstruction

A proper method of reconstructing ecliptical coordinates of Hipparchan stars from the information in the Commentary must not follow Vogt's method.[58]

The idea for the following method is rather simple. One searches for the ecliptical coordinates λ_i, β_i which *explain* Hipparchus values in the Commentary best, i.e. the differences between the corresponding calculated phenomena and the values of the text should be as small as possible.

---

[58] The major disadvantage of Vogt's method consists in first combining two phenomena for the determination of the position of the ecliptic in the horizon system and then adding some other data to take means, which are then used for further calculations. This method sometimes overemphasizes the impact of one value over the other in the final result.

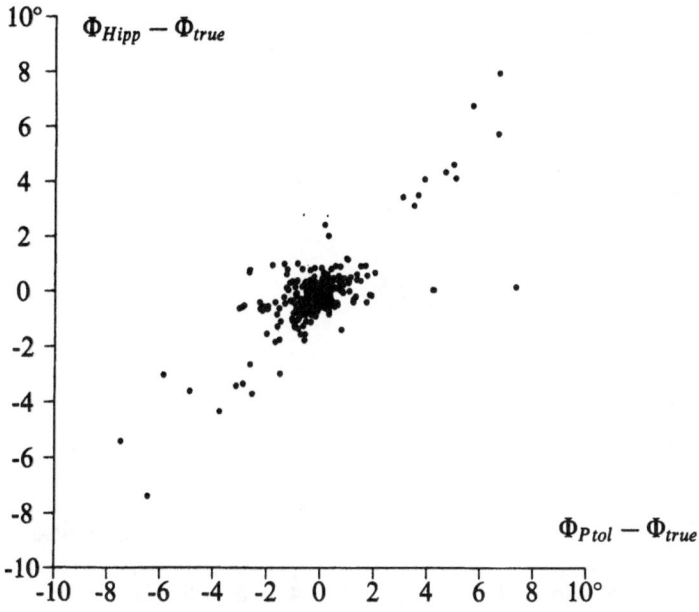

Figure 5.32: Correlation of errors without culminating stars.

As was true before in the determination of the reference time of the Commentary, the condition of a minimal sum of differences is satisfied when the function *fit* is minimal:

$$fit = \sum_{i=1}^{n} |\Phi_{Commentary} - \Phi_{Calc}(\lambda, \beta)| \qquad (5.33)$$

The Commentary gives five phenomena for the star π Hydrae[59]

Phenomenon	degree
1	195.5
2	107
3	160
4	258.5
5	183

Table 5.17: Phenomena of π Hya.

---

[59]The additional description on page 244 is not taken into account because that would require the further information of the size of the vagueness of the phrase "a little west of the meridian".

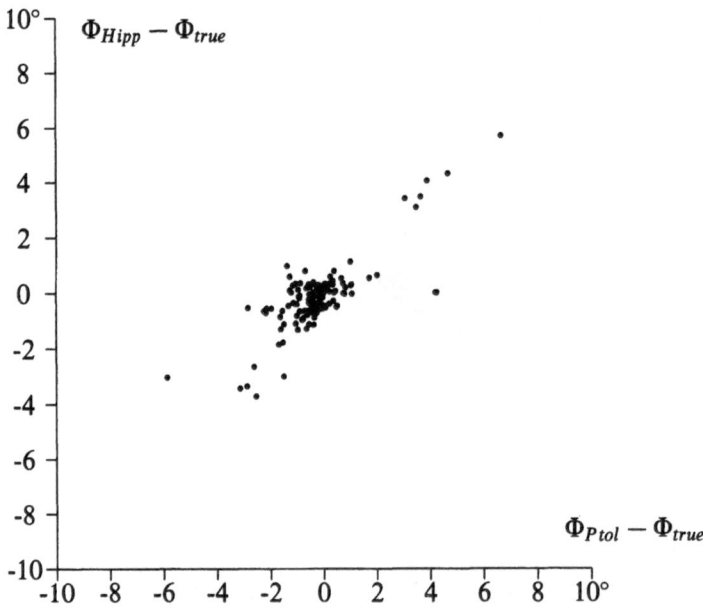

Figure 5.33: Correlation of errors for rising stars.

As an initial value one could assume any position λ_i, β_i in the sky and compare its fit value with any surrounding position. The second position is taken as a better initial guess if the sum of errors for that position is smaller than the previous one. This search is continued until the minimum is in a well defined region surrounding the position under consideration. Any algorithm finding the minimum of a two-dimensional function could be used for the search for best fit. In the following minimizing procedures a latitude of $36°$ and an obliquity of the ecliptic of $23°\!.86$ has been assumed.

From Book II of the Commentary the data for valid stars suffices for the reconstruction of 55 ecliptical coordinates.

The next figure shows longitudinal errors of the Hipparchan stars and the corresponding coordinates from the Almagest. The small bars parallel to the x-axis give the values of the estimated double standard deviation of the reconstructed Hipparchan values. They might serve as a rough orientation for the accuracy of the data.

The errors in latitude show a much more compact distribution. Again significant is the orientation to the diagonal line, as was shown in the simulated picture with the tilted error box.

All these correlations of errors support the thesis of a common observational origin, although in the case of the reconstructed coordinates the influence of systematic errors cannot be excluded as easily as before in the direct test of the phenomena.

The next two figures display only errors of stars with a very deep minimum of the error sum. The coordinates of such cases are better determined than the ones

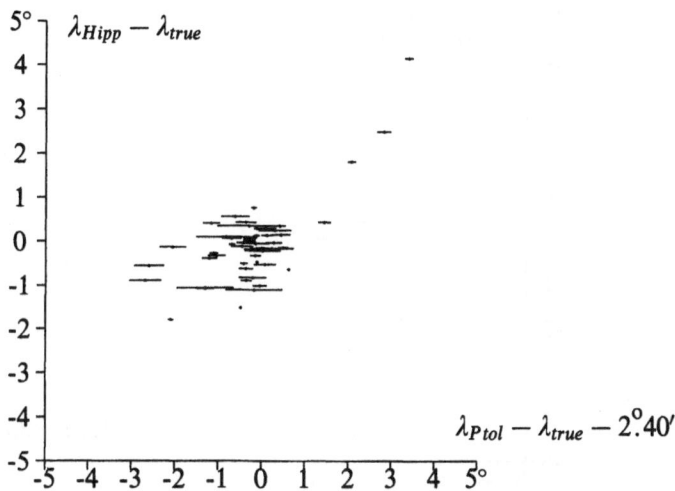

Figure 5.34: Correlation of errors of reconstructed longitudes.

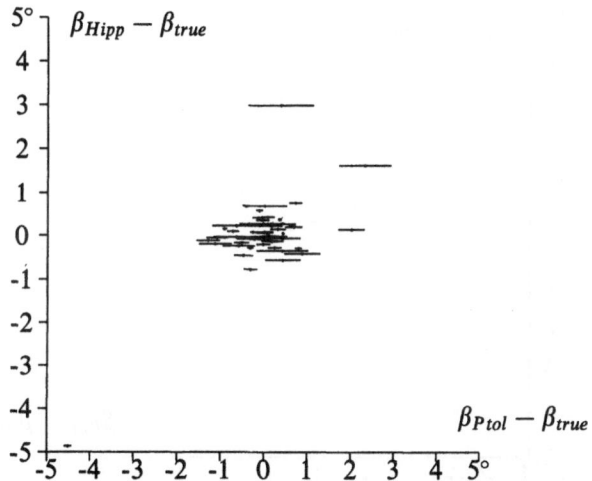

Figure 5.35: Correlation of errors of the reconstructed latitudes.

with a very flat fit function. Longitudinal errors for standard deviations smaller than 0.8:

Latitudinal errors for standard deviations smaller than 0.8 show a very small difference from the actual, i.e. calculated latitude. Nevertheless the inclination of the error distribution towards the right is very obvious.

In the final figure the latitudinal error of the star is marked with a number. This number stands for the first decimal digit of the fraction of degree, as it is reported

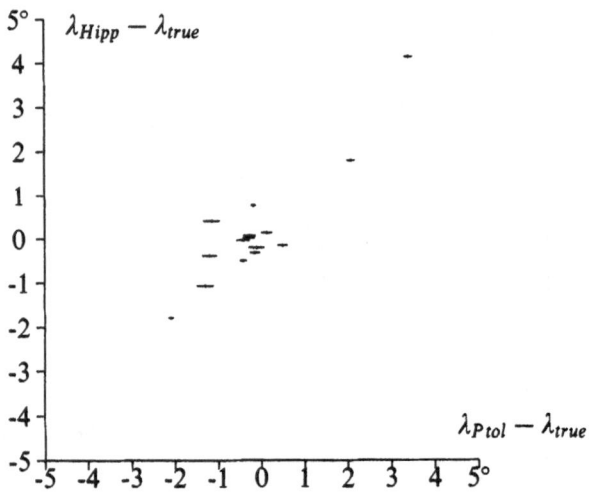

Figure 5.36: Correlation of well approximated longitudes.

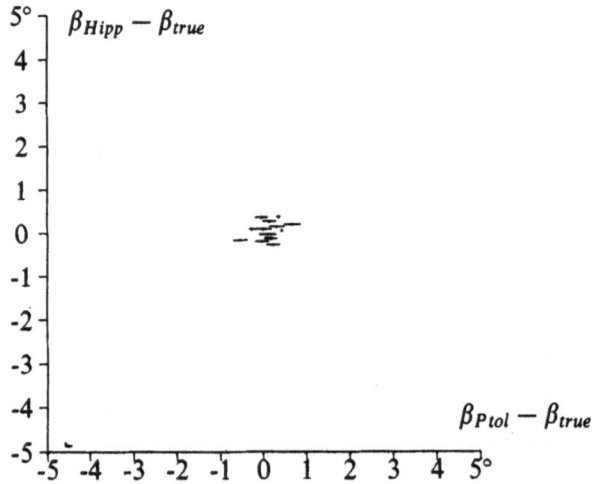

Figure 5.37: Correlation of well approximates latitudes.

in the Almagest. If a star has the latitude of $13°.25$, then its error point would be represented by a '2'. If the fraction of a degree is $0°.5$, the number would be '5' and so forth.

It is surprising, that some areas in the error distribution show a significant preference for some fractions of a degree. No satisfactory explanation has been found. It might be the key to the definitive explanation of Hipparchus Commentary: the description of the coordinate system and the positional accuracy he employed in his "star catalogue".

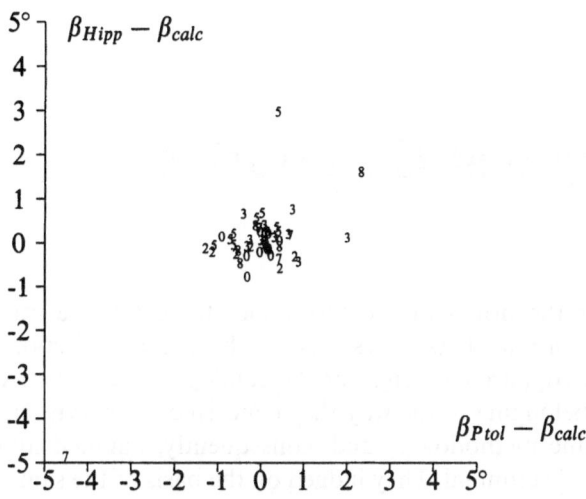

Figure 5.38: Correlation of fraction numbers.

As a speculation, one could imagine a list of star coordinates expressed as declination of the star and the degree on the ecliptic simultaneously culminating with the star. These types of coordinates could easily be observed with a meridian instrument, whereas the ecliptical longitude is merely an abstract concept that can be derived only by an intermediate step, for example from a table which shows the culminating degree on the ecliptic after a certain time in the night.

It has been proven that Ptolemy's stellar coordinates were at least to a large extent observed by Hipparchus or a contemporary astronomer. This does not imply that Hipparchus used a catalogue like the one in the Almagest. Any type of Hipparchan coordinates could be transformed to ecliptical coordinates later on: they would still share the ϵ error in the observational generation path. For example, one could imagine that Ptolemy used a precession globe with all stars plotted on it and derived the ecliptical coordinates by an appropriate rotation.

The key to a final understanding of the type of Hipparchus' star catalogue might still be hidden in the Commentary. One would approach very close to this if he were to succeed in deriving all numbers in the Commentary from the coordinates in the Almagest. For a successful reconstruction one has to provide a full description for all four types of errors in the transformation of positions.

6. Theory and Observation

The debate over the origin of the Ptolemaic star catalogue was accompanied by a moralistic evaluation of its genesis. After the accusing faction had offered their evidence for the Hipparchan origin of the catalogue, they proceeded to impose the ideal standards belonging to the way they conceived their own scientific activities on Ptolemy's scientific methodology and, consequently, damned him as a plagiarizer, forger and scientific criminal. They judged on the basis of the same historical impetus with which the theories of the modern era replaced the traditional Ptolemaic world-view and, in general, made the empirical fact, as experiment or as observation, the real foundation of a correct understanding of nature.

The defenders of Ptolemaic morals in this historical autodafé saw themselves forced, in face of the reproaches that were raised, to resurrect the methodological reputation of the Almagest and to supply evidence that the author had indeed carried out his work with authentic observations understood according to the modern standards of scientific practice.

Both of these idealizing positions, however, served to conceal the historical situation in which Ptolemy was dependent not only on the astronomical knowledge available to him at the time, but on the ancient methodology of science as well.

6.1   The Aristotelian Heritage

In the foreword to the Almagest Ptolemy situates himself in the tradition of Aristotelian methodology. He reviews the Aristotelian diviison into theoretical and practical philosophy and the three main genres of theoretical philosophy, namely, physics, mathematics and theology.[1] For Ptolemy only mathematics promises a way to attain knowledge. The two other genres necessarily included speculative elements, "theology because of its completely invisible and ungraspable nature, physics because of the unstable and unclear nature of matter; hence there is no hope that philosophers will ever be agreed about them."[2]

Two elements, the visible and the coherently structurable, are indispensable preconditions for obtaining scientific knowledge. Theology could not provide the groundwork for knowledge, for it does not deal with visible objects; and neither

---

[1]Ptolemy, C. (1984), pp. 35f.
[2]Ptolemy, C. (1984), p. 36.

could physics, for it deals exclusively with the empirical, neglecting to recognize the underlying principles concealed from the appearances.

These epistemological deficiencies are alien to astronomy:[3]

> "Hence we were drawn to the investigation of that part of theoretical philosophy, as far as we were able to the whole of it, but especially to the theory concerning divine and heavenly things. For that alone is devoted to the investigation of the eternally unchanging. For that reason it too can be eternal and unchanging (which is a proper attribute of knowledge) in its own domain, which is neither unclear nor disorderly;"

The science of astronomy is for Ptolemy distinguished by the quality of the knowledge thereby obtained, for it is at the mercy of the permanent tension between the phenomena observed and the recognizable principles underlying them which are surely derivable on the strength of the certainty of a clear and straightforward mathematical line of reasoning. The certainty of astronomical knowledge is not grounded in the evidence of empirical data, but rather in the rigorous mathematical deduction of the statements explaining the empirical data.

The twin hallmarks of Ptolemaic science are the clear priority of a theoretically legitimate knowledge and the rather sharp epistemological cleft between mathematical knowledge and the speculative character of physics. While Aristotle understands astronomy as an integral part of physics, Ptolemy gives it the rank of a "first" science whose laws differ qualitatively from those of physics.[4] Only in his confrontation with the heliocentric world view does Ptolemy resort to arguments taken from physics to demonstrate the "impossibility" of the hypothetical alternative,[5] expanding on Aristotle's argumentative line in "De Caelo".

Ptolemy's self-conception as scientist has little in common with that of a modern scientific researcher after Galileo, for whom empirical data makes up the true source of scientific knowledge. For the latter, pure theory is a delicate and dubious construct that must at some time stand the test of observable and evident relationships. Instead, the Ptolemaic reality is to be found beyond the visible façade of the phenomena in the mathematical structures that generate them. All doubt directed towards the empirical statements is justified by the methodological situation of astronomical science in antiquity.

6.2 The Uncertainty of Empirical Data

Plato categorically denies that perception leads to knowledge. The Aristotelian conception of science is more pronouncedly shaped by empirical elements. Aristotle opens the first book of his Metaphysics with rather poetical sentences: "All men

---

[3]Ptolemy, C. (1984), p. 36.

[4]Mittelstrass, J. (1962), *Die Rettung der Phänomene in der Geschichte der Astronomie*, Göttingen, pp. 164ff; Kanitschneider, B. (1974), *Philosophisch-historische Grundlagen der physikalischen Kosmologie*, Stuttgart, pp. 52ff; Bourgey, J. (1955), *Observation et expérience chez Aristote*, Paris, pp. 35ff; Wieland, W. (1962), *Die aristotelische Physik*, Göttingen, pp. 231ff; Wieland, W. (1982), *Platon und die Formen des Wissens*, Göttingen, pp. 150ff.

[5]Ptolemy, C. (1984), pp. 38ff.

by nature desire to know. An indication of this is the delight we take in our senses."[6] One readily understands that perception alone does not yet constitute for Aristotle the empirical point of departure from which the underlying principles can be discovered. Only through the act of memory and critical reflection can perceptions become experience which, as an empirical given, can then and only then enter into a logical relationship with the theories of science.[7]

> "But the human race lives also by art and reasonings. And from memory experience is produced in men; for many memories of the same thing produce finally the capacity for a single experience. Experience seems to be very similar to science and art, but really science and art come to men through experience; for 'experience made art', as Polus says, 'but inexperience luck'."

The experience of the observer, then, is the first guarantee that in the context of astronomy perceptions of celestial events are converted into a scientifically relevant experience.[8]

> "And art arises, when from many notions gained by experience one universal judgement about similar objects is produced."

Yet even more so than those endowed with experience, the wise men stand closer to knowledge because they grasp the causes:[9]

> "For men of experience know that the thing is so, but do not know why, while the others know the 'why' and the cause. Hence we think that the master-workers in each craft are more honorable and know in a truer sense and are wiser than the manual workers, because they know the causes of the things that are done."

Finally, "that is why ... the theoretical kinds of knowledge [are held] to be more of the nature of wisdom than the productive. Clearly then wisdom is knowledge about certain causes and principles."[10]

Now the remarkable accomplishments of science consist precisely in being able to deduce the empirical statements from the original principles with deductive certainty.

In contrast to Plato scientific knowledge does not find its certainty in abstract ideas which are uncontaminated with any sensual experience, but grounds itself first of all in the concrete and empirical. Relying on this as a point of departure, the original causes can be discovered only with the assistance of reason in a scrupulous and methodical process of abstraction. At the same time the validity of theoretical hypotheses together with their empirical prognoses can be checked and controlled when compared with actual observations.

---

[6] Aristotle (1984), Metaphysics, in: *The Complete Works of Aristotle*, ed. J. Barnes, 2 vols., Princeton, 980a21.

[7] Aristotle (1984), 980b27.

[8] Aristotle (1984), 981a1.

[9] Aristotle (1984), 981a29.

[10] Aristotle (1984), 981b30.

The central task of astronomy as Ptolemy saw it was the synthesis of a coherent theory from the empirical material of the gifted observer and astronomical theoretician Hipparchus – a theory capable of describing the empirical phenomena that were known with sufficient accuracy. The problem for the historian here, however, is to examine what kind of observations Ptolemy carried out and their relation to the astronomical theory. In connection with the star catalogue the question unavoidably arises: are Ptolemy's observations to be taken as experience which, when they contradict the hypotheses intended to explain them, could falsify them according to simple logical syllogisms? In other words, what relevance do the Ptolemaic observations actually have in view of the competing observations of Hipparchus?

For the ancient astronomer, one basic problem of making an observation was to differentiate between an unqualified perception on the one hand, and the sole theoretically relevant empirical data on the other. He could develop techniques of measuring which set the conditions of the experiments in such a way that the astronomical principles or parameters under examination became apparent or directly measurable, preferably through a series of reproducible observations.

In astronomy, where the possible variation in the experimental arrangements is limited, rather simple numerical strategies such as the use of mean values from repeated observations or a set of theoretical parameters based on different empirical data could lead to an "improved", i.e. more qualified numerical representation. Then, even the reliability of the observation could be estimated on the basis of standard deviation of the data set.

However, this methodological aid was not available to Ptolemy. Consequently there was no criterion besides the intuitive experience of the observer against which the quality of an observation could be gauged.

Ptolemy was well aware of the errors inherent in astronomical observation. In the Almagest he names several examples to criticize the results handed down by the astronomers before the time of Hipparchus. In the context of planetary theory, he complains openly about the almost useless empirical data the older astronomers had provided:[11]

> "[It is] also [confusing] that most of the ancient [planetary] obser-
> vations have been recorded in a way which is difficult to evaluate, and
> crude. For [1] the more continuous series of observations concern sta-
> tions and phases [i.e. first and last visibilities]. But detection of both of
> these particular phenomena is fraught with uncertainty: stations cannot
> be fixed at an exact moment, since the local motion of the planet for
> several days both before and after the actual station is too small to be
> observable; in the case of the phases, not only do the places [in which
> the planets are located] immediately become invisible together with the
> bodies which are undergoing their first or last visibility, but the times too
> can be in error, both because of atmospherical differences and because
> of differences in the [sharpness of] vision of the observers."

It is not only the astronomical situation that serves to disqualify certain types of observations for a theoretical appraisal, such as the planetary phases during the

---

[11]Ptolemy, C. (1984), pp. 420f.

time of opposition or the heliacal risings. Only a very experienced observer can rule out the corrupting influence of eye irritation during a measurement:[12]

> "In general, observations [of planets] with respect to one of the fixed stars, when taken over a comparatively great distance, involve difficult computations and an element of guesswork in the quantity measured, unless one carries them out in a manner which is thoroughly competent and knowledgeable."

An ignorant or unskilled observer arrives at only approximately correct numerical values which are not worthy of theoretical consideration.[13] The length of the year, for example, was determined by Ptolemy only through the observation of the equinoxes, "choosing amongst them, for the sake of accuracy, those which Hipparchus especially noted as very securely determined by him, and those which we ourselves have made with the greatest accuracy using the instruments for such purposes described at the beginning of our treatise."[14]

In many similar places in the Almagest Ptolemy discusses the possibilities of error for an observation in detail.[15]

Variations in the numerical values derived from measurements confronted the astronomer of antiquity with considerable difficulties. For a series of measurements with n values p_i, from which the value of a astronomical constant – for example, the precession constant – is to be determined, the modern astronomer would simply calculate the mean value $\mu(p)$. With this numerical value, which need not agree with any of the individual measurements, the hypothetically assumed or theoretically derived value of the parameters can be checked. The hypothesis H, from which the observable numerical parameter O can be derived, is then considered as confirmed when O lies in a statistically defined interval with variation S around the mean value μ, or:

$$(H \longrightarrow O) \quad \text{and} \quad |O - \mu| < S, \text{ for } S > 0 \tag{6.1}$$

The hypothesis stands in contradiction to the result of the series of measurements when the calculated mean value is found outside of the interval of variation around O, in other words when:

$$(H \longrightarrow O) \quad \text{and} \quad |O - \mu| \leq S, \text{ for } S > 0 \tag{6.2}$$

In such a case the observer has either to examine the observation series for a systematic error or has to adjust the hypothesis so that condition (6.1) is fulfilled again. Both criteria define a simple hypothesis test. They presuppose that the empirical statement O contains one single numerical value representing the whole measurement series along with a value for the range of possible variation, as is done

---

[12]Ptolemy, C. (1984), p. 421.
[13]Ptolemy, C. (1984), p. 421.
[14]Ptolemy, C. (1984), p. 137.
[15]Viz. Ptolemy, C. (1984), pp. 131ff; pp. 173ff; pp. 190ff; pp. 416f; pp. 419ff; pp. 453f.

in modern science. The tolerable fluctuation, for example, could be calculated from the standard deviation of the measurement values. In cases like these, the empirical statement O is not dependent on the hypothesis H.[16] A constructed example can clarify the methodology of this test:

Suppose an astronomer wants to check whether the ecliptical latitudes of the stars remain constant for a short period of time. Older measurements are available which can be compared with the measurements now carried out. The older measurement of a stellar position obtained an ecliptical latitude β_a (O). The hypothesis of constant latitudes is now tested with a series of repeated new measurements with a mean value μ and the fluctuation S.

In this rather simplified situation, the hypothesis will be rejected when the prediction for the latitude of the star (O) lies outside the calculated variation interval S around the mean value μ. The hypothesis would be confirmed if the mean value μ lies within the limits of accuracy S around the old value O. The modern astronomer could use this method to check a simple numerical hypothesis concerning the precession constant.

Now Ptolemy was in an entirely different situation. He knew of neither the method of using the mean value nor any method for calculating the variation intervals for a series of measurements within which the expected value must lie for a confirmation of the hypothesis. This lack of methodological knowledge not only impaired the accuracy of empirical statements, it even made it impossible to test a hypothesis according to (6.1) and (6.2). To be sure, at that time the arithmetical, geometrical and harmonic means of two numbers were known.[17] But no astronomer used them to enhance the precision of empirical statements about the results of a measurement series. The Islamic astronomers of Baghdad, seven hundred years after Ptolemy, were the first ones to have a method at their disposal similar to that of the construction of the mean value. Here the astronomer Ibn Yunus recommended that the average speeds of the planetary motions be determined by splitting the interval between two reliable observations.[18] Even an extremely simplified application of the mean value construction like this is nowhere to be found in the Almagest. Instead, every single observation has a direct relationship to an astronomical hypothesis, providing its numerical result is qualified as validated empirical data.

However, to call an observation a "qualified" measurement presents problems. When each numerical result of a comprehensive series of observations is understood as an exact result, it is impossible to use the measurements for evaluation. As a rule, the various results contradict themselves. If the calculated results of repeated observations for the size of the precession motion were the values $2°40'$, $4°0'$ and $2°45'$, then, among other things, the first value contradicts the hypothesis that the precession had increased the longitudes by $3°40'$. For another, the third value contradicts the hypothesis that the precession had increased the longitudes by $2°40'$ during the period of time under consideration. One readily sees that in this case no series of measurements, no matter how reliable, can lead to making any kind

---

[16]I.e. the interval of variation and the mean value of the measurement are not dependent on the parameters of the hypotheses H.

[17]Heath, Th. (1921), *A History of Greek Mathematics*, repr. 1981, New York, vol. 1, p. 85.

[18]Hartner, W. (1977), The Rôle of Observations in Ancient and Medieval Astronomy, *Journal for the History of Astronomy* **8**, p. 9.

of empirical statements about the precession motion when all numerical results are understood as rigorous results. What is needed here is a methodological procedure with which an astronomical hypothesis can be empirically tested. Two opposing methodological approaches are of relevance for an assessment of the Almagest.

6.3 Radical Empiricism

Methodology: Any theoretical formulation has to be directly confirmed by measurements. There is no method of selection that is capable of determining the true value among equally legitimate qualified measurements.

From the quantity of divergent measurement results, only one number standing for the measurement can be true. A proponent of a radical empiricism concedes in fact the susceptibility of the measurements to error – for only one single value can be correct in a strictly logical sense – but for him no criterion exists which would permit the correct number to be recognized. The empirical values and the theoretical entities stand in a reversible relationship of representation.

Methodology :

(T) Exactly one value of the empirical measurements corresponds to each theoretical parameter

and, conversely,

(E) a theoretical parameter corresponds to a single distinct empirical value which can be tested by measurement.

The problems of testing theories with this methodology have already been described. Astronomical theories can only be developed to the extent that empirically derived statements do not contradict the totality of the relevant measurements. Consequently, it is in principle impossible, for instance, to advance an exact value for the precession constant. On the basis of the constructed series of measurements discussed above, one could only conclude that the precession has a minimal value of 2°40′ over the examined period of time. It is not permissible to say whether the precession increases the longitudes by 2°40′ or by 3°40′.

Several of Ptolemy's comments in the Almagest indicate that Hipparchus attempted to construct his astronomical theory from his observations in accordance with this methodological rule. Ptolemy cites Hipparchus when trying to prove the precession motion:[19]

> "For if the solstices and equinoxes were moving, from that cause, not less than $\frac{1}{100}$th of a degree in advance [i.e. in the reverse order] of the signs, in the 300 years they should have moved not less than 3°."

The Hipparchan evaluation of the observations corresponds exactly to the empiricist model. The smallest value measured by Hipparchus allows a precession

---

[19]Ptolemy, C. (1984), p. 328.

motion of one degree in one hundred years, whereas the calculated mean value of all of his measurements, when judged by today's standards, lies around the substantially larger value of about $1°.4$ per century.

Ptolemy's determining of an unambiguously defined value of the precession constant could easily be understood from the modern point of view when Ptolemy, in contrast to Hipparchus, had at his disposal sufficient empirical material proving the Hipparchan minimal value as an exact parameter. The historical situation exhibits the curiosity that the empirical material of the declination measurements in the Almagest does not confirm the small minimal value of Hipparchus, but the accurate value. It is the basis for accusations of forgery being levelled at Ptolemy.

Such an unhistorical perspective fails to consider that modern methods of evaluation were not available to Ptolemy and that he had to free himself from the empiricistic strategy of Hipparchus if he wanted to obtain richer and more significant theoretical results.

In many passages of the Almagest the critique of Hipparchan astronomy does not concern the accuracy of the measurements nor the logical validity of the theoretical inferences, but rather the concealed, overly strict empirical attitude. If Ptolemy had intended to include a star catalogue in the Almagest with which the following generations of astronomers could work and be able to convert the coordinates to their time with the inclusion of the precession motion, then one had to calculate with one definite numerical value for the precession. The Hipparchan stellar coordinates are transferable to the star catalogue of the Almagest only under the condition that a constant precession value is added to all Hipparchan longitudes. For this conversion the Hipparchan minimal value is unsuitable.

In the long first chapter of the third book, Ptolemy mentions in great detail the methodological weaknesses of a Hipparchan empiricism focusing on the determination of the length of the year as one of the key parameters for all other astronomical theories. Hipparchus succeeded in distinguishing the sidereal year – the time the sun needs for returning to the same place relative to the stars – from the tropical year as the time the sun needs to reach the spring equinox once again.[20] This distinction marked the first time that the effect of the precession motion was included in the description of the sun's motion. The measuring of the length of the year placed Hipparchus in a position to make this conceptual distinction, though his empirical model excluded inferring a constant length of the tropical year from the slightly diverging figures. Each measuring of the length of the year involves fluctuations that are either a consequence of observational errors or point to a further irregularity regarding the tropical year. In line with the empirical approach Hipparchus could not derive from this material the constant length of the tropical year, but could at

---

[20]Ptolemy, C.(1984), 131: "The very first of the theorems concerning the sun is the determination of the length of the year. The ancients were in disagreement and confusion in their pronouncements on this topic, as can be seen from their treatises, especially those of Hipparchus, who was both industrious and a lover of truth. The main cause of the confusion on this topic which even he displayed is the fact that, when one examines the apparent returns [of the sun] to [the same] equinox or solstice, one finds that the length of the year exceeds 365 days by less than $\frac{1}{4}$-day, but when one examines its return to [one of] the fixed stars it is greater [than $365\frac{1}{4}$ days]. Hence Hipparchus comes to the idea that the sphere of the fixed stars too has a very slow motion, which, just like that of the planets, is towards the rear with respect to the revolution producing the first [daily] motion, which is that of a [great] circle drawn through the poles of both equator and ecliptic."

best describe the limits within which the length of the year can vary.

Precisely this outlook is cited by Ptolemy in the Almagest:[21]

> "Now since Hipparchus is somewhat disturbed by the suspicion, derived from a series of observations which he made in close succession, that this same revolution [of the sun] is not of constant length, we shall try to show succinctly that there is nothing to be disturbed about here."

In contrast to the theoretically more cautious Hipparchus, Ptolemy assumes that an error in observation must absolutely be the culprit here.[22]

> "But we also guess from Hipparchus' own calculations that his suspicion concerning the irregularity [in the length of the tropical year] is an error due mainly to the observations he used."

Hipparchus, who like Ptolemy recognizes the possibility of error at observations, although he cannot methodologically distinguish fallacious observations from correct ones, restricts himself to fixing empirically sound limits of the hypothesis, as he does in the case of the precession constant.[23]

> "However, Hipparchus himself does not think that there is anything in the above observations which provides convincing support for his suspicion that there is an irregularity in the length of the year. Instead he makes computations on the basis of certain lunar eclipses, and declares that he finds that the variation in the length of the year, with respect to the mean value, is no more than $\frac{3}{4}$ of a day. This would be sufficiently great to take some account of, if it were indeed so; but it can be seen to be false from the very considerations which he adduces [to support it]."

The way Ptolemy corrects Hipparchus' mistake reveals very much of the two fundamentally different methodological viewpoints. The citation continues:[24]

> "For he uses certain lunar eclipses which were observed to take place near [specific] fixed stars to compare the distance of the star called Spica in advance of the autumnal equinox at each [eclipse]. By this means he thinks he finds, on one occasion, a distance of $6\frac{1}{2}°$, the maximum in his time, and on another a distance of $5\frac{1}{4}°$, the minimum [in his time]. Thence he concludes that, since it is impossible for Spica [itself] to move so much in such a short time, it is plausible to suppose that the sun, which Hipparchus uses to determine the positions of the fixed stars, does not have a constant period of revolution."

Hipparchus' analysis of the eclipses considers no astronomical error nor any error in calculation. Solely due to these two observations, he had to keep the theoretical

---

[21] Ptolemy, C. (1984), p. 132.

[22] Ptolemy, C. (1984), p. 132.

[23] Ptolemy, C. (1984), p. 135.

[24] Ptolemy, C. (1984), p. 135.

possibility open that the length of the tropical year could be subject to small fluctuations. Starting the mathematical demonstrations from the measurements, Hipparchus could only narrow down the possible frame of hypotheses; in particular he was able to refute unsuitable theories. According to this methodology, the possibility of an observational error is not in itself sufficient to disqualify the observation and allow one to assume a constant year. Hipparchus had indeed carried out, independently of his eclipse observations, parallel measurements of the spring equinox from which he came to the conclusion of a constant tropical year.[25] Here, however, an error could be concealed in the measurements, which would conceal a small variation in the length of the year.

The absence of any criterion which could enable Hipparchus to recognize a corrupt measurement forces him to consider the intervals of fluctuation in all theoretical hypotheses, "his love of truth led him not to suppress anything which might in any way lead some people to suspect [such an anomaly]."[26] It prevents more advanced constructions of theories. In addition a methodical rule similar to (E) compelled Hipparchus to prove the constancy of the tropical year with the direct help of his observations.

In one case there is a time difference of 11 years between two position measurements by eclipse observations.[27] The observations of the close conjunctions of Spica with the moon seemed to imply an increase of the ecliptical longitude of Spica by more than 10′, and Hipparchus felt justified by the observation in conceding the possibility of a variable tropical year.

Here Ptolemy criticizes Hipparchus as inconsistent. Although he does not describe the Hipparchan method of calculation in detail, it can nevertheless be assumed (just as Ptolemy relates) that Hipparchus calculated the position of the moon during the eclipse by means of the position of the sun derived from the solar theory, and found the position of Spica from its distance from the eclipsed moon. Since for Hipparchus too the solar theory assumes a constant tropical year, one cannot, according to Ptolemy, doubt the truth of that constancy afterwards: "but this kind of computation cannot be made without using the sun's position at the eclipse as a basis."[28]

Ptolemy considers Hipparchus' methodology inadmissible. First of all the truth of the hypothesis regarding the constancy of the length of the year must be presupposed to obtain any calculated results whatsoever and then, following the analysis of the observations, it serves precisely to undermine the truth of the hypothesis advanced in the first place:[29]

> "Since, then, the sun has been shown to complete its revolution (as measured with respect to those equinoxes) in a time neither greater nor less than the $[365]\frac{1}{4}$-day interval, and since it is impossible for Spica to move $1\frac{1}{4}^{\circ}$ in such a small number of years, surely it is perverse to use calculations based on the above foundations to impugn the very

[25] Ptolemy, C. (1984), p. 136.
[26] Ptolemy, C. (1984), p. 136.
[27] Ptolemy, C. (1984), p. 136.
[28] Ptolemy, C. (1984), p. 136.
[29] Ptolemy, C. (1984), p. 136.

foundations on which they were based. It is perverse to ascribe the reason for such an impossibly large motion of Spica solely to the equinoxes on which the calculations are based (which entails the simultaneous assumptions, both that they are accurately observed, and that they have been inaccurately observed), when there are several possible causes for so great an error."

Hipparchus could not have let the Ptolemaic objection stand. Precisely because of the assumptions implicitly contained in the solar theory, the positional measuring of Spica would have to have shown the same value, besides a very small increase attributable to the precession. The considerable change in the observed position of Spica might at least make the doubt cast on the correctness of the premises seem legitimate. At this point we see different methodological positions taken by Hipparchus and Ptolemy.

Before presenting the Ptolemaic position more precisely, the weaknesses of the Hipparchan approach to the matter will be elucidated with another example.

Despite his methodological doubts about the empirical foundation of the constancy of the year, Hipparchus went ahead and developed the solar theory on the basis of this precondition. In the case of this most fundamental of all astronomical theories, Hipparchus had no alternative but to ignore his methodological misgivings and to construct a usable theory through a different approach to the empirical material, mainly via the period relationships inherent in Babylonian astronomy.[30] The data in most of these tables, however, show no observational error because of their entirely theoretical nature.

In another case, Hipparchus did not take this step. For the theory of the planets, all he did was to take the observations he had carefully carried out while waiving any progressive theoretical evaluation. In the Almagest Ptolemy describes Hipparchus' troubles in detail:[31]

> "Hence it was, I think, that Hipparchus, being a great lover of truth, for all the above reasons, and especially because he did not yet have in his possession such a groundwork of resources in the form of accurate observations from earlier times as he himself has provided to us, although he investigated the theories of the sun and moon, and, to the best of his ability, demonstrated with every means at his command that they are represented by uniform circular motions, did not even make a beginning in establishing theories for the five planets, not at least in his writings which have come down to us. All that he did was to make a compilation of the planetary observations arranged in a more useful way, and to show by means of these that the phenomena were not in agreement with the hypotheses of the astronomers of that time."

It is clear from the quotation that Hipparchus did not describe the reasons of his failure, though Hipparchus' requirements for a proper planetary theory must be evident for Ptolemy. Ptolemy also relates that the theoretical cautiousness shown by

---

[30] Viz. Toomer, G. J. (1978).
[31] Ptolemy, C. (1984), p. 421.

Hipparchus cannot be explained by a shortage of older observations. In accordance with (E) Hipparchus – or what Ptolemy presumed of Hipparchus – had set as his goal,[32]

> "that anyone who was to convince himself and his future audience must demonstrate the size and the period of each of the two anomalies by means of well-attested phenomena which everyone agrees on, must then combine both anomalies, and discover the position and order of the circles by which they are brought about, and the type of their motion; and finally must make practically all the phenomena fit the particular character of the arrangement of circles in his hypothesis. And this, I suspect, appeared difficult even to him."

These three steps in the development of an appropriate model require a demonstration of the hypothetical components from the empirical data, and 'demonstration' has the meaning of a strict mathematical proof. Such a rigorous limitation to the astronomical argument cannot lead to the construction of a Ptolemaic planetary model even when only those observations are evaluated which particularly suit the determination of certain parameters and geometrical arrangements.

Hipparchus was denied the chance to develop a theory of planetary motion because the variation of the measurements could not be condensed to clear and unambiguous phenomena which would permit a sound demonstration of the parameters of the orbits. Ptolemy, whose goal was to develop a comprehensive theory of celestial phenomena, formulated the relationship between observation and theory anew.

6.4 Holistic Rationalism

The novelty of his strategy and the shift away from the stringent empiricism prompted Ptolemy to justify the new method in detail after he had described the reasons for the missing Hipparchan planetary theory:[33]

> "The point of the above remarks was not to boast [of our own achievement]. Rather, if we are at any point compelled by the nature of our subject to use a procedure not in strict accordance with theory (for instance, when we carry out proofs using without further qualification the circles described in the planetary spheres by the movement [of the body, i.e.] assuming that these circles lie in the plane of the ecliptic, to simplify the course of the proof); or [if we are compelled] to make some basic assumptions which we arrived at not from some readily apparent principle, but from a long period of trial and application, or to assume a type of motion or inclination of the circles which is not the same and unchanged for all planets; we may [be allowed to] accede [to this compulsion], since we know that this kind of inexact procedure will not affect the end desired, provided that it is not going to

---

[32]Ptolemy, C. (1984), p. 422.

[33]Ptolemy, C. (1984), pp. 422f.

result in any noticeable error; and we know too that assumptions made without proof, provided only that they are found to be in agreement with the phenomena, could not have been found without some careful methodological procedure, even if it is difficult to explain how one came to conceive them (for, in general, the cause of first principles is, by nature, either non-existent or hard to describe); we know, finally, that some variety in the type of hypotheses associated with the circles [of the planets] cannot plausibly be considered strange or contrary to reason (especially since the phenomena exhibited by the actual planets are not alike [for all]); for, when uniform circular motion is preserved for all without exception, the individual phenomena are demonstrated in accordance with a principle which is more basic and more generally applicable than that of similarity of the hypotheses [for all planets]."

Starting from three reasons justifying the use of "inexact procedures", Ptolemy develops a method of theory tests which has nothing in common with the radical empiricism. The assertion of theoretical hypotheses cannot be limited by

(i) radical empirical demonstration. A hypothesis – or one of its parameters – requires no direct demonstration through an observable phenomenon. The equants of the planetary orbits, for example, can be introduced without having to be directly observable. Ptolemy does not leave many clues in the Almagest how he arrived at the bisection of the eccentricity. Though there might be strong astronomical reasons to modify the preliminary planetary models and introduce this unique feature, it cannot be directly demonstrated from a set of observations without reference to the other hypotheses developed so far.

(ii) radical methodical demonstration. The theoretical way leading to the formulation of an hypothesis has, similarly, no influence on its validity.

(iii) theoretical demonstration. The choice of hypotheses does not have to be oriented on considerations of similarity to direct visual impressions, such as that of the apparent paths of planets in the sky compared to the mathematical form of their orbits. In other words, the geometric structures of astronomical models require no direct confirmation through perception.

All three modifications guarantee a large degree of freedom in the theoretical constructions. Ptolemy counters the disadvantage of Hipparchan empiricism, namely, that it prevents theoretical development when one is confronted by a wealth of empirical material, with a methodological concept that, at first, places no empirical restrictions on theoretical proposals. Only after the complete theory is developed is its quality reviewed in face of the empirical data.

Methodology:
(B) A single observation is accepted as "qualified" if it satisfies the theoretically required circumstances that allows a direct mathematical deduction from one of the tested parameters to the observations. Furthermore it should agree with the theoretical prediction, because otherwise one has to suspect an inaccuracy in the observation.

(T) Individual theories or hypotheses will be no longer tested on single observations; rather, all the theoretical formulations will be tested on observations selected according to (B).

This new methodological approach made it possible for Ptolemy to first of all develop far-reaching theoretical assumptions and to test these hypotheses on the totality of the observational material after the complete theoretical model had been constructed.

This "holistic" approach calls for an empirical test of the entire theory. In the next passage Ptolemy points out that there are rarely observations which would allow a direct demonstration of the hypotheses, hence one has to save the phenomena in general.[34]

> "We have made the above remarks, not to disparage the preceding method of determining the periodic returns, but to show that, while it can achieve its goal if applied with due care and the appropriate kind of calculations, if any of the conditions we set out above are omitted from considerations, even the least of them, it can fail utterly in its intended effect; and that, if one does use the proper criteria in making one's selection of observational material, it is difficult to find corresponding [pairs of eclipse] observations which precisely fulfil all the required conditions."

Although the limits within which a proposal for a hypothesis is permitted according these three criteria are generous, these regulations are not in themselves sufficient to allow the construction of an astronomical theory going beyond the Hipparchan achievements. The holistic test requires that the entire theory be checked against the observations. Every theory with observable predictions would be refuted if practically all observations varying in their numerical results would have to be considered as being in principle equally sound. Ptolemy was forced to find criteria to choose between those empirical data that can either refute or confirm a theoretical proposal. A single numerical value can fulfil this task only if it represents the entire series of measurements. In what follows, this value will be called by the specific name "measurement value" or "observation".

We have said before that Ptolemy had no access to a method of finding the mean value. Therefore, he could not deduce solely from the set of observations a single datum which would test the theory in place of the observations. Such a test criterion could only be a selection of the presumably best empirical result from the total set of observations.

As the first step only those types of observations are considered whose method of measurement produces particularly reliable results. This could be verified if a repetition of the measurement yields similar results. Ptolemy remarks in the context of the planetary theory:[35]

> "The observations which we use for the various demonstrations are those which are most likely to be reliable, namely [1] those in which there

---

[34]Ptolemy, C. (1984), p. 178. Holistic considerations can be found in Ptolemy, C. (1984), p. 136.
[35]Ptolemy, C. (1984), p. 423.

is observed actual contact or very close approach to a star or the moon, and especially [2] those made by means of the astrolabe instruments."

However, the selection of the observations according to the standpoint of reliable measuring methods is in general not sufficient to obtain identical results when a measurement is repeated, as we have already seen, for instance, regarding the determination of the precession constant.

Ptolemy needed an additional criterion which would allow a special value to emerge from the series of similar observations which could be used as the "true" observed value even if the "truth" of observational results can be revised by later analyses. The only way to select this measurement requires the astronomical theory under test.

Methodology:
(B') From a set of measurements obtained with a qualified procedure, that value will be chosen as "sound result of a measurement" which coincides with the theoretical prediction.

It appears paradoxical to test a theory empirically with a measurement that was selected from a data set on the basis of its best accordance with the theory. Indeed, is it possible at all to reject theories with such a methodology?

Two factors allow Ptolemy to employ a fruitful theory test:

(i) An hypothesis or theory determines the selection of a measured value from the set of the results. However, a positive test presupposes that there exists a qualified measurement that can then be selected. If, for example, a precession of 6° per century were to be hypothetically assumed and no single observation with a qualified method leads to this result, the hypothesis is then refuted.[36]

> "And in general, we consider it a good principle to explain the phenomena by the simplest hypotheses possible, in so far as there is nothing in the observations to provide a significant objection to such a procedure."

(ii) Ptolemy stresses repeatedly that an hypothesis is not tested on a measured value in isolation, but has to demonstrate its legitimacy in face of the totality of phenomena in the theoretical net of all hypotheses. This implies that theoretical parameters, whose values cannot be measured directly and would according to (i) always be verified, could be empirically controlled through the fit of the theory to the totality of phenomena.

With this methodological precondition of Ptolemaic astronomy one can see why the observations reported in the Almagest by and large agree with those derived from Ptolemy's theory, while they sometimes significantly deviate from the accurate values. The best example of this procedure is the measuring of the

---

[36]Ptolemy, C. (1984), p. 136.

precession constant.[37] According to Ptolemy's methodology, a correct observational result requires its theoretical derivability. If the theoretically expected value agrees with the observed value, this value can be interpreted as a confirmation of the theory and, correspondingly, can be cited in the Almagest as an observation with which, conversely, the values of the theoretical parameters are "derived". There is no such thing as an observation without an explanatory deduction from theory. Only in this way is it understandable that a position measurement of Spica carried out by Timocharis in the year -293, whose observations Ptolemy had previously branded as "not trustworthy, having been made very crudely", is used together with a measurement of the same astronomer from the year -282, in order to utilize a usually negligible displacement of $10'$ as successful confirmation of the faulty precession constant of one degree per century.

Not only did Ptolemy analyze the observations according to the standpoint of correspondence to the theoretical prediction, but he even corrected the reports of the observations where he saw fit:[38]

> "In the 48th year of the same [First Kallippic] Cycle, he says that on the sixth day from the end of the last third of Pyanepsion, with is Thoth 7, when as much as half an hour of the tenth hour had gone by, and the moon had risen above the horizon, Spica appeared exactly touching the northern point on the moon.
>
> This moment is in the 466th year from Nabonassar, Thoth 7/8 in the Egyptian calendar [-282 Nov. 8/9]; [the hour is], according to Timocharis himself, $3\frac{1}{2}$ seasonal hours after midnight, or approximately $3\frac{1}{8}$ equinoctial hours, since the sun was near the middle of Scorpius; but, according to logical reasoning, [it must have been] $2\frac{1}{2}$ hours after midnight. For that is the time when $82°30'$ is culminating, and $172°5'$ (approximately) is rising: and that was the longitude of the moon at that moment when, as he says, it was rising. Reckoning with respect to mean solar days, we find that only 2 equinoctial hours had passed since midnight. At this time the positions of the centre of the moon were as follows:

> true [longitude]:   distance from the summer solstice: 81;30°
> true [latitude]: $2\frac{1}{6}°$ south of the ecliptic
> apparent longitude: $82\frac{1}{2}°$ [from the summer solstice]
> apparent [latitude]: $2\frac{1}{4}°$ south [of the ecliptic]

> Therefore, according to this observation too, Spica was the same distance of about 2° south of the ecliptic, and was $82\frac{1}{2}°$ from the summer solstice."

Together with the first positional measurements of Timocharis yielding a distance of Spica of $82°20'$ from the summer solstice, Ptolemy can conclude: "So in the 12

---

[37] A series of observations in the Almagest are presumably selected or constructed according to theoretical considerations: equinox observations (second and third book), the obliquity of the ecliptic (first book), the eclipses used for the determination of the lunar parameter (fourth book), the parallax measurements of the moon (fifth book).

[38] Ptolemy, C. (1984), p. 336.

years between the two observations it moved about $\frac{1}{6}^{\circ}$ towards the rear from the summer solstice."[39]

One equinoctial hour divides the day into exactly identical 24 time intervals. In the summer with its long days, the seasonal day hours are for that reason longer than the night hours. Only at the beginning of spring and fall are all hours equally long. Both time concepts refer to the actual rising or meridian transit of the sun on a given day. As the earth moves on an ellipse around the sun, the apparent motion of the sun is quicker the nearer it is to the earth. For this reason the actual sun sometimes rises a bit later or earlier than the theoretical mean sun – presuming a uniform motion. This time difference is called "the equation of time". For the time of the observation the equation of time is about 30 minutes, and Ptolemy subtracts it from the corrected $2\frac{1}{2}$ equinoctial hours in order to derive the theoretical position of the moon for the time of 2 equinoctial hours.

In his translation, Toomer draws particular attention to a Ptolemaic error regarding the conversion of the times.[40] The $3\frac{1}{2}$ seasonal hours do not tally with the $3\frac{1}{8}$ equinoctial hours given in the Almagest.

On the night of observation the sky rotates with 16;38° per seasonal hour.[41] One seasonal hour corresponds therefore to 1;6 equinoctial hours. In view of this ratio, the observation of Timocharis lasting $3\frac{1}{2}$ seasonal hours would have to be converted to 3.88 equinoctial hours. The Ptolemaic time difference of $3\frac{1}{8}$ equinoctial hours is not larger but smaller than the seasonal time. Toomer points out that Ptolemy mistakenly calculated with the shorter day-hour instead of the longer night-hour: one day-hour is at the date of observation only 53 equinoctial minutes long, therefore $3\frac{1}{2}$ seasonal hours would correspond exactly to the $3\frac{1}{8}$ hours mentioned in the Almagest.

Considering this obvious error, Ptolemy proceeded to correct the time in the observational report of Timocharis by more than one hour. What reasons could Ptolemy have had for justification?

The "logical reasoning" could be based on two points:

(i) The time of observation is changed to the moment when the moon rises as it calculated by the lunar theory. Ptolemy would have successfully eliminated an inconsistency between the lunar theory and the observation of Timocharis, but he had still derived the position of Spica from the empirical evidence that the star had the same ecliptical longitude as the moon at rising.

(ii) The basis of the correction is not the rising time of the moon, but the theoretically expected position of Spica. In this case, the position of Spica is first of all derived from the precession formula, and then the time from when the moon reaches the same longitude.[42] In such a procedure the rising of the moon does not determine the observation time.

The control calculation shows that in reality Spica rose at 2;50 equinoctial hours. That means that 40 minutes must have elapsed after the time assumed by Ptolemy

---

[39] Ptolemy, C. (1984), p. 336.

[40] Ptolemy, C. (1984), p. 336.

[41] Ptolemy, C. (1984), p. 336.

[42] Toomer's calculation of the lunar position with Ptolemy's theory agrees well with the time of 2 equinoctial hours after midnight. Cf. Ptolemy, C. (1984), p. 336 and p. 652.

before the moon and Spica were visible at all above the horizon.[43] This conspicuous difference cannot be explained if Ptolemy employed the rising time according to (i) as the basis of his calculations. Thus, the correction of the observation is oriented exclusively on the theoretical requirements of the lunar theory and the precession motion, and is in full agreement with the revised methodological procedure of developing the theories.

The example makes clear that Ptolemy not only selected a measurement from a group of data according to theoretical requirements, but shows himself ready to correct the observational reports to conform to fundamental parameters.

Hartner examined the similar case of the measurement of the lunar parallax and again pointed out that the parameters obtained thereby could not possibly have been deduced from authentic observations.[44] However, the consequence which he draws: "I call this procedure wishful thinking rather than hoax, as does Newton, and claim that the history of science is full of examples of this kind,"[45] obscures the actual historical context of Ptolemy's astronomy.

The Ptolemaic procedure should not be misinterpreted as forgery or wishful thinking, judged from the perspective of the modern scientific self-conception; rather, it is essentially an expression of the constricting predicament which Ptolemy could escape only by means of a theoretically directed selection and correction of observations for the purpose of constructing a comprehensive astronomical theory. Moreover, in this situation it shows methodological progress to control the results of observations by theoretical hypotheses. Otherwise an essential stage in the development of astronomical theories would not have taken place.

In this context, then, it becomes clear why, in spite of the statement in the Almagest that the stellar positions in the catalogue had been observed by Ptolemy, a considerable number of coordinates stem from Hipparchan observations. The stellar positions – converted with the precession constant – represented for Ptolemy a theoretical proposition with a certainty vastly greater than his own contingent measurements done with the astrolabe. Even if Ptolemy had really observed all the stars in the interests of control, the results of the theoretical positions, that is to say, the Hipparchan positions, must have been correspondingly selected or corrected.

When Ptolemy was testing the precession motion discovered by Hipparchus by comparing his own observations, he says:[46]

> "For when we observe the latitudinal distance of any star with respect to the ecliptic, as measured along the great circle through the poles of the ecliptic, we find that it is practically the same as that computed from the records of Hipparchus, or if there is a discrepancy, it is of very small size, such as can be accounted for by small observational errors."

Now these differences are so small that they serve to confirm the Hipparchan values. The inclusion of Hipparchan coordinates with larger errors like π Hydrae,

---

[43]Coordinates of Spica (-282): $\alpha = 172^{\circ}.06, \delta = 1^{\circ}.40$, star time $t_s=2.89551$h, equation of time $t_{equ}=0.37$h; $\varphi=31^{\circ}.2$. Newton calculates the rising time as 2;47 equinoctial hours. Ptolemy's error in the position of Spica leads only to an error in the rising time of about 2 minutes.

[44]Hartner, W. (1977), pp. 3f.

[45]Hartner, W. (1977), p. 4.

[46]Ptolemy, C. (1984), pp. 329f.

α Centauri, and θ Eridani in the catalogue of the Almagest shows that Ptolemy in each case considered the Hipparchan coordinates with the theoretical conversions more sound than his own. What is more, the errors in his solar theory would not have put Ptolemy in a position to eliminate the systematic error of one degree in his stellar coordinates.

From the methodological point of view Ptolemy calls theoretical statements "observations", which in principle could be controlled by measurements, though the details in their formulation are worked out theoretically.

Previously, Ptolemy quoted Hipparchus:[47]

> "Now Hipparchus agrees with [the idea of] the motion taking place about the poles of th ecliptic. For in 'On the displacement of the solsticial and equinoctial points' he deduces from the observations of Timocharis and himself that Spica (again) has maintained the same distance in latitude, not with respect to the equator but with respect to the ecliptic, being 2° south of the ecliptic at both earlier and later periods."

This passage offers evidence that Hipparchus had already calculated and argued with stellar positions in the ecliptical coordinate system. It must have been immediately clear to him that a star catalogue should be compiled in ecliptical coordinates due to the precession motion. Whether or not Hipparchus had ever carried out such a project cannot be judged on the basis of presently available data. However, it is certain that Hipparchan measurements of stellar positions found their way into the star catalogue of the Almagest.

Against this backdrop Ptolemy emerges as a most circumspect astronomer who wanted to develop far-reaching theories of celestial phenomena proceeding from the rich observational material of Hipparchus or his Babylonian sources. To do so, he only had to abandon the strictures of a radical Hipparchan empiricism.

---

[47]Ptolemy, C. (1984), p. 329.

7. Appendix A

7.1 Stars and Constellations

A correct identification of the Ptolemaic stars is essential for the analysis of the catalogue. In the majority of cases the traditional assignments can be accepted.

A successful identification accounts for the whole set of data in the Almagest: the ecliptical longitude and latitude, the magnitude class and the arrangement of the star within the constellation. Furthermore one has to realize that the coordinates of the stars are not independent of each other. Relative positions of stars within the constellation are described by reference to neighboring stars and systematic errors are often shared by a group of stars.

A graphical representation of the sky together with the Ptolemaic catalogue stars has proven its value in many critical cases of identification. The maps of all 48 Ptolemaic constellations show the stars of the Bright Star Catalogue and the bright clusters in the environment of the Ptolemaic stars in the Cartesian coordinate system of the ecliptical coordinates. The Ptolemaic stars are plotted with their catalogue number and a dotted line connects them to their reference star. The positions of the reference stars are calculated for the year -128 to display a better separation of the positions of the two catalogues. Consequently the number of the Ptolemaic star is usually placed left of the reference star with an average shift in longitude of about $2°40'$. Therefore the charts do not show the deviation of Ptolemy's values from the actual positions at his epoch. A catalogue position free of errors would be connected to the reference star by a dotted line of the length of $3°40'$.

The scale of projection varies with the constellation and with it the scale of the connection lines between the Ptolemaic coordinates and the position of the reference star. The reference stars and clusters are plotted by different symbols depending on their magnitude:[1]

object	$> 1.5^m$	$1.5^m - 2.5^m$	$2.5^m - 3.5^m$	$3.5^m - 4.5^m$	$4.5^m - 5.5^m$	$5.5^m - 6.0^m$	$< 6.0^m$
star	●	●	●	●	•	•	·
nebula	○	○	○	○	○	○	○

---

[1] It should be noted that the symbols differentiate between the rounded values of magnitude and not the truncated ones as in standard star maps. This is justified by the assumption that Ptolemy's magnitudes represent rounded values.

7.2 Identifications

In the majority of cases the identifications of stars are the same as Toomer's.[2]
Difficulties in finding the correct reference star can be demonstrated with the stars
No. 250–251. Toomer designates both stars as 51 Oph without further justification.
The Almagest describes star 250 as "the last and rearmost of the 4 (stars on the
right foot)" and No. 251 as "the star to the rear of these, which touches the heel".[3]
Since both coordinates of the stars differ as well, one can exclude the possibility of
a double entry of the same star. Apparently the designation of the star 251 must be
changed.

Peters-Knobel consider three major alternatives.[4]

- Baily assigns the star to 52 Oph (HR 6545) with $\lambda_{-128} = 234\overset{\circ}{.}68$ and $\beta_{-128} = 1\overset{\circ}{.}56$, but a small magnitude of $6\overset{m}{.}57$. The coordinates fit well to the analogue displacements of the neighboring stars. Only, the star is so faint that it is questionable whether it was visible at all.

- Schjellerup and Manitius favour 58 Oph (HR 6595) with $\lambda_{-128} = 236\overset{\circ}{.}60$, $\beta_{-128} = 2\overset{\circ}{.}01$ and a magnitude of 4.87^m. Here the magnitude fits the catalogue, but the longitude is shifted by $2°$ and the latitude by $1°$.

- Peters also considers 2 Sge (HR 7369) as a possible candidate, though the coordinates would differ even more from the listed position.

Difficulties in the assignment might reflect larger typing or copying errors of figures
in the edition of the Almagest. Not all of them can be traced down by comparing
different manuscripts. In such cases one has to assume an authentic catalogue value.
The designation to a star is decided by the least difference of the calculated "true"
values to Ptolemy's position, magnitude and arrangement in the constellation with
respect to the average error of the surrounding Ptolemaic stars.

In general the critical identifications concern stars No. 540–543, 728–732, 785–788
and 1023–1028.

Modifications of Toomer's edition:

No. 802–804: The identification of Peters-Knobel is favoured, because it suits
better the coordinates. Toomer follows the identification of aṣ-Ṣūfī.

No. 836: Changed to HR 2648, compare the finder chart (fig. 4.1).

[2]Toomer, G. J. (1984), Ptolemy's Almagest, London.
[3]Toomer, G. J. (1984), p. 354.
[4]Peters, C. H. F., Knobel, E. B. (1915), p. 114.

No. 870: The difference in latitude is too large for HR 3439 (brightness 5.2^m), Toomer's identification. The star is approximately in the middle of the chart with the long dotted line to the reference star. At about half of the distance one finds the star HR 3535 with a brightness of 5.8^m. It fits the Almagest much better, especially when one considers similar errors of the stars No. 871ff. The following map shows Toomer's identification.

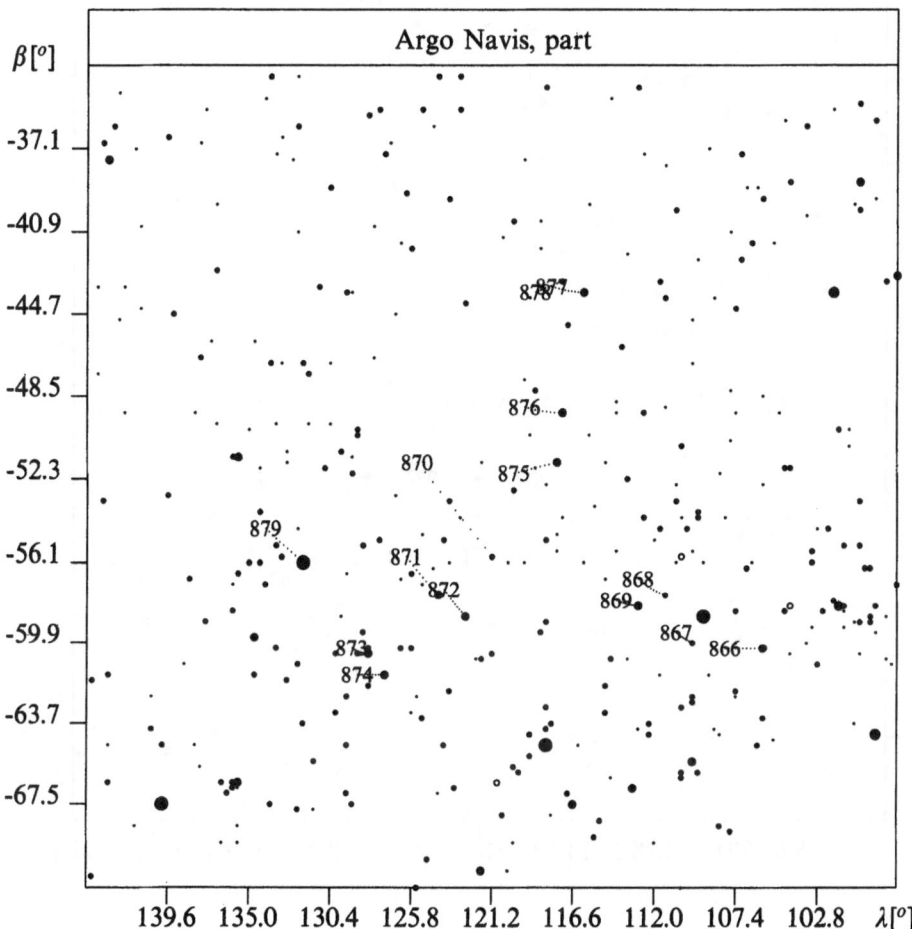

No. 987–988: We follow the identification of Peters-Knobel, otherwise the difference in longitude would be too large. Toomer supports his identification by the better relative positions of the stars, though the pair No. 989–990 has an analogous error in the latitude of the northern star.

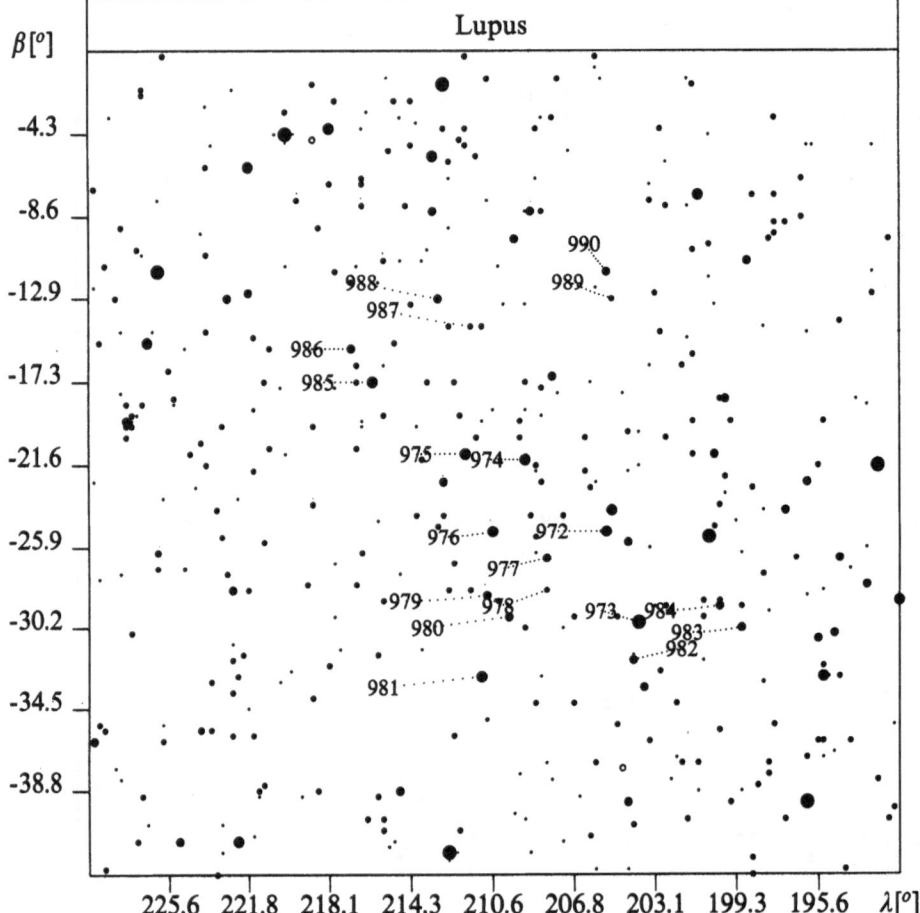

No. 1027-1028: The last stars of the Almagest are notoriously difficult to identify. Ptolemy describes No. 1028 as the "northernmost" and No. 1027 as southernmost. This requires changing the identification of both stars.

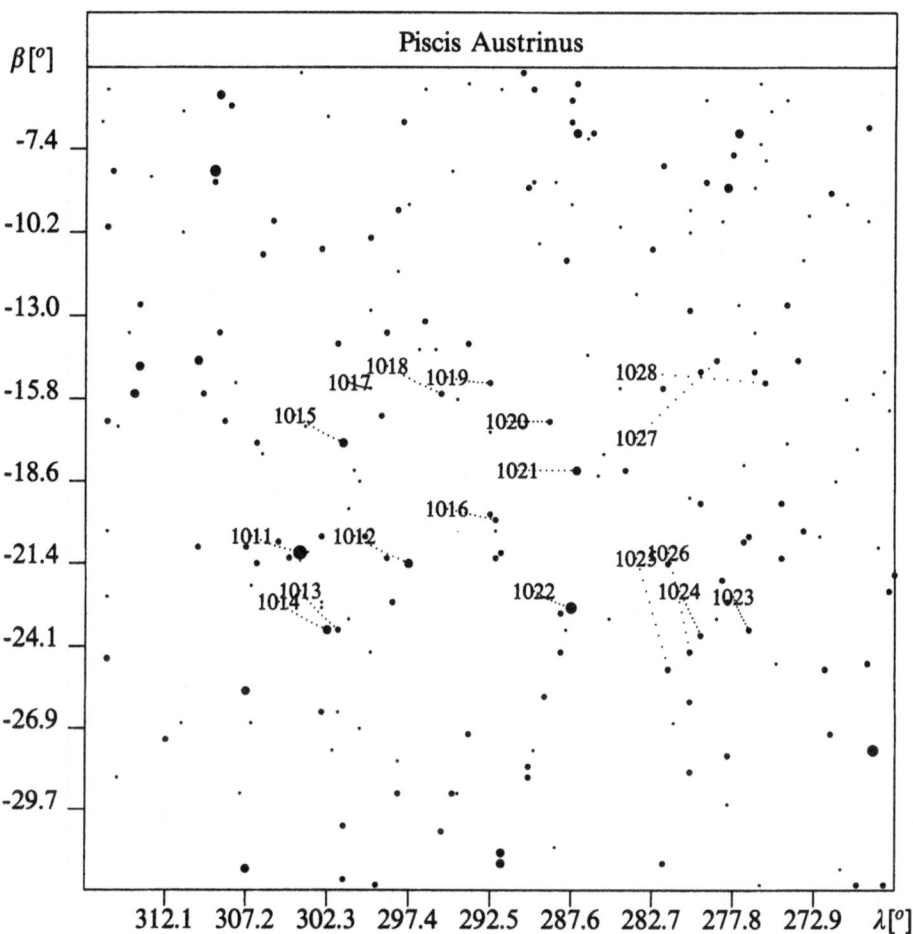

The following charts are printed in the sequence of constellations in the Almagest.

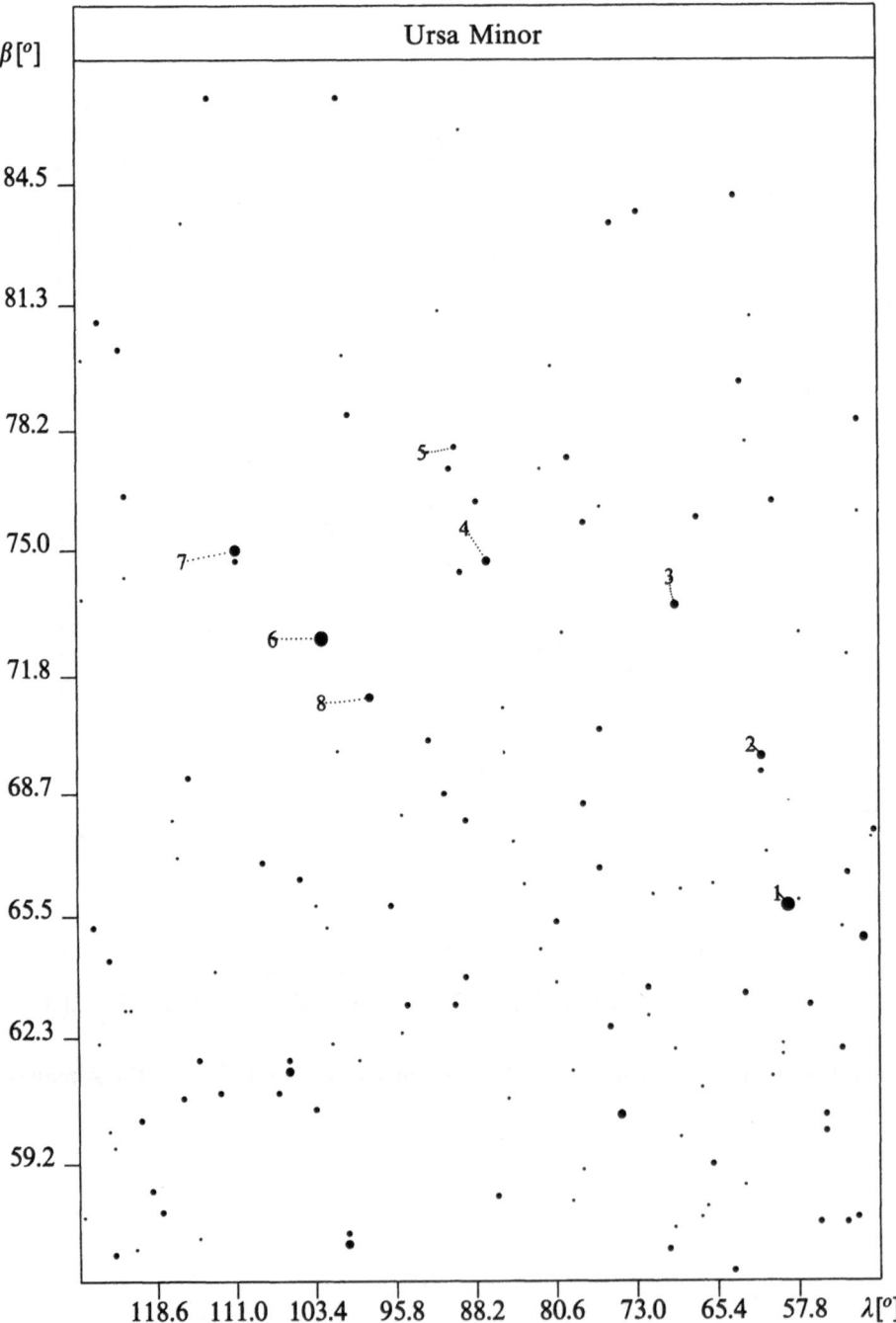

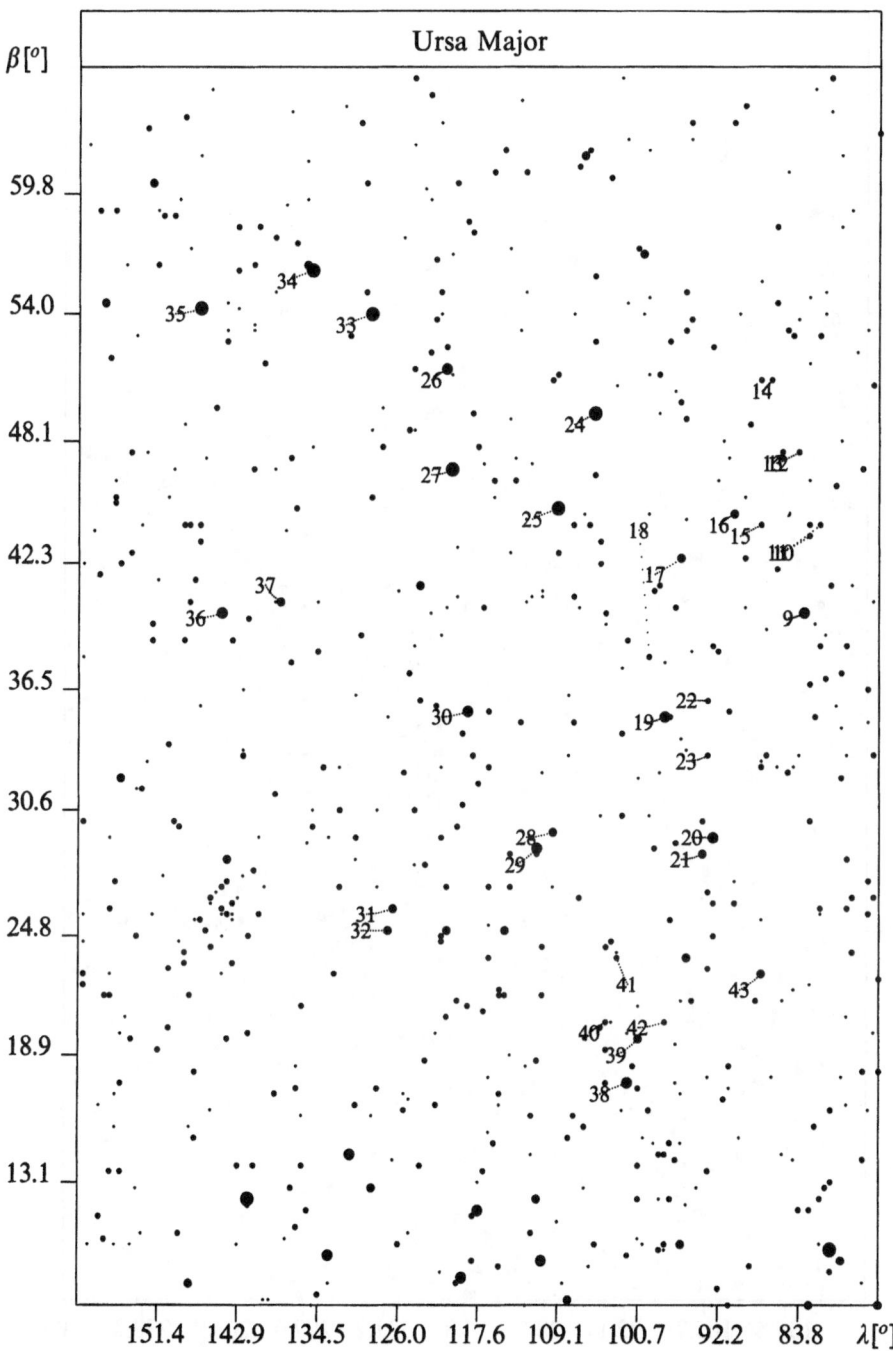

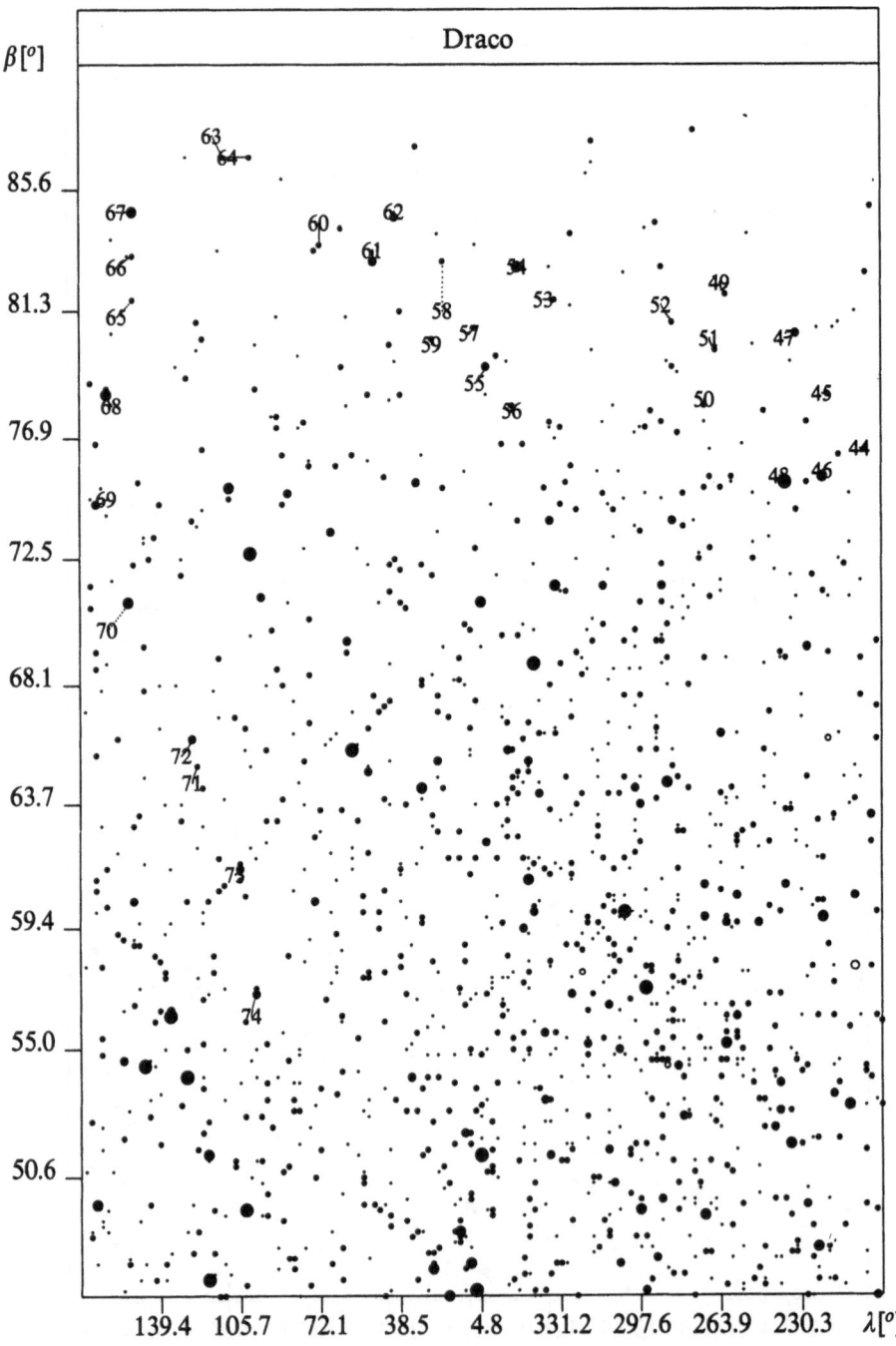

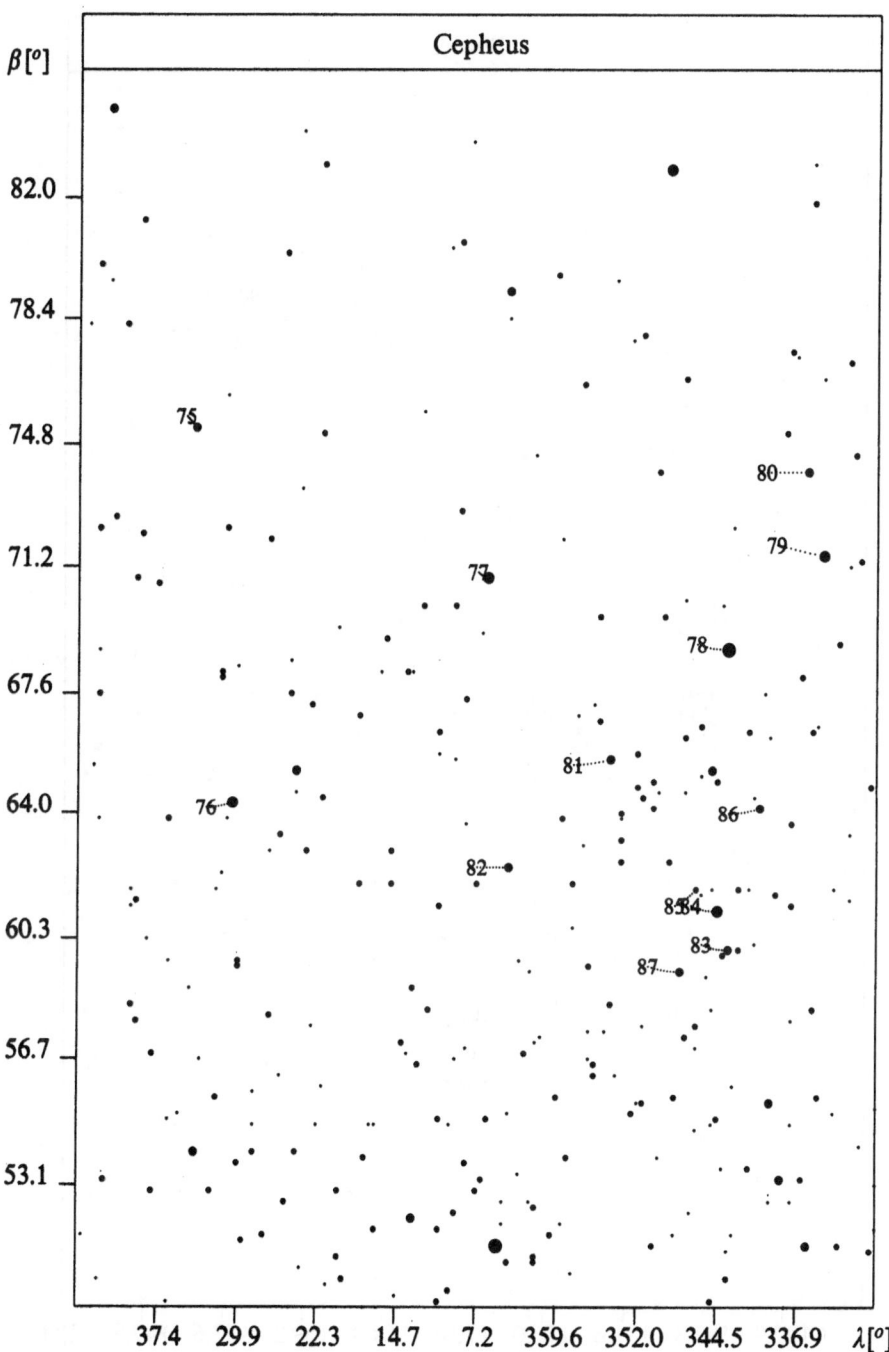

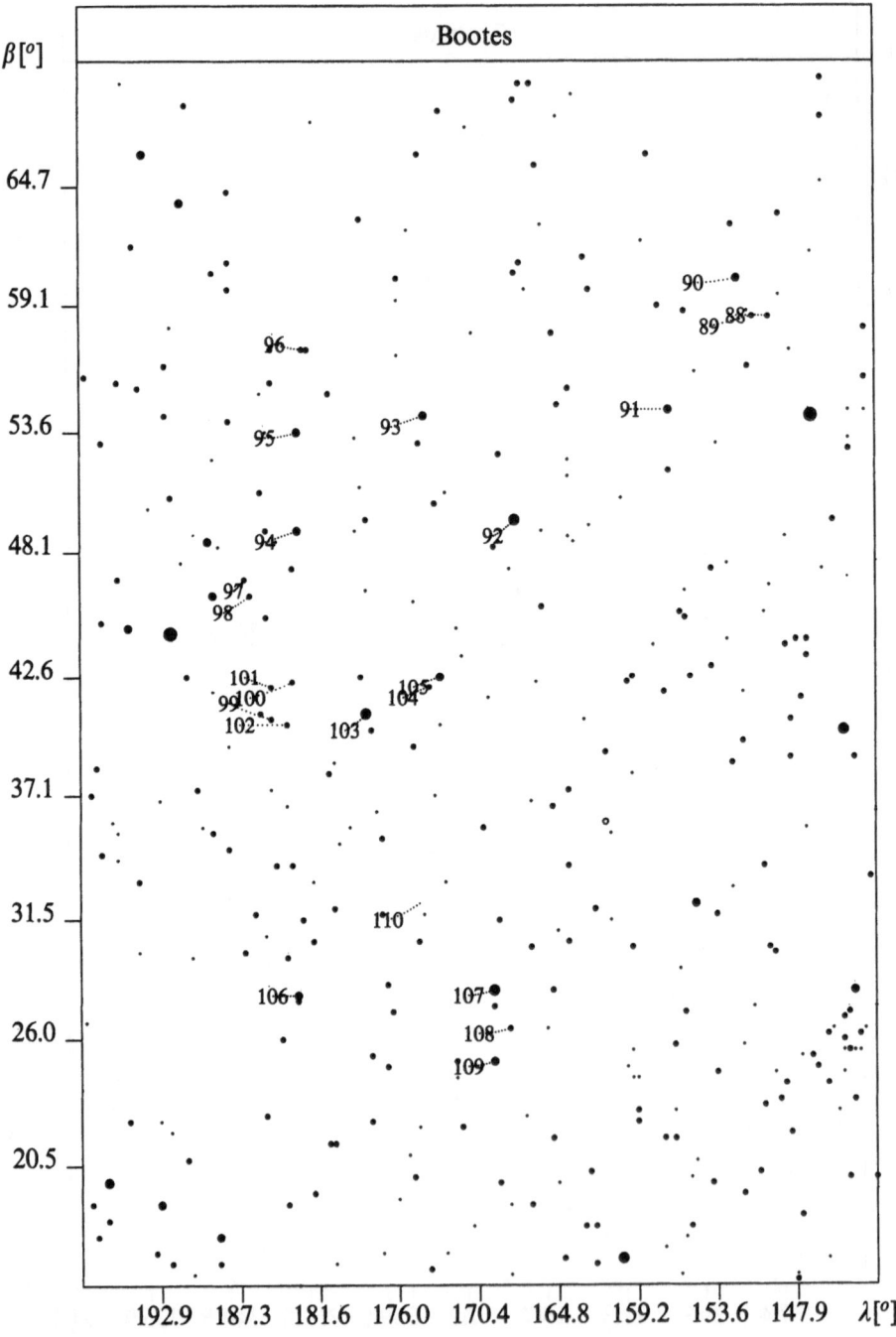

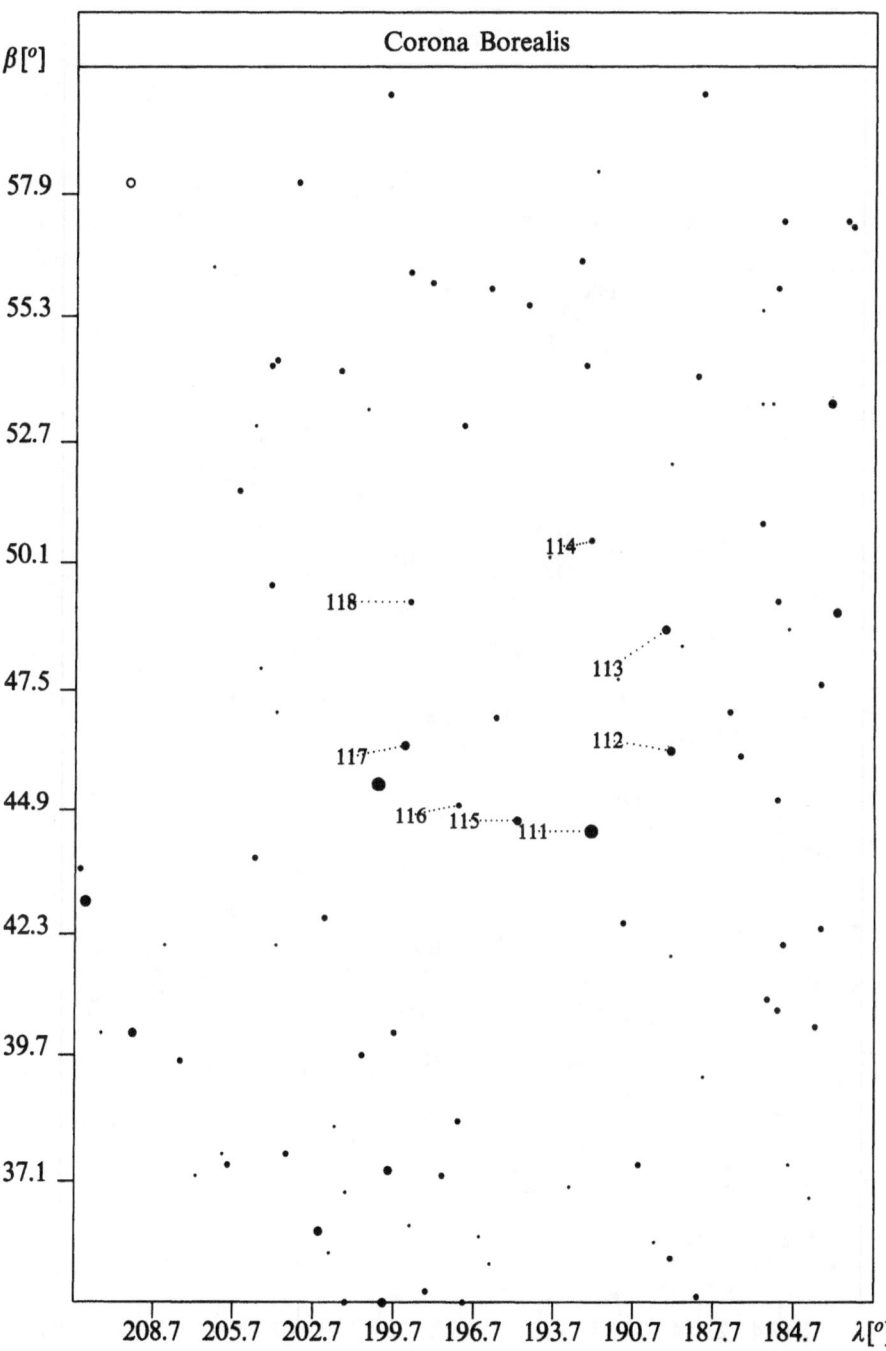

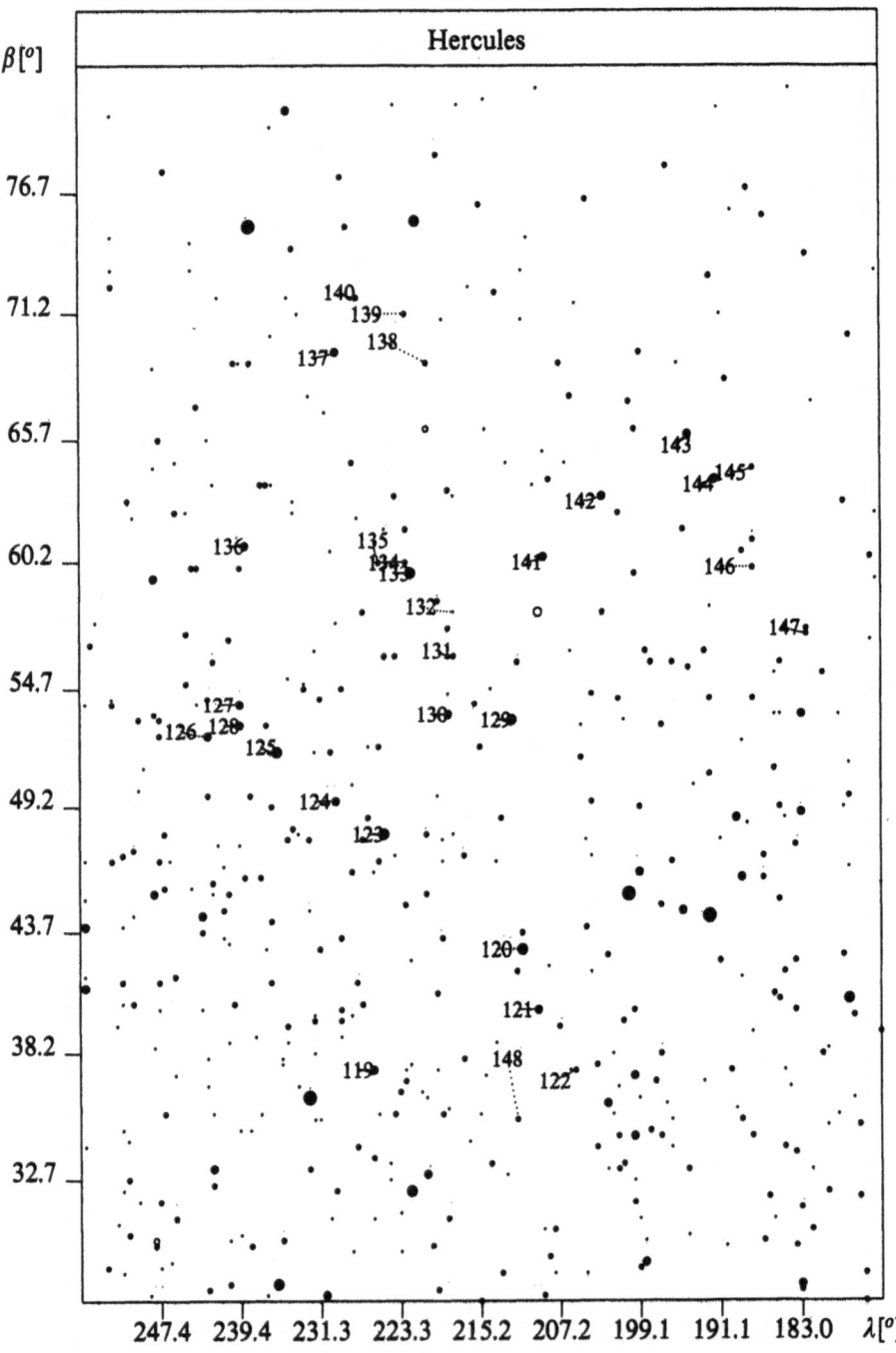

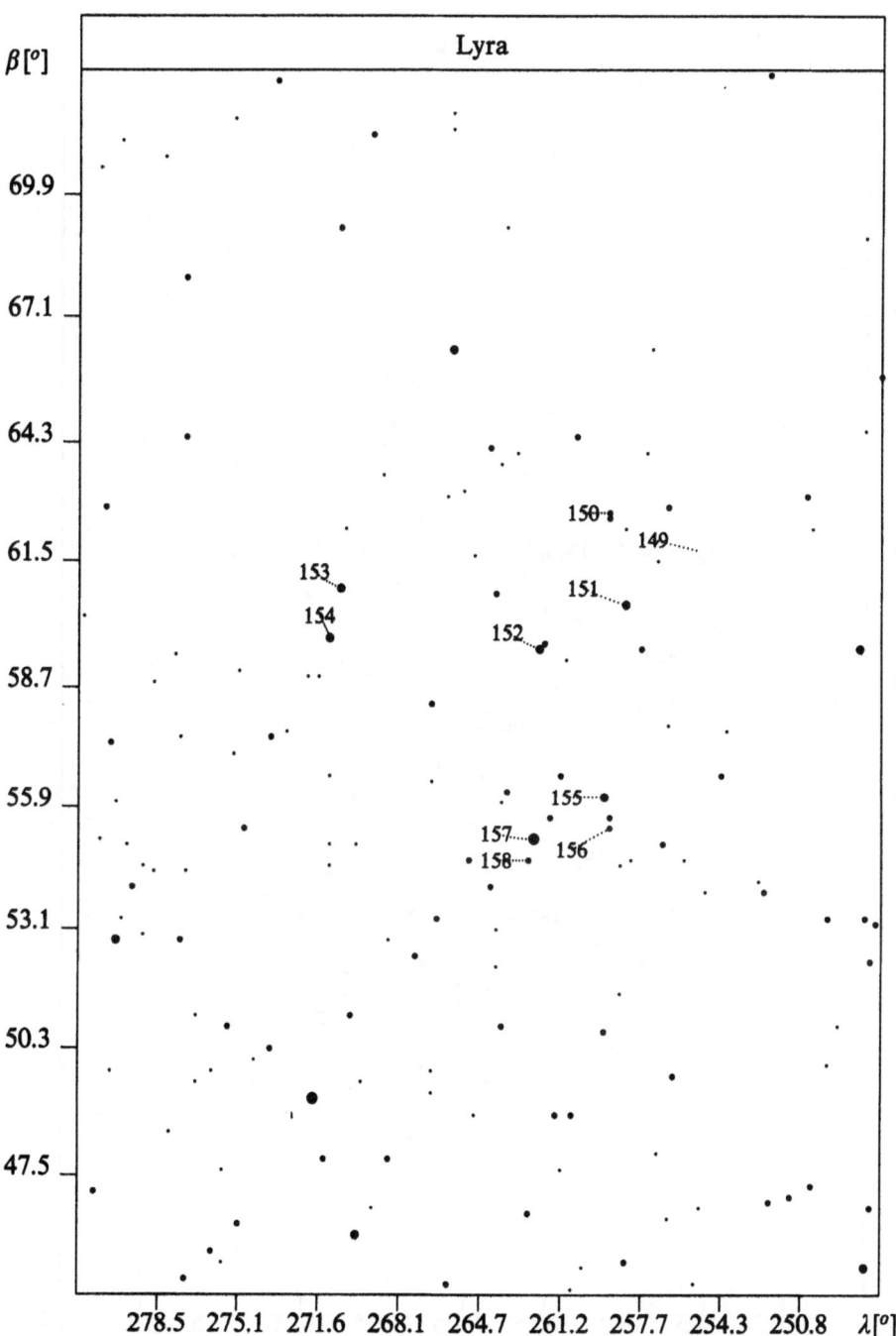

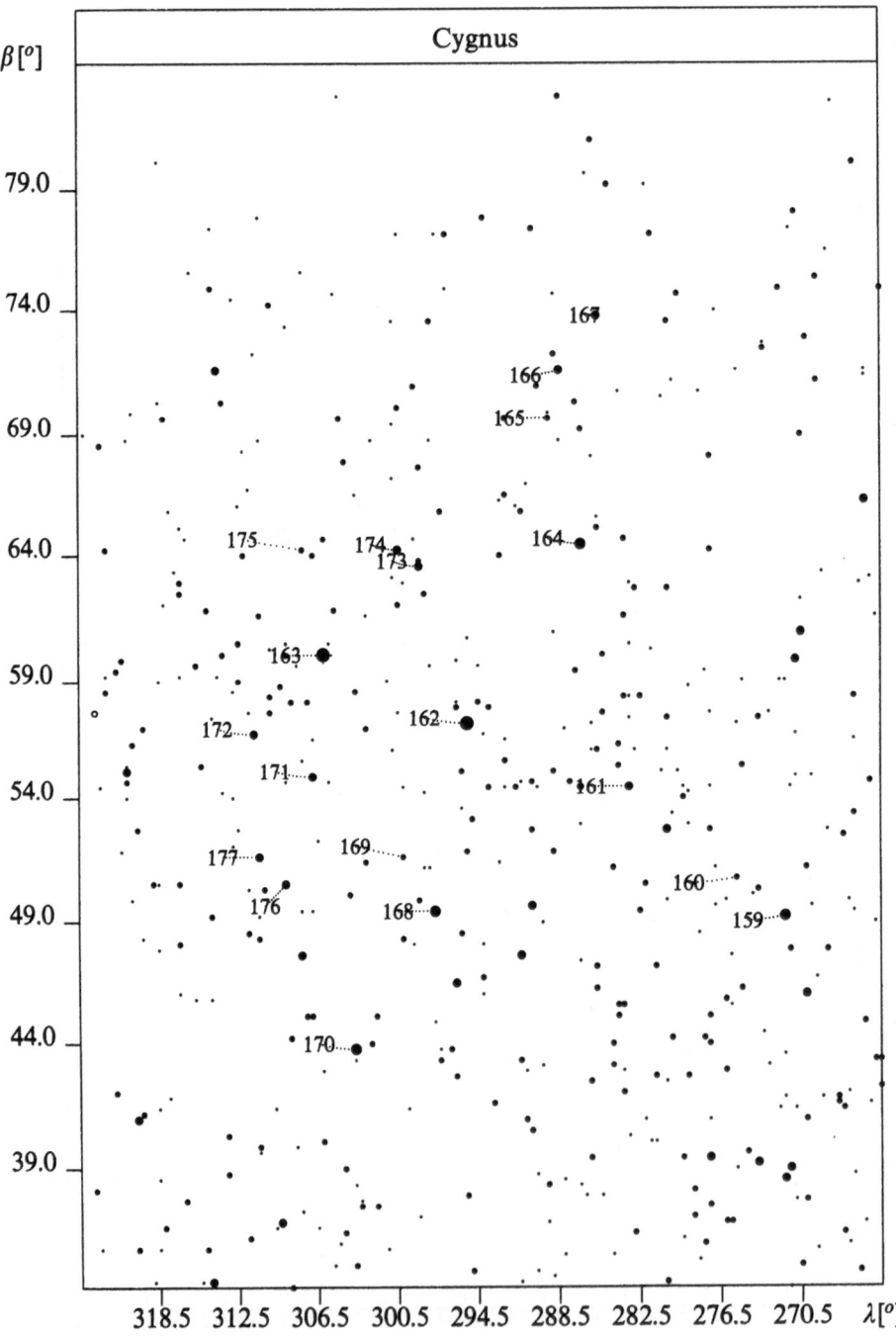

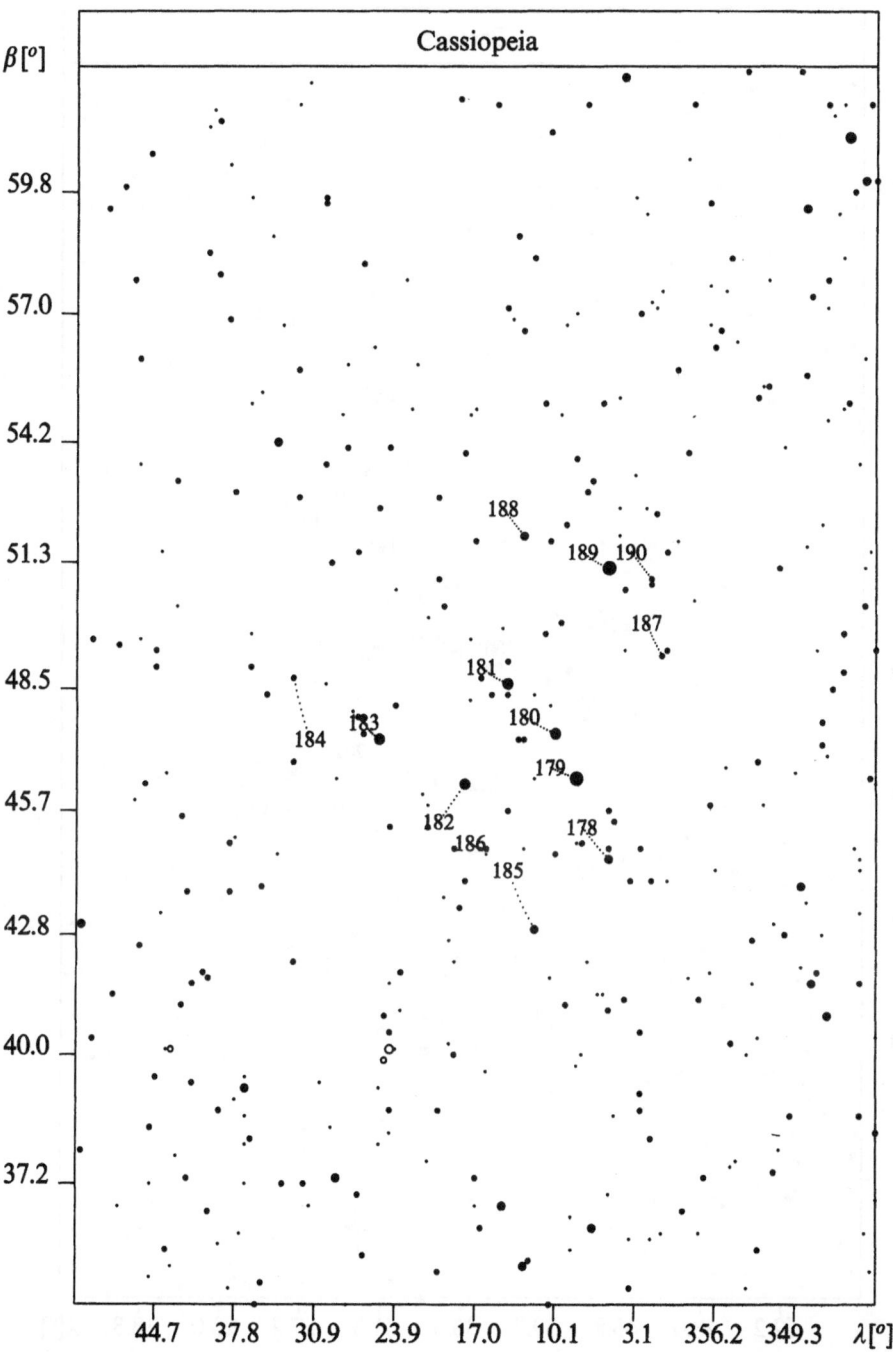

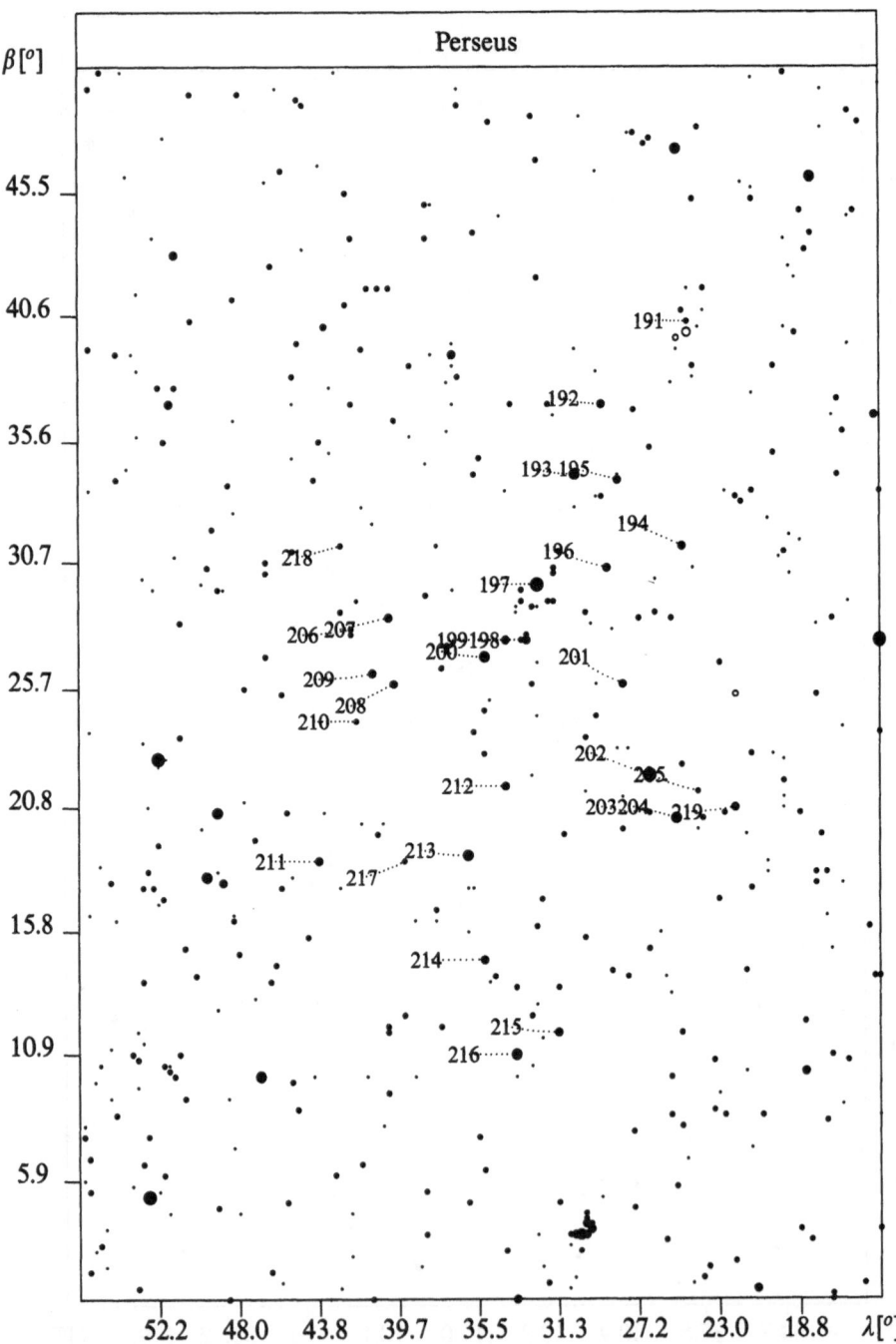

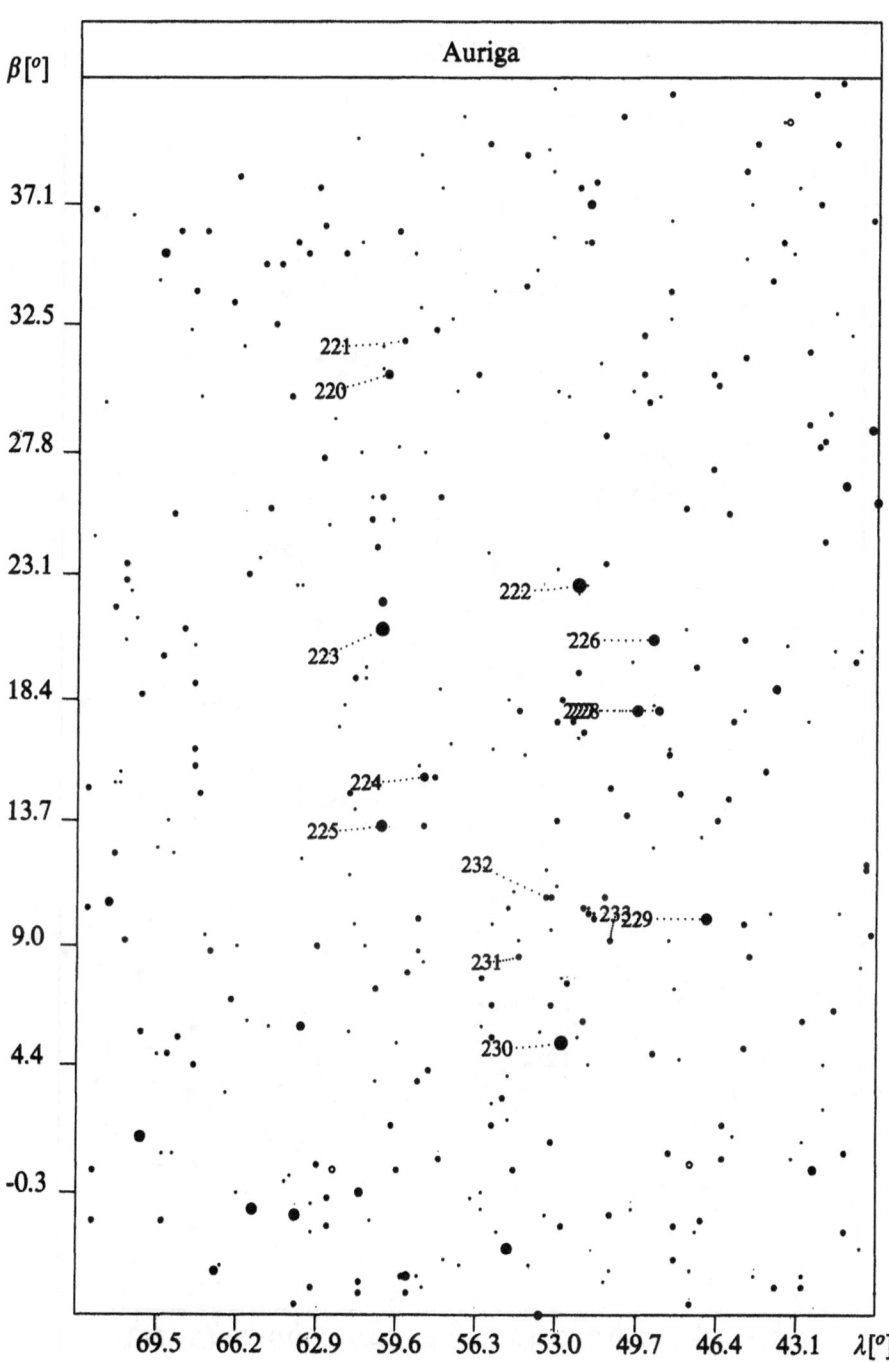

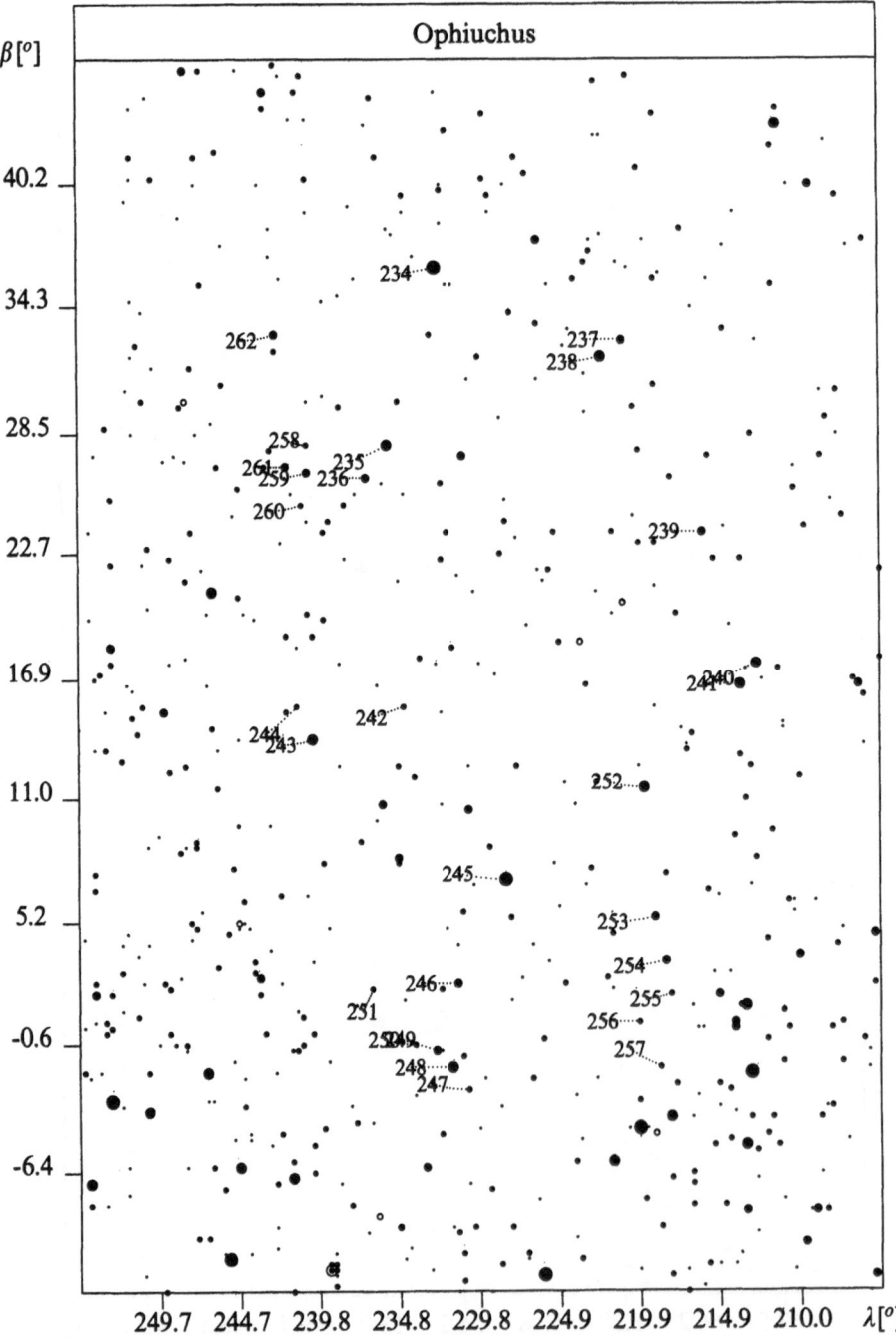

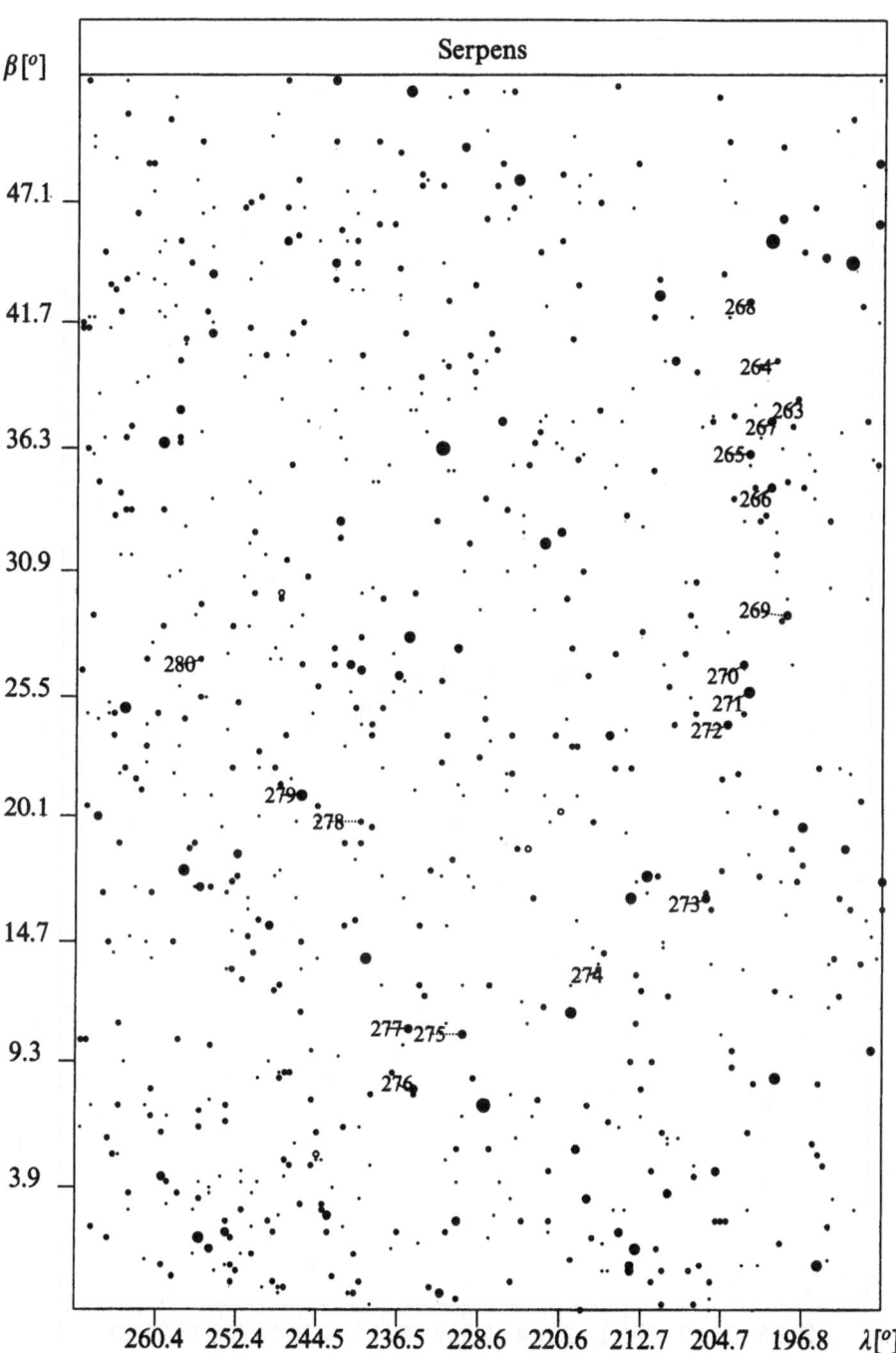

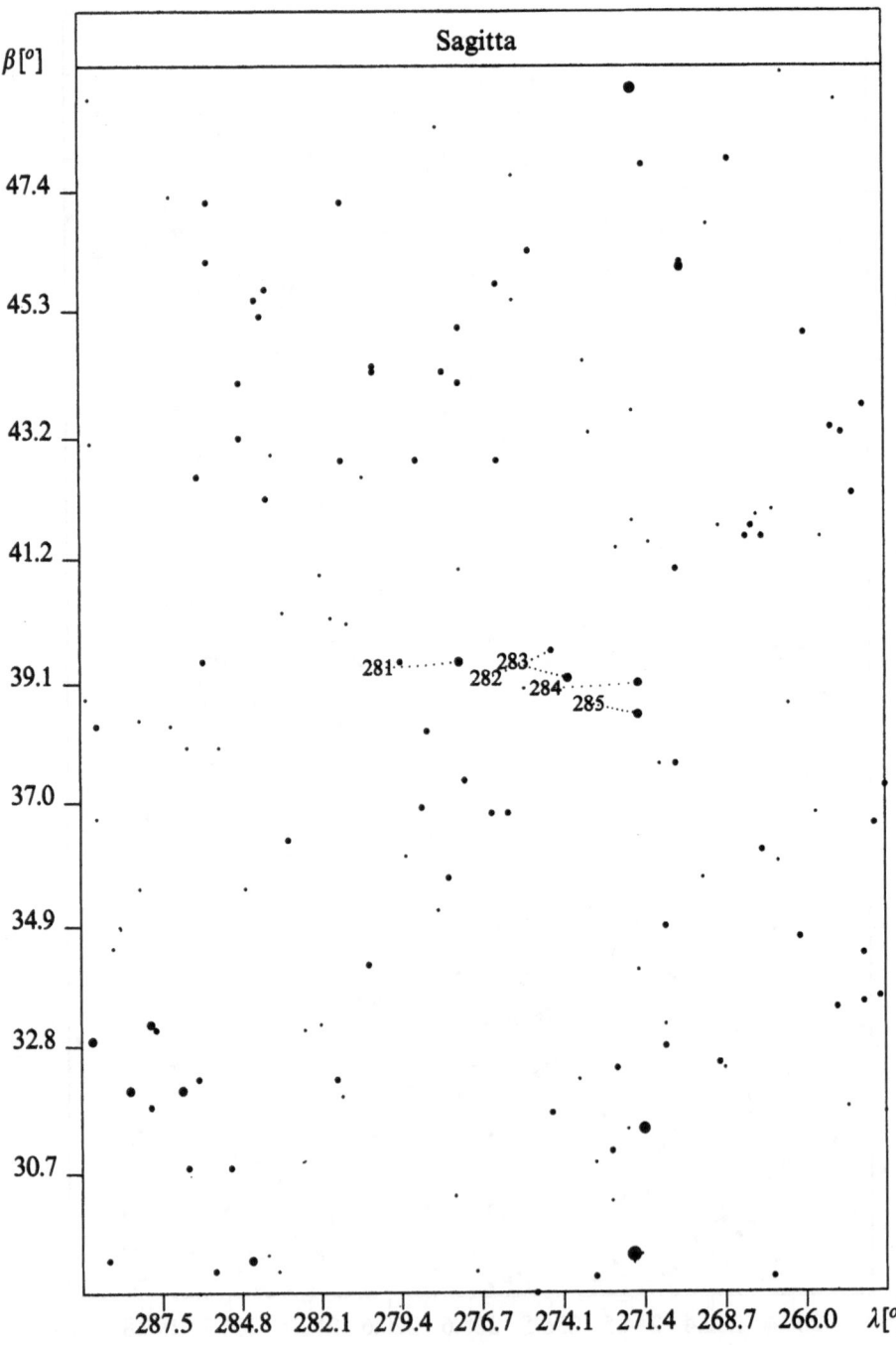

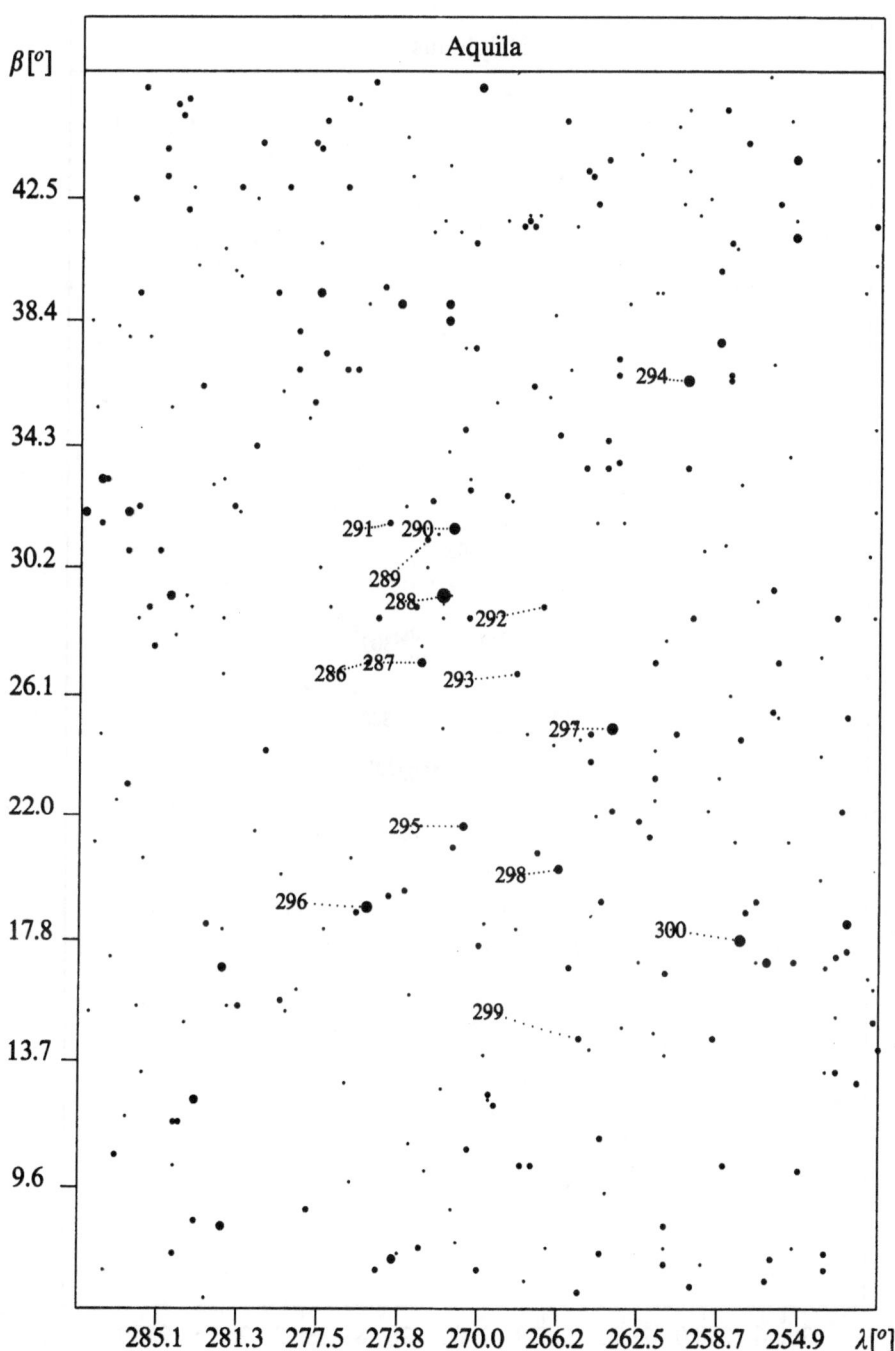

238

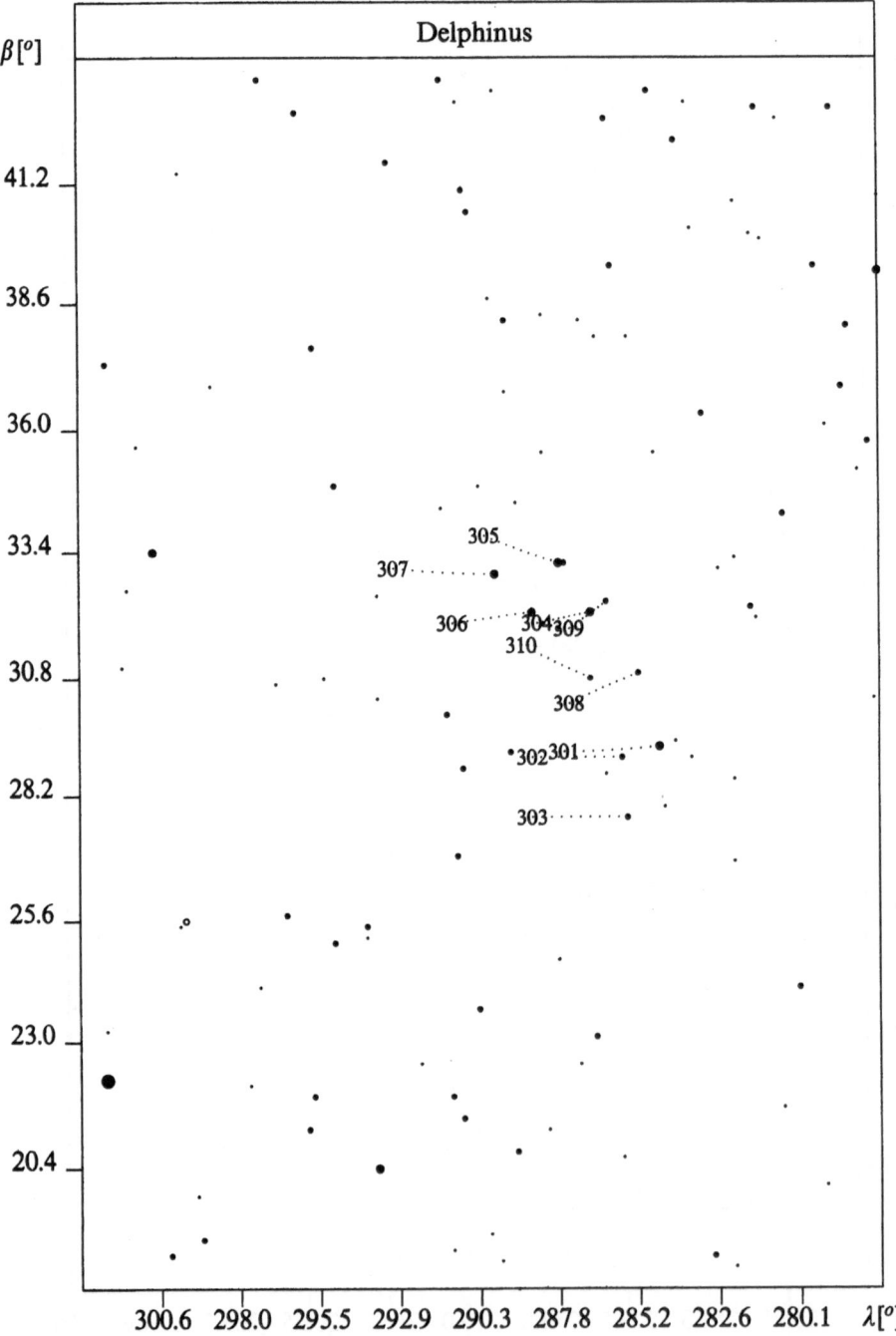

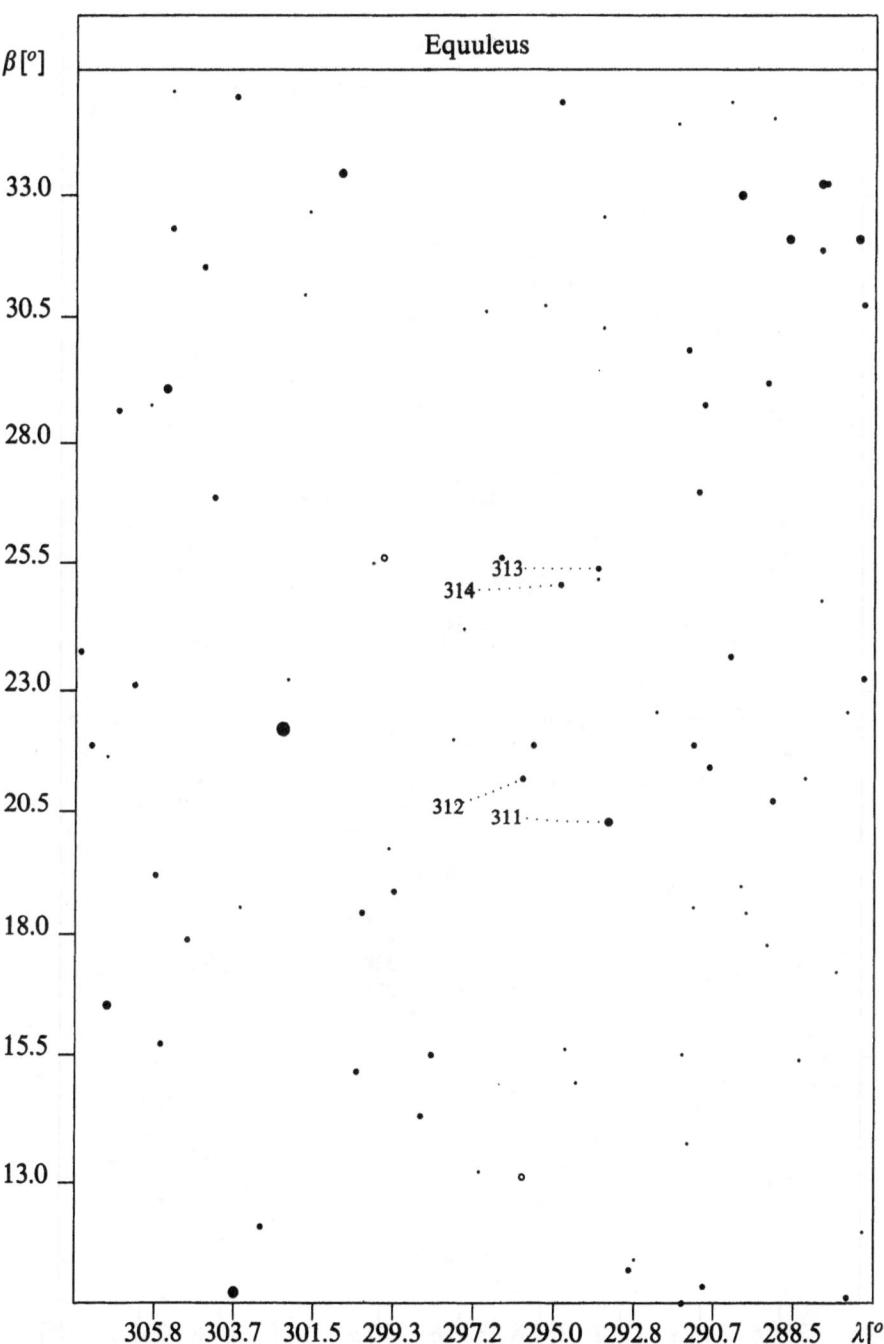

240

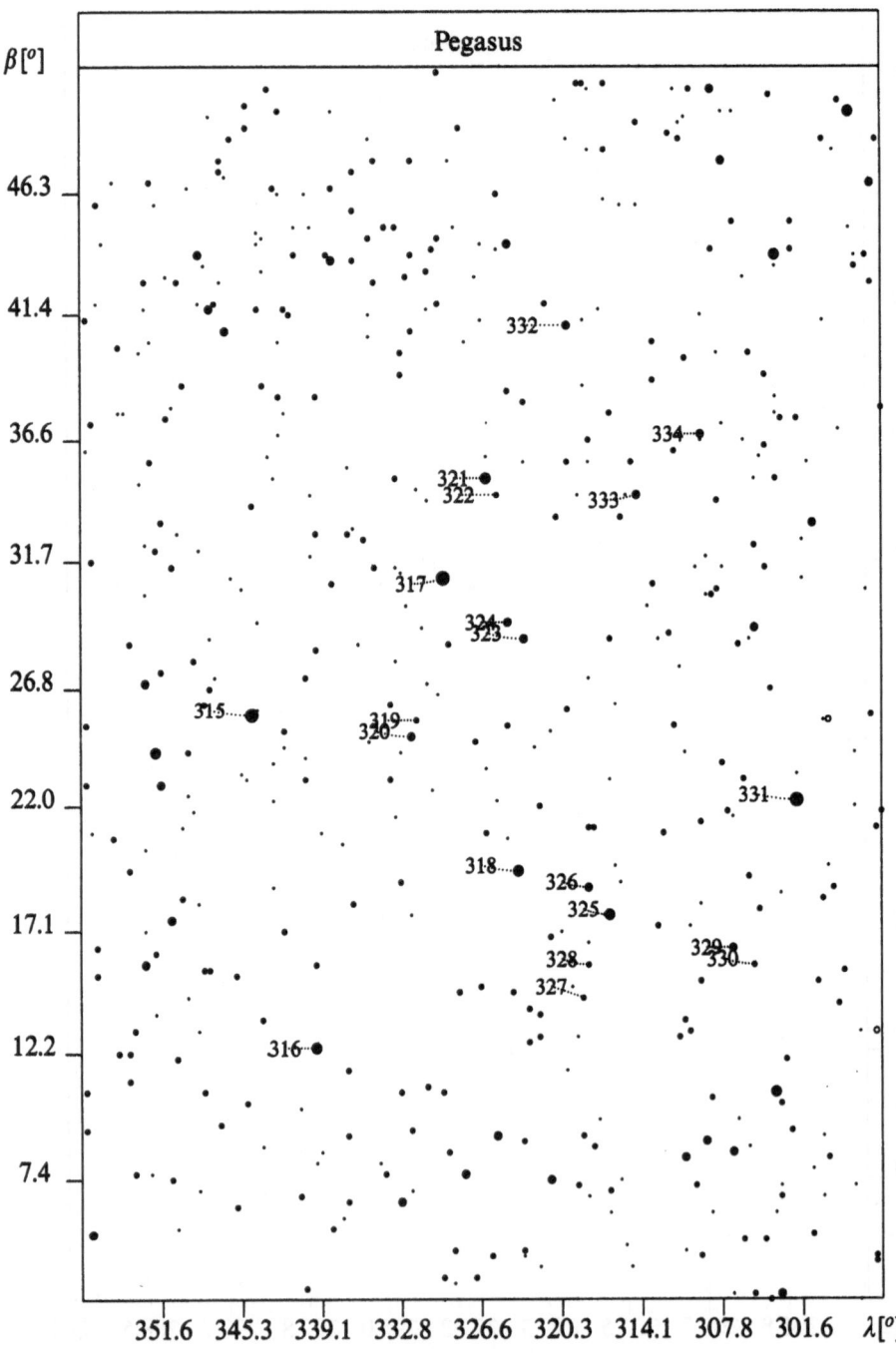

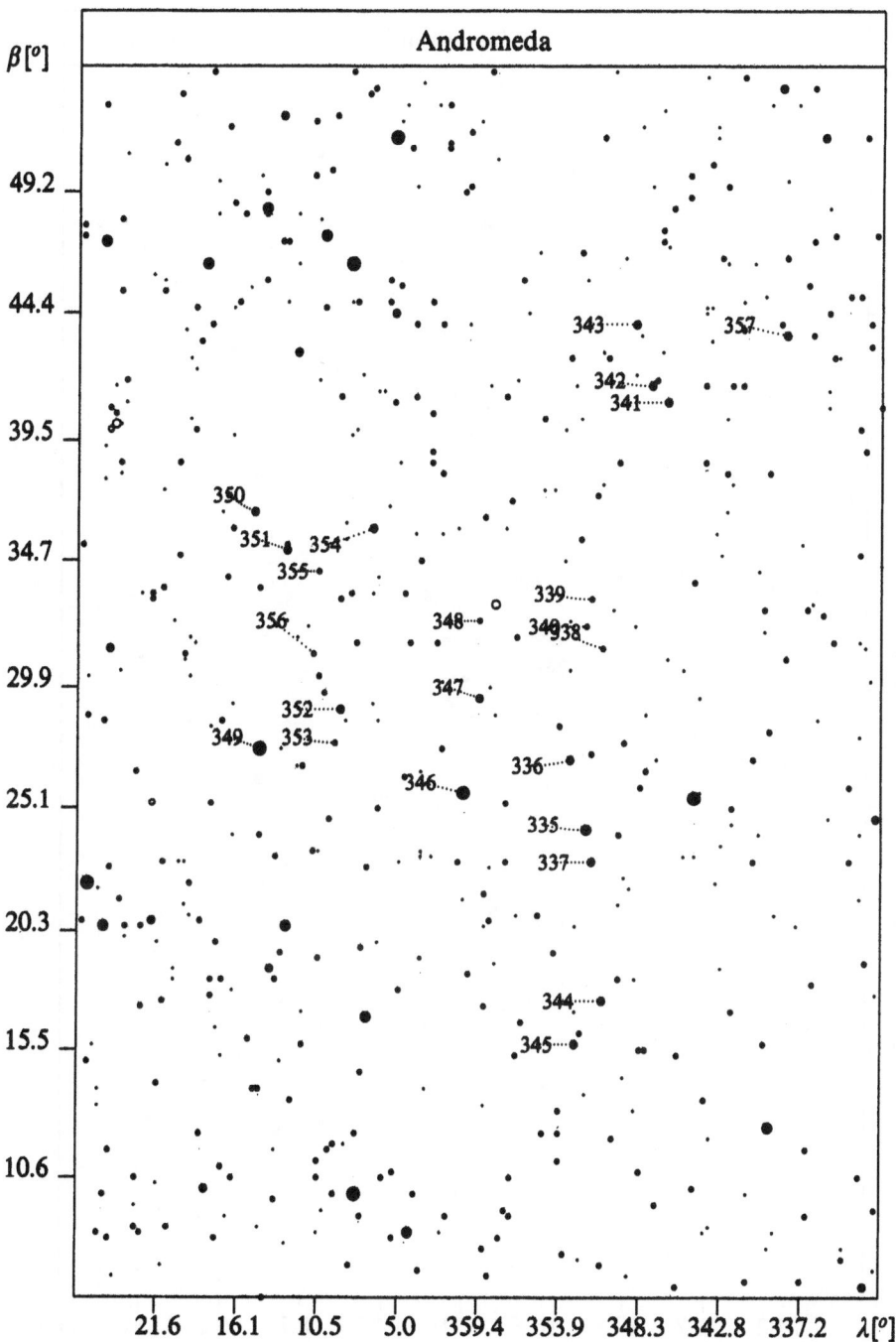

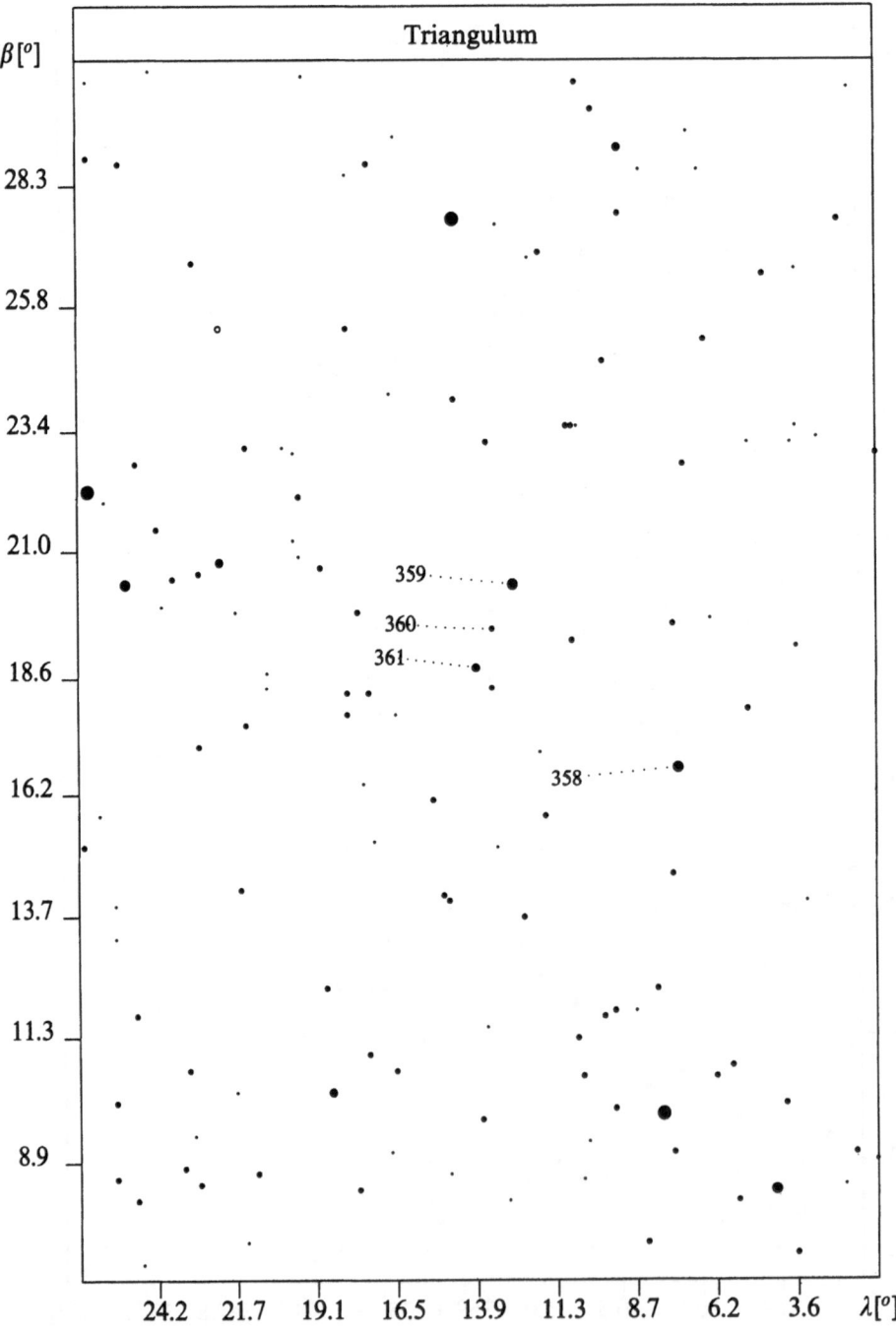

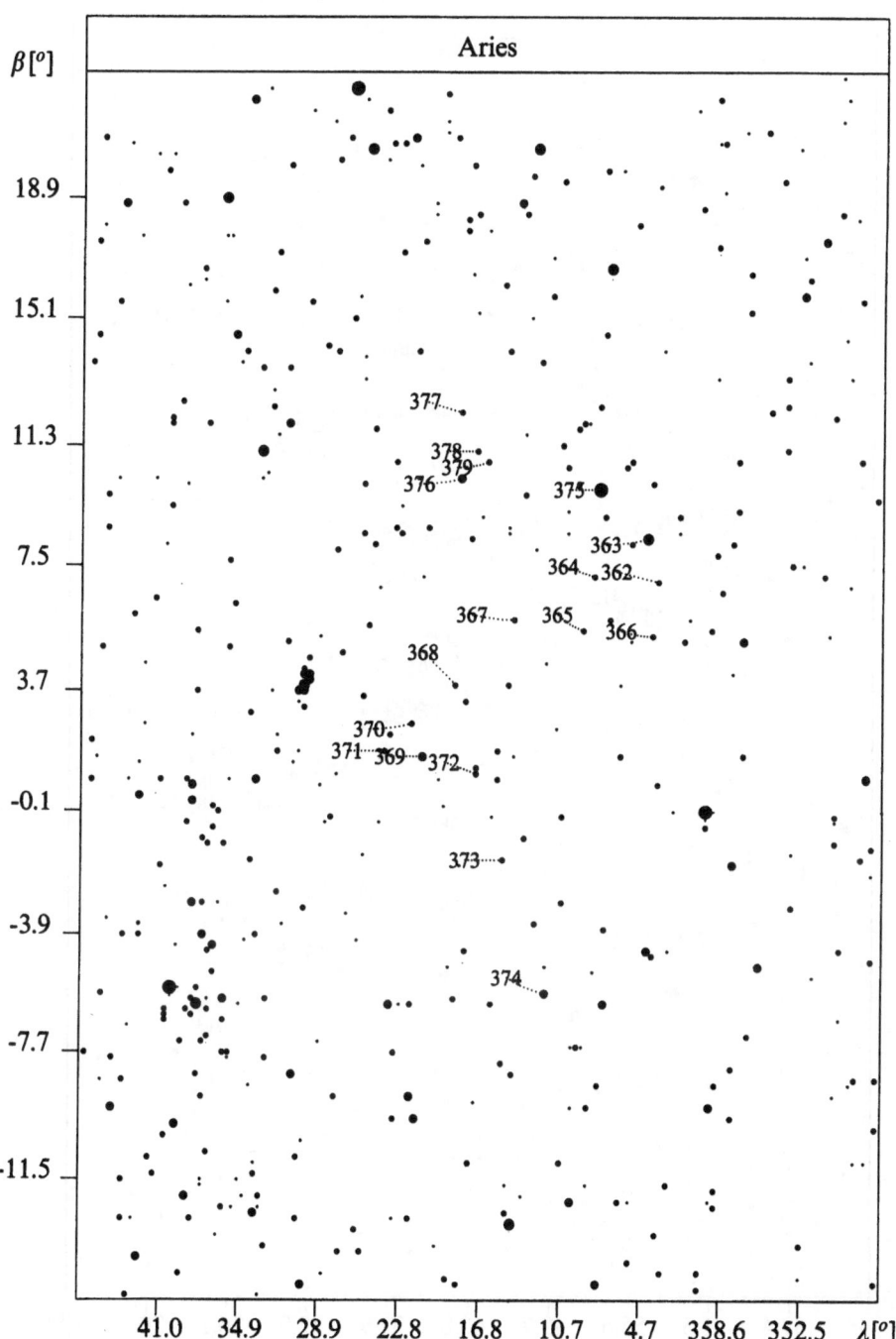

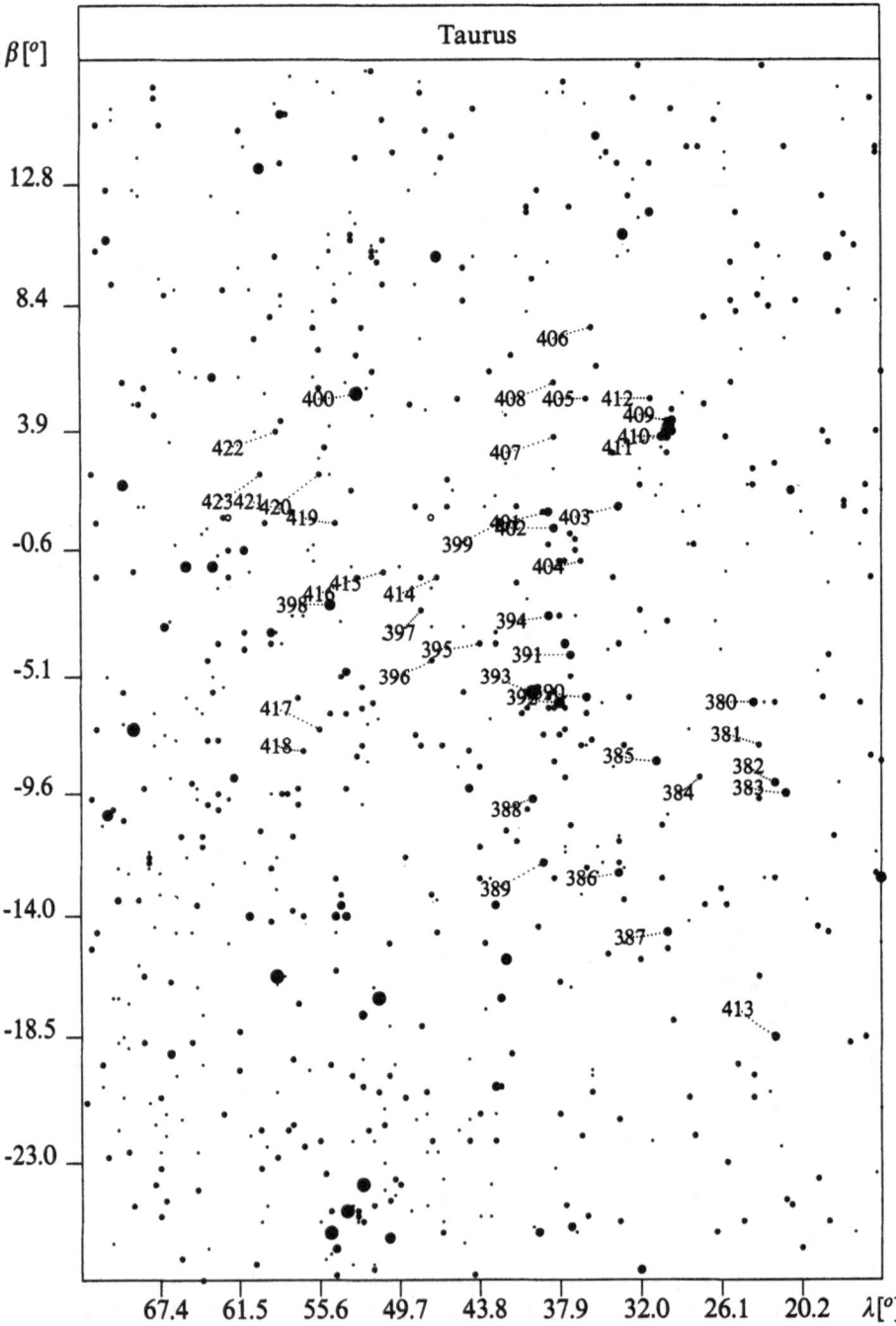

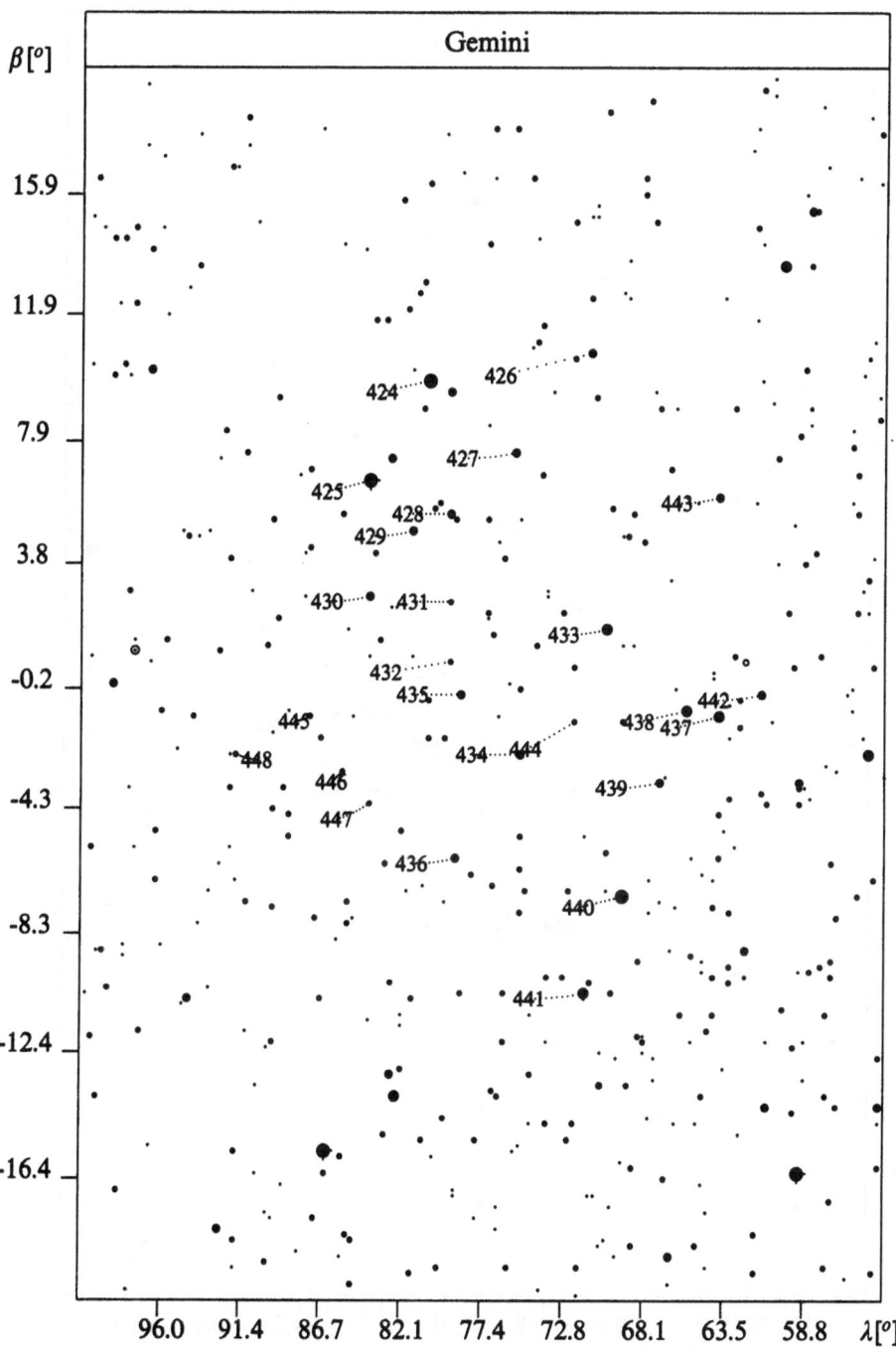

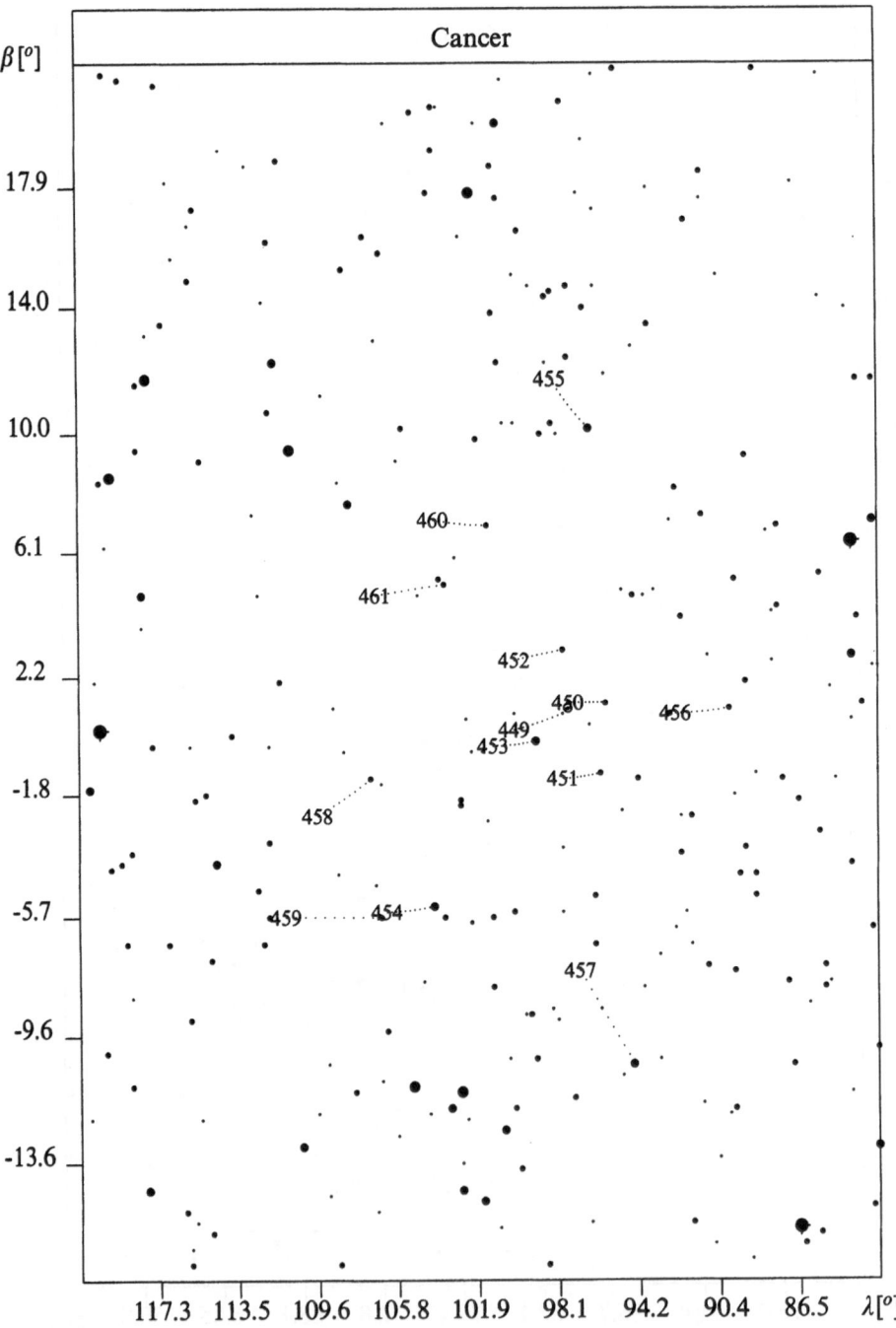

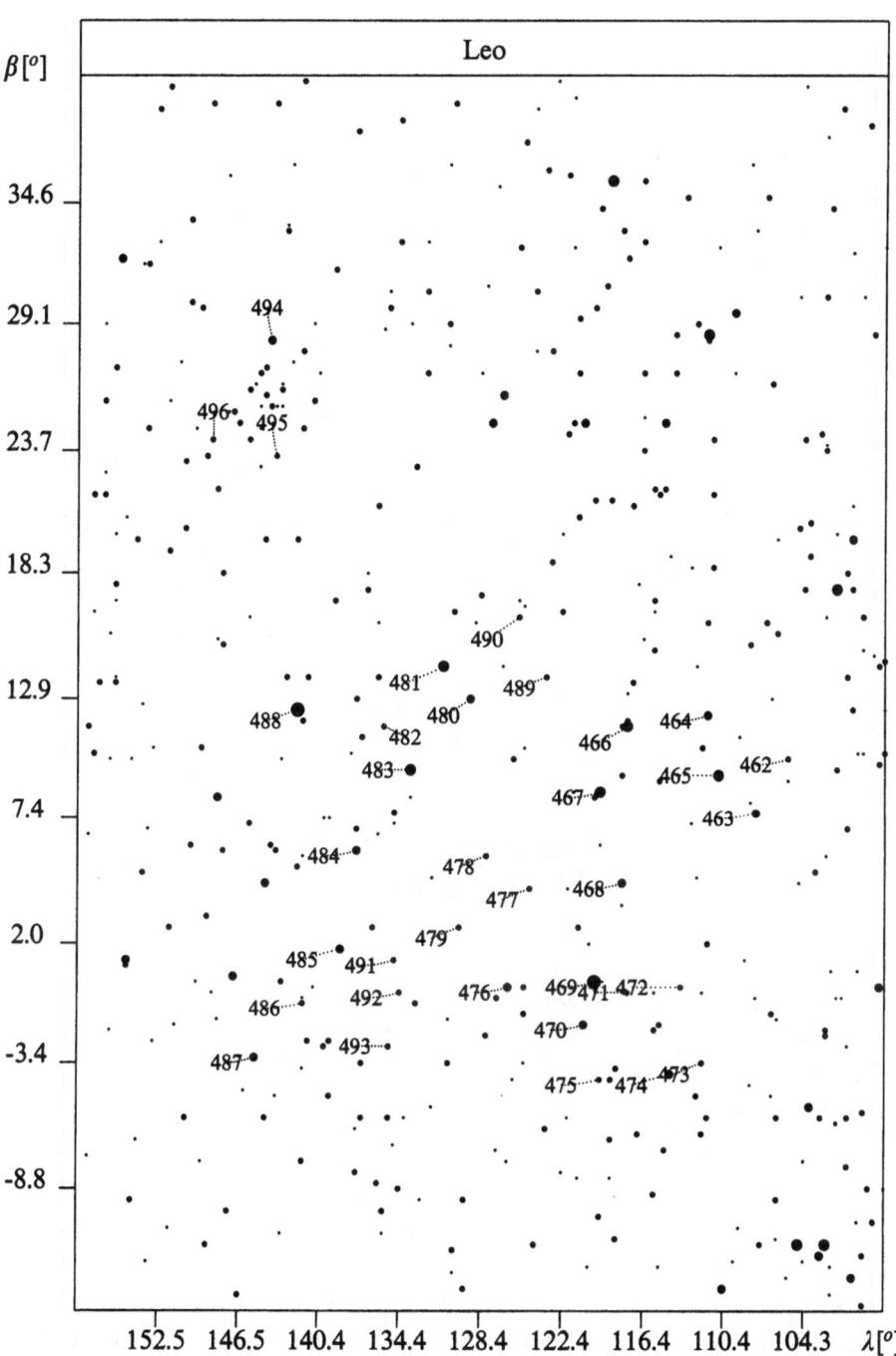

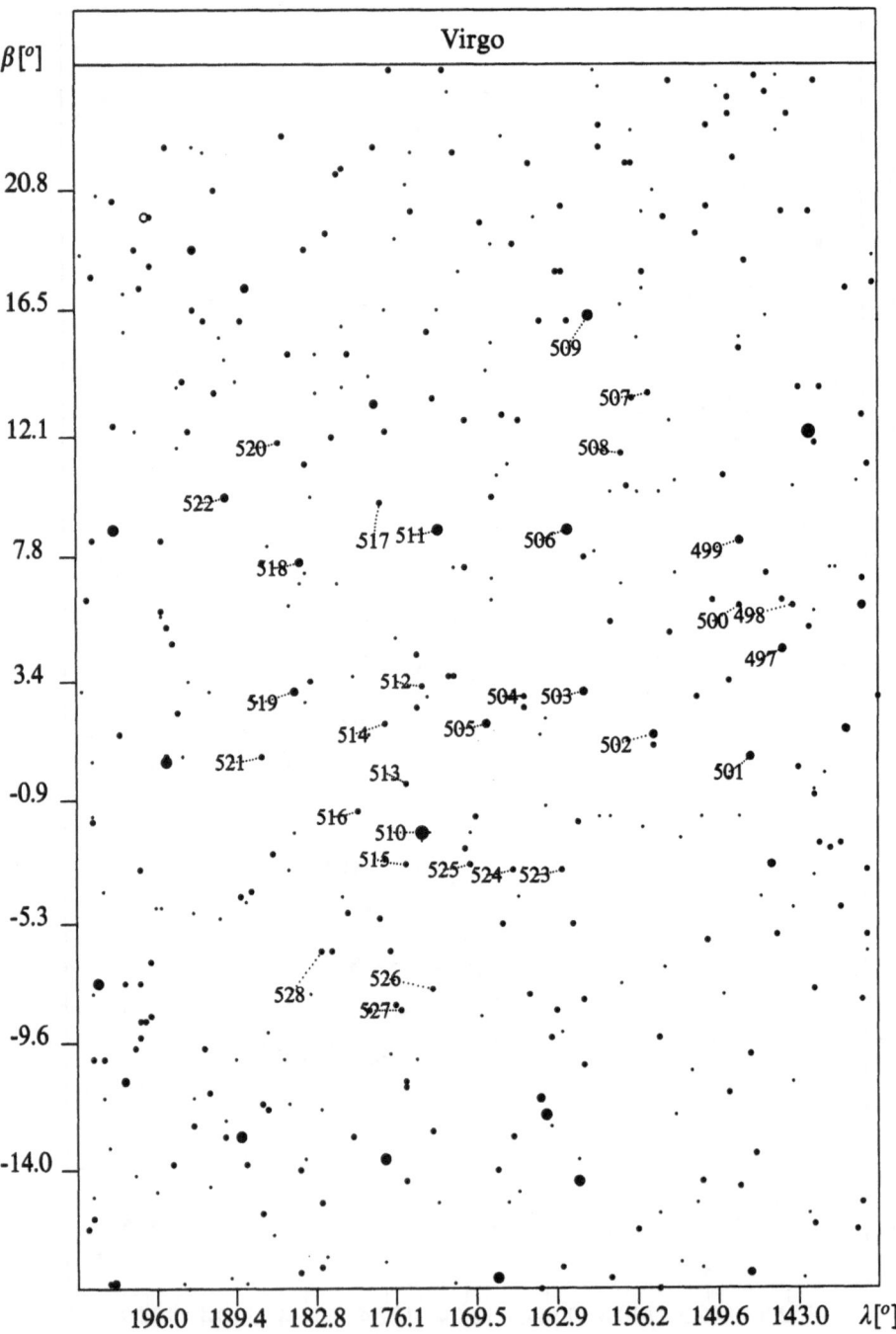

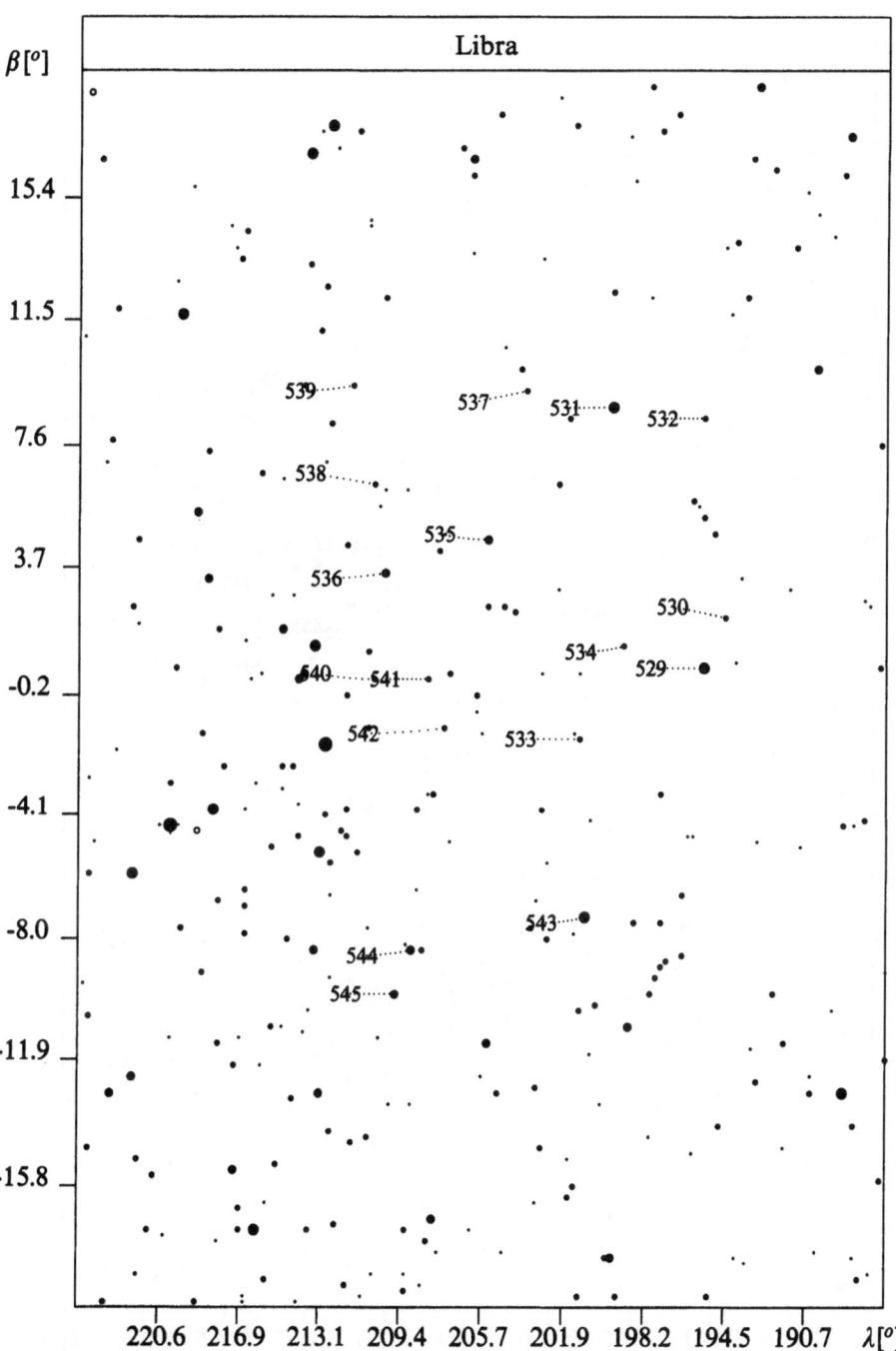

250

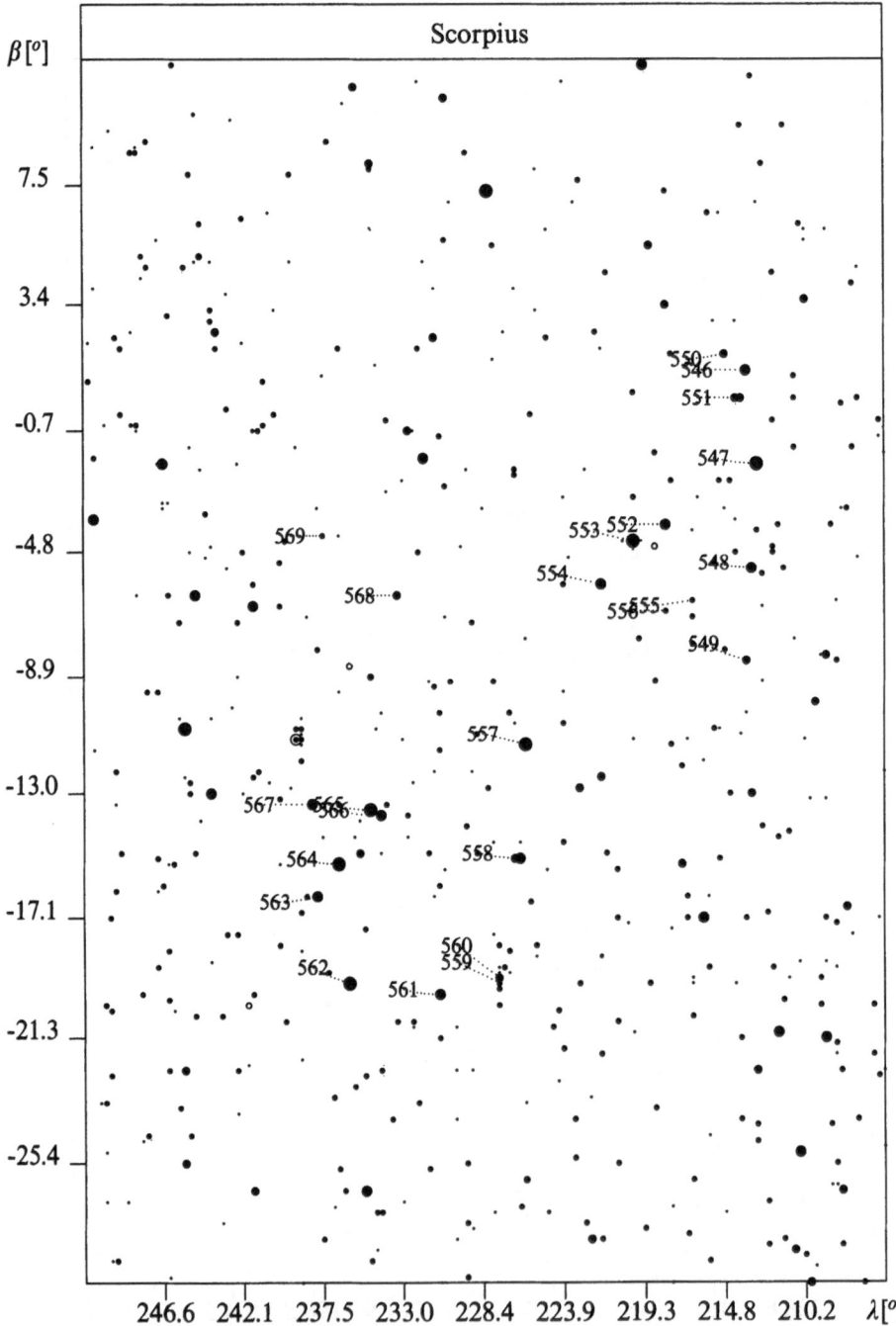

Scorpius

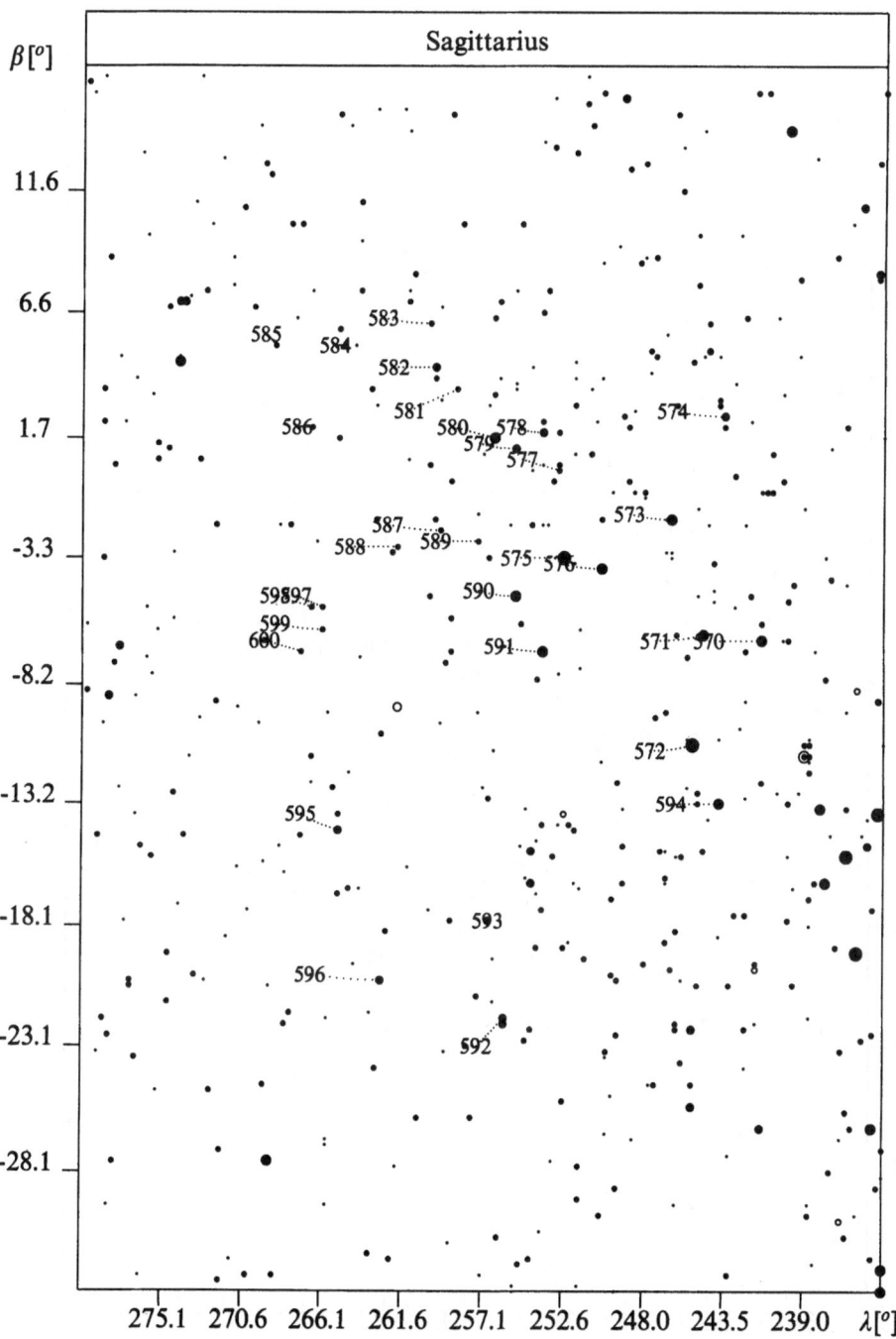

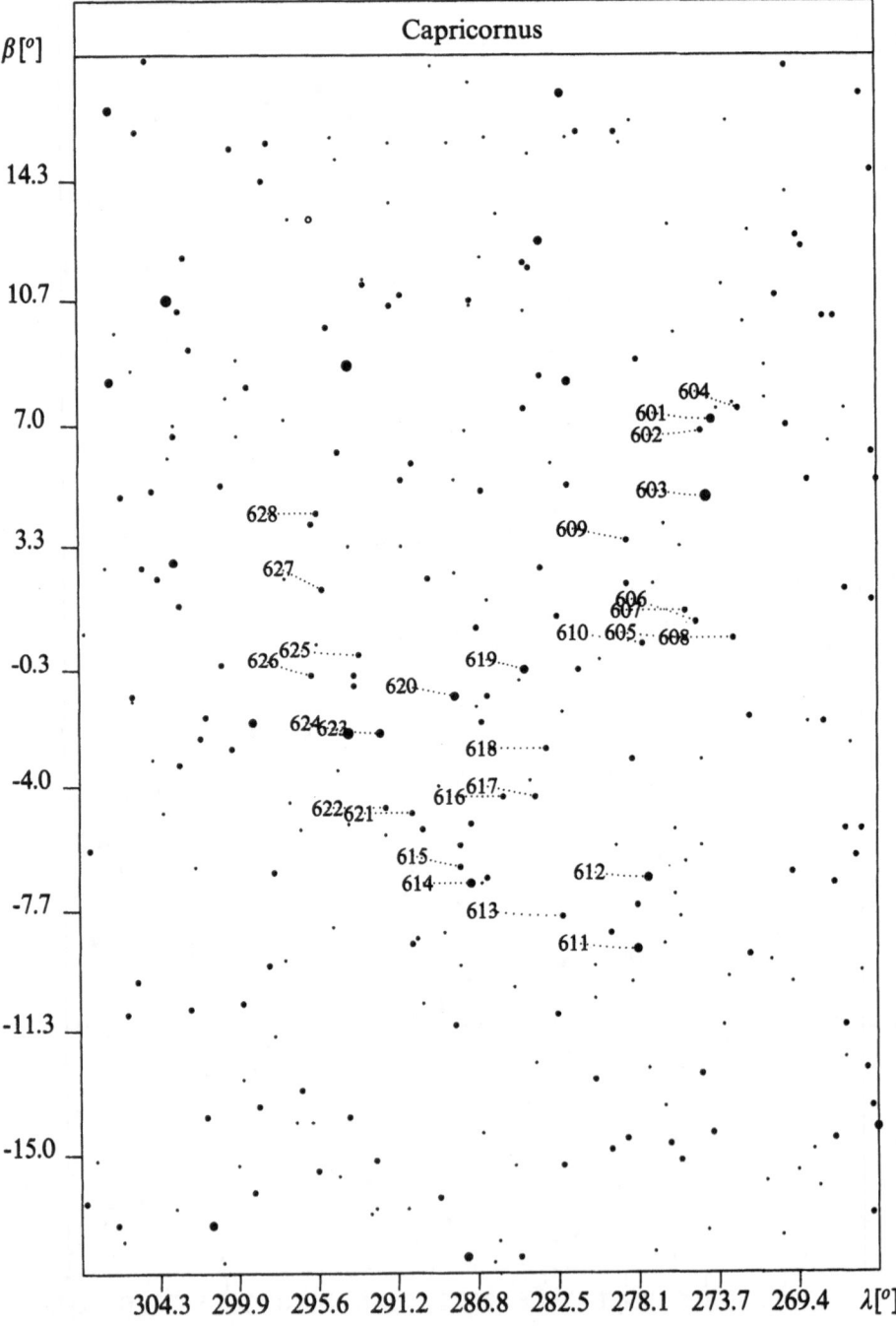

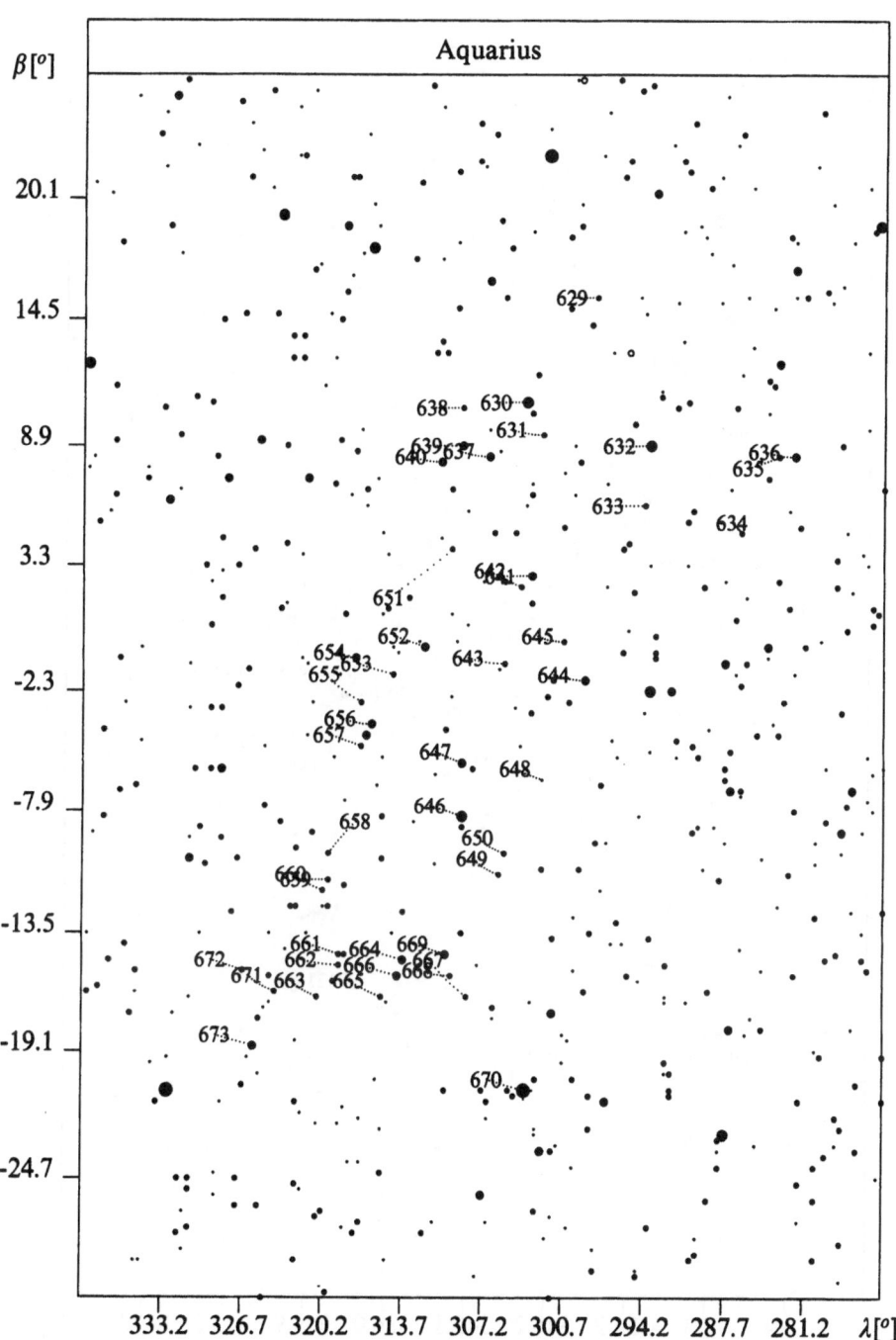

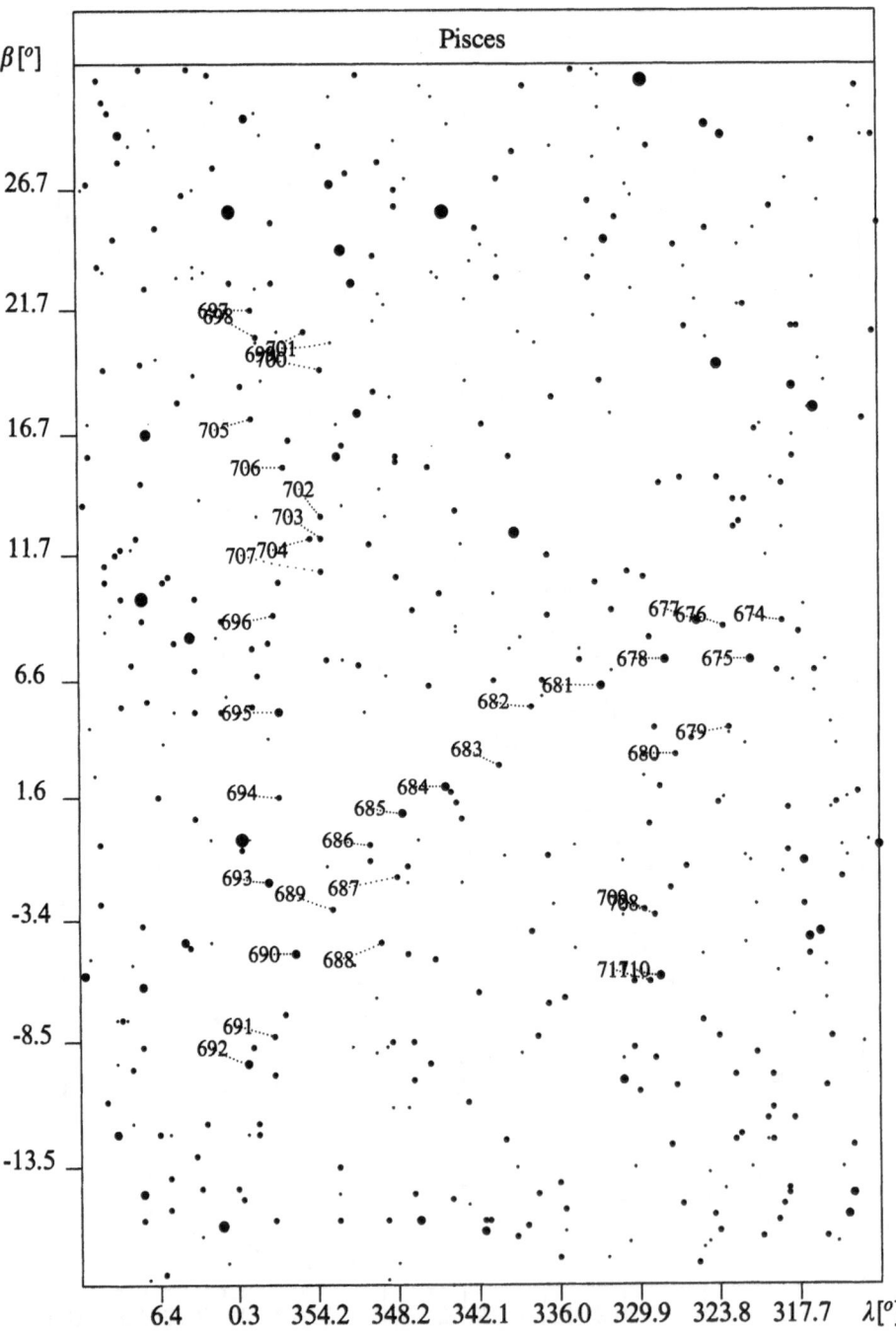

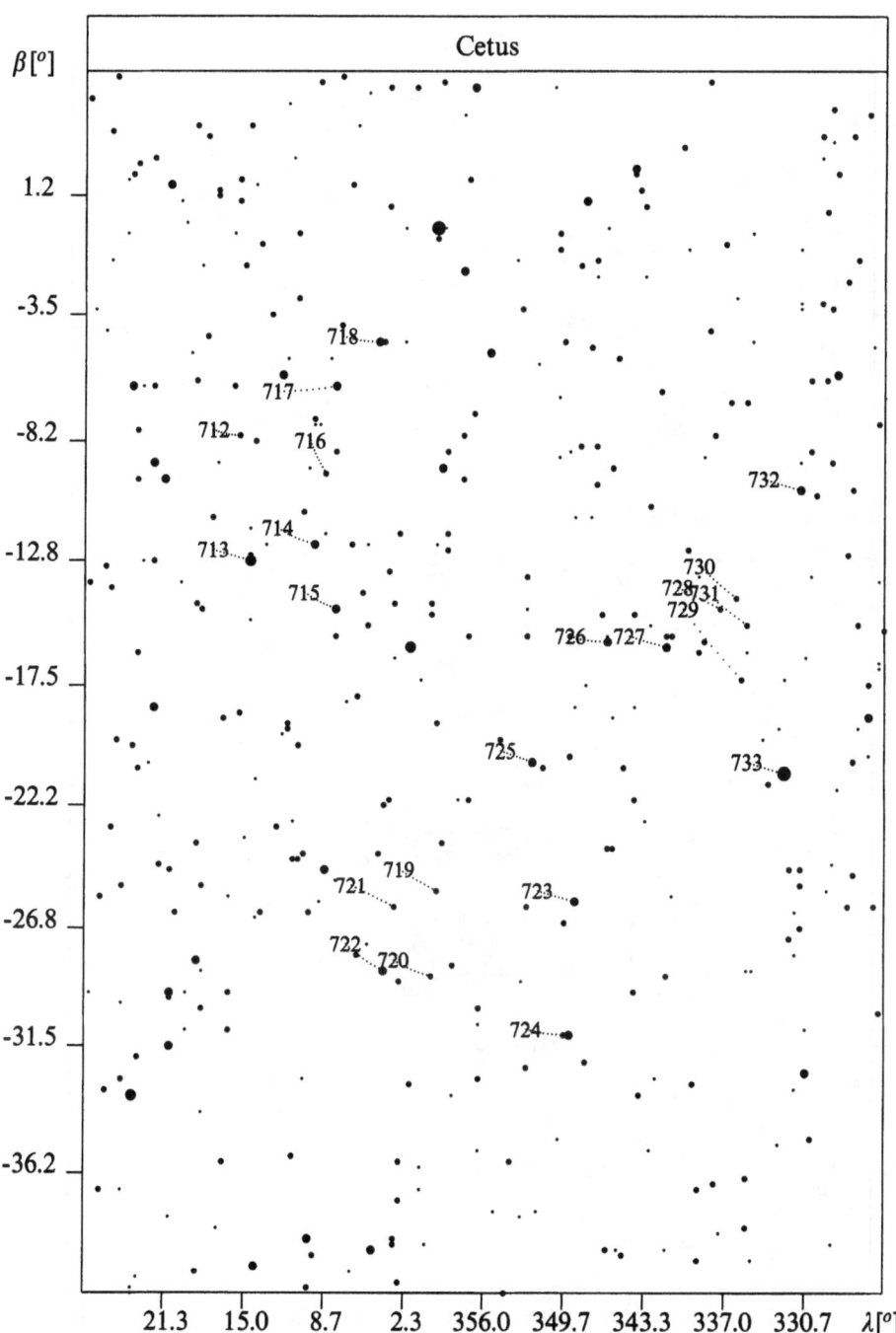

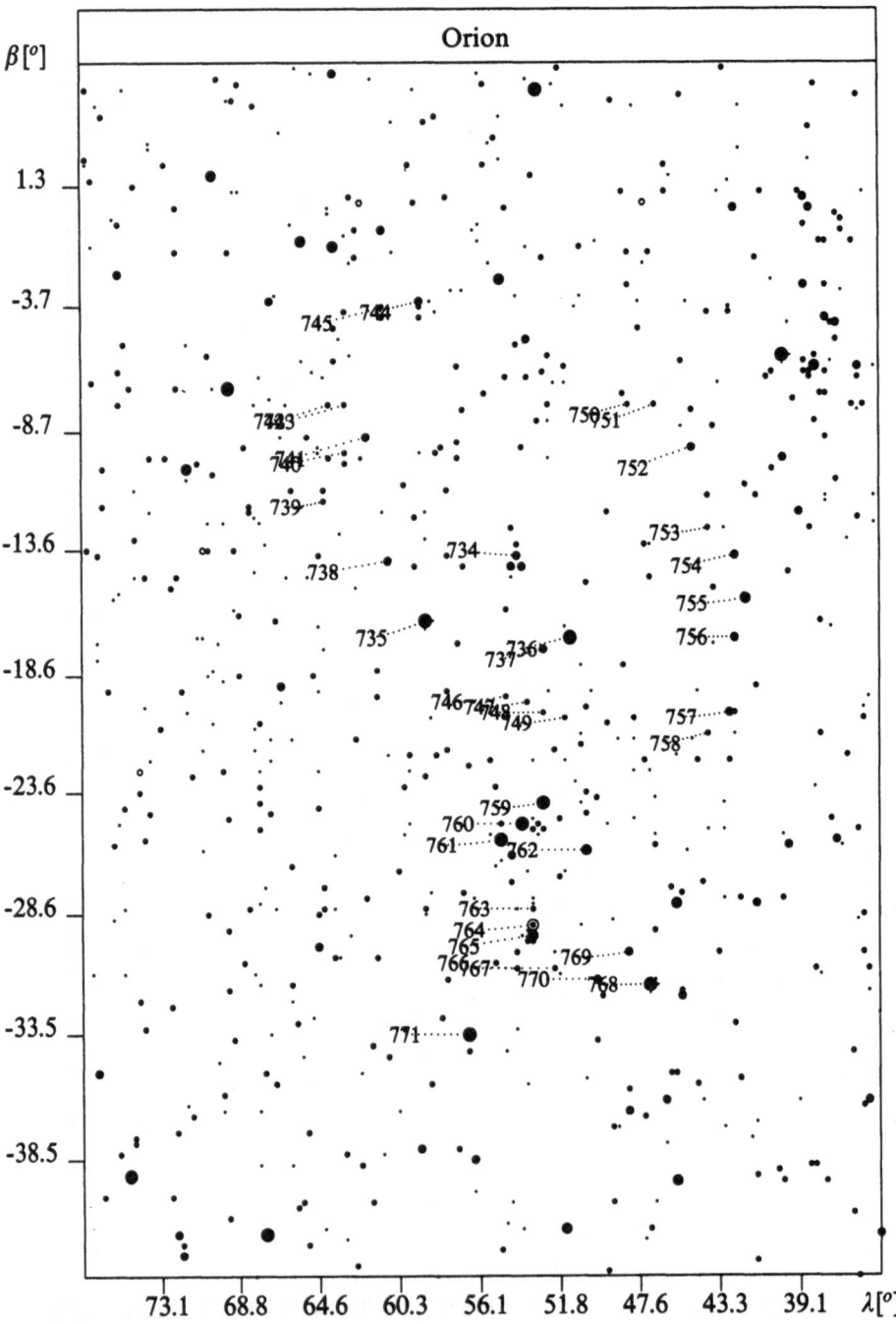

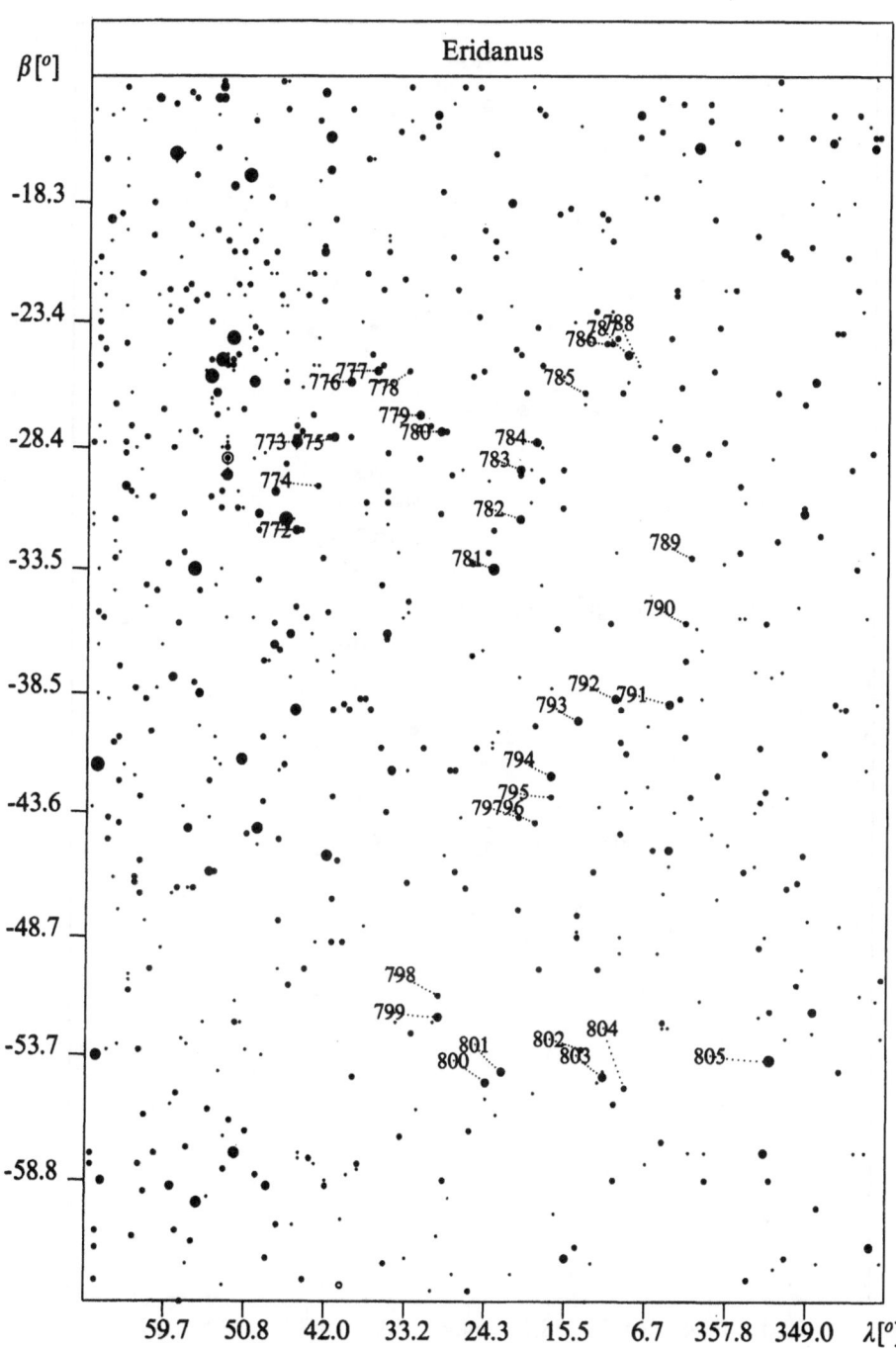

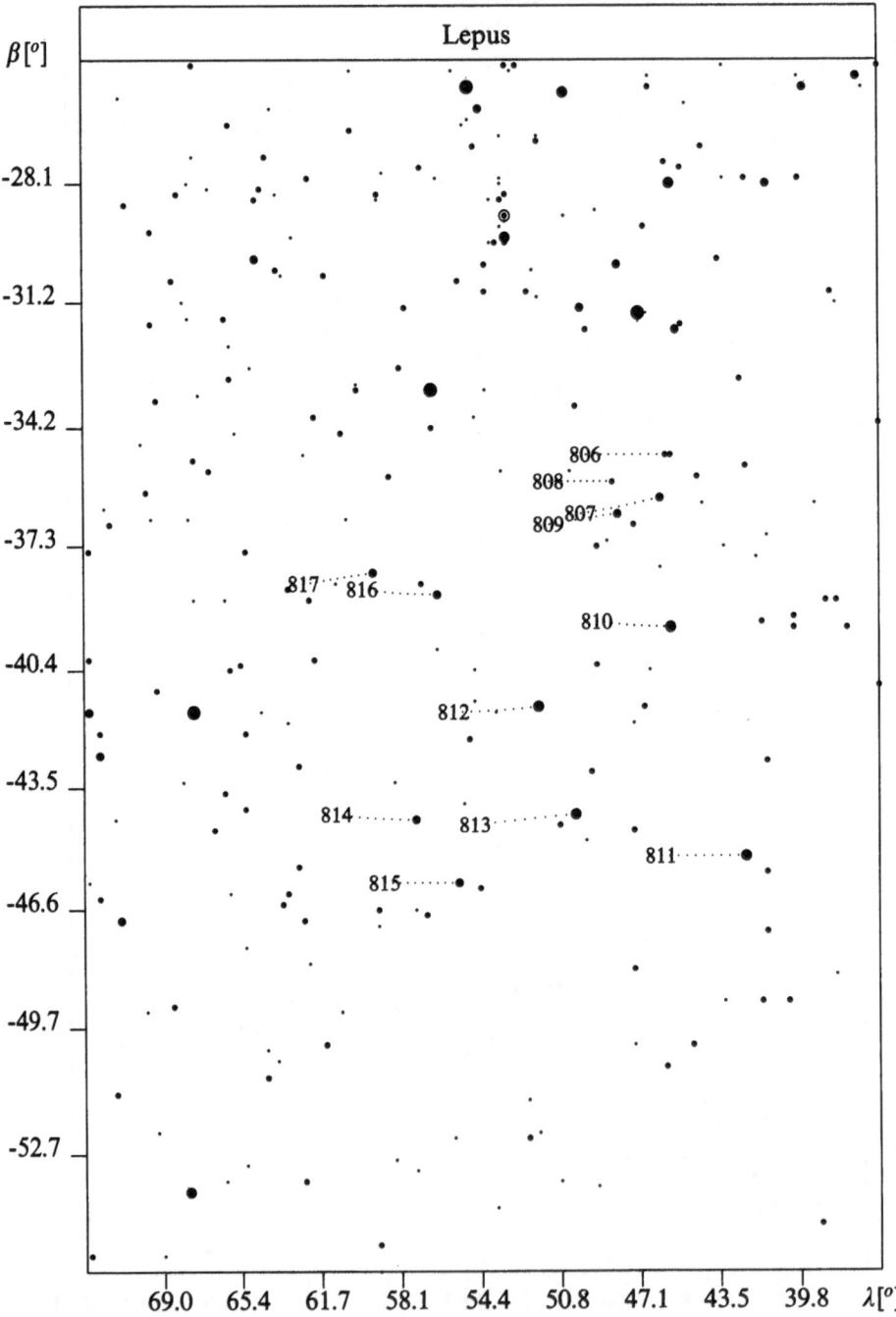

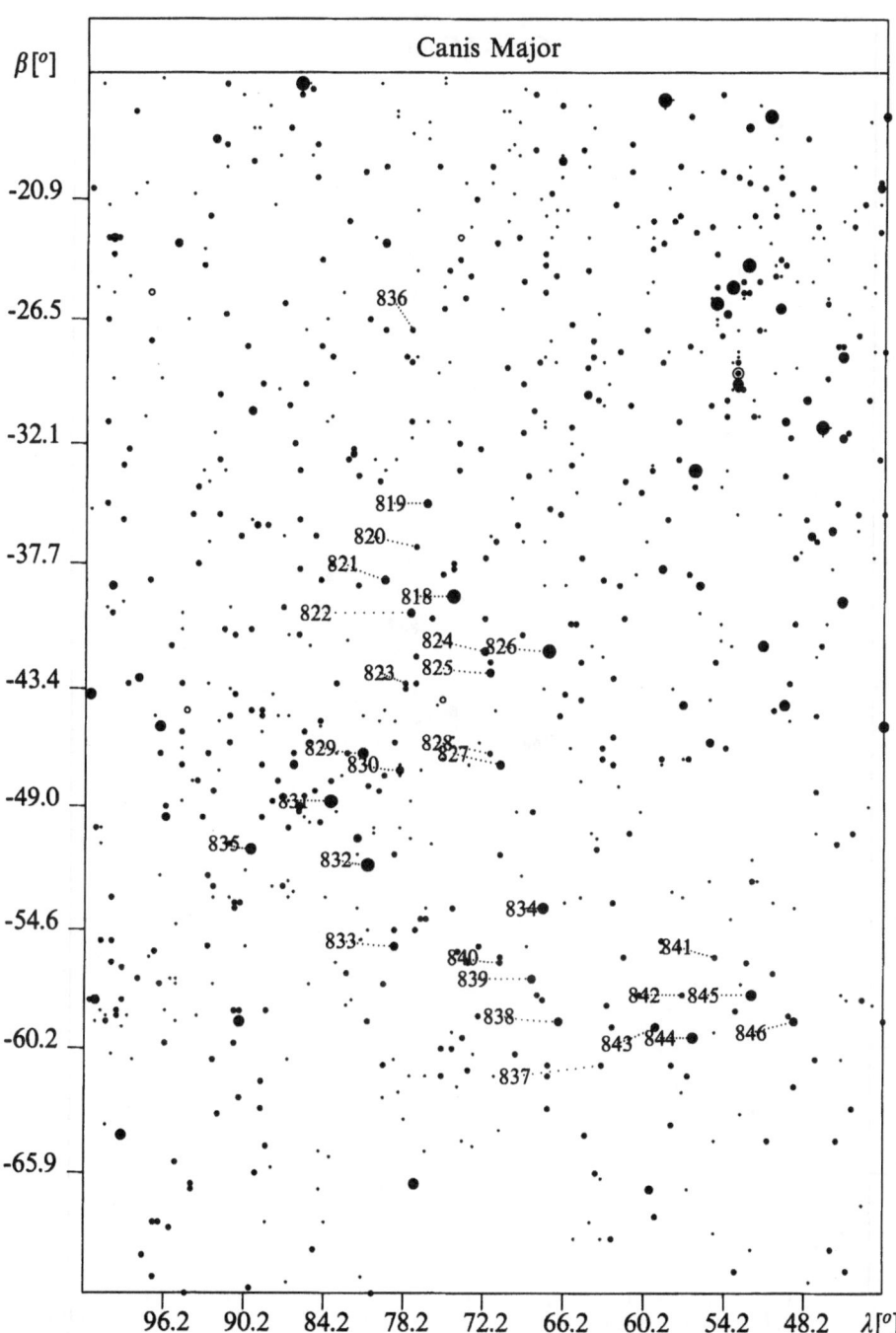

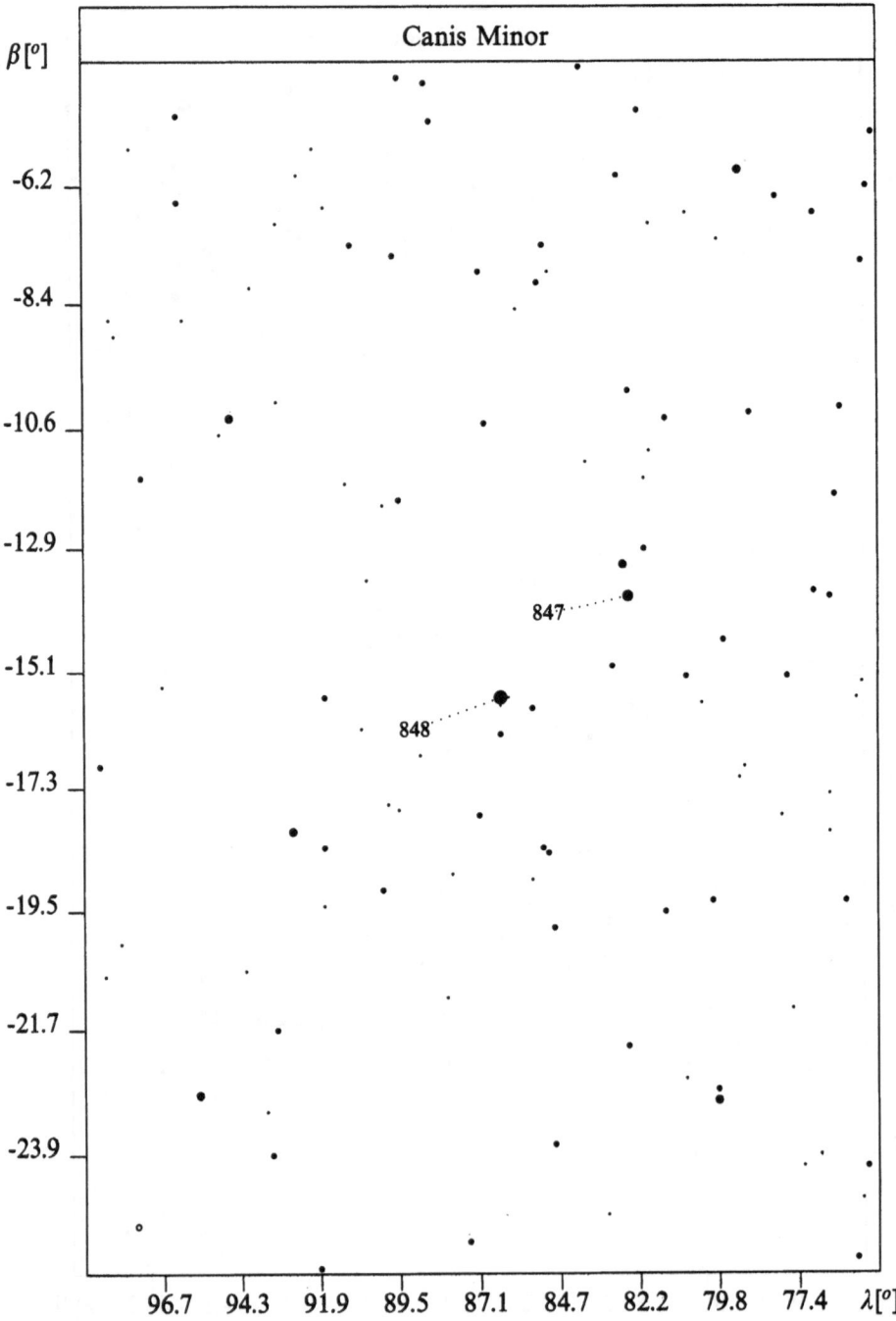

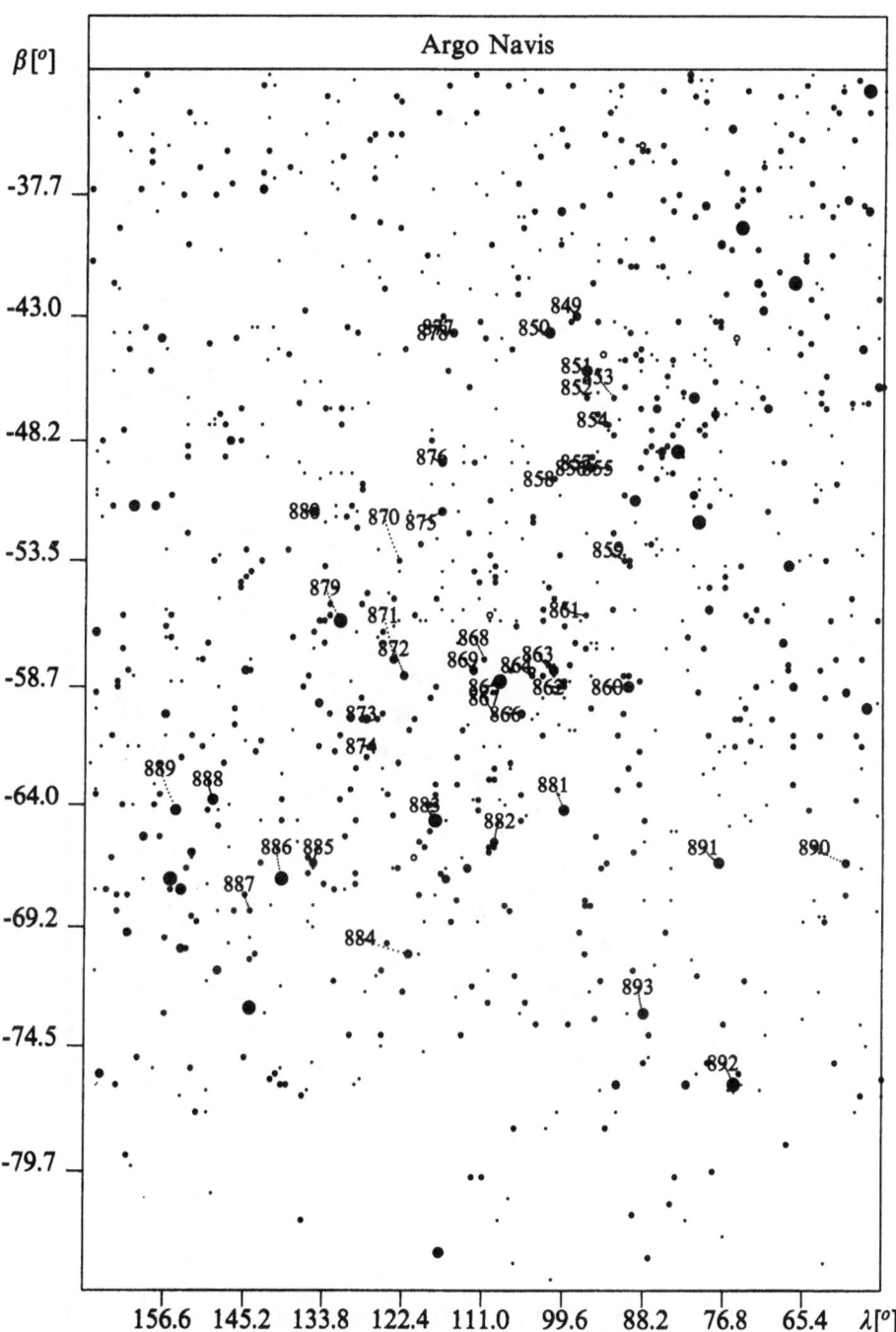

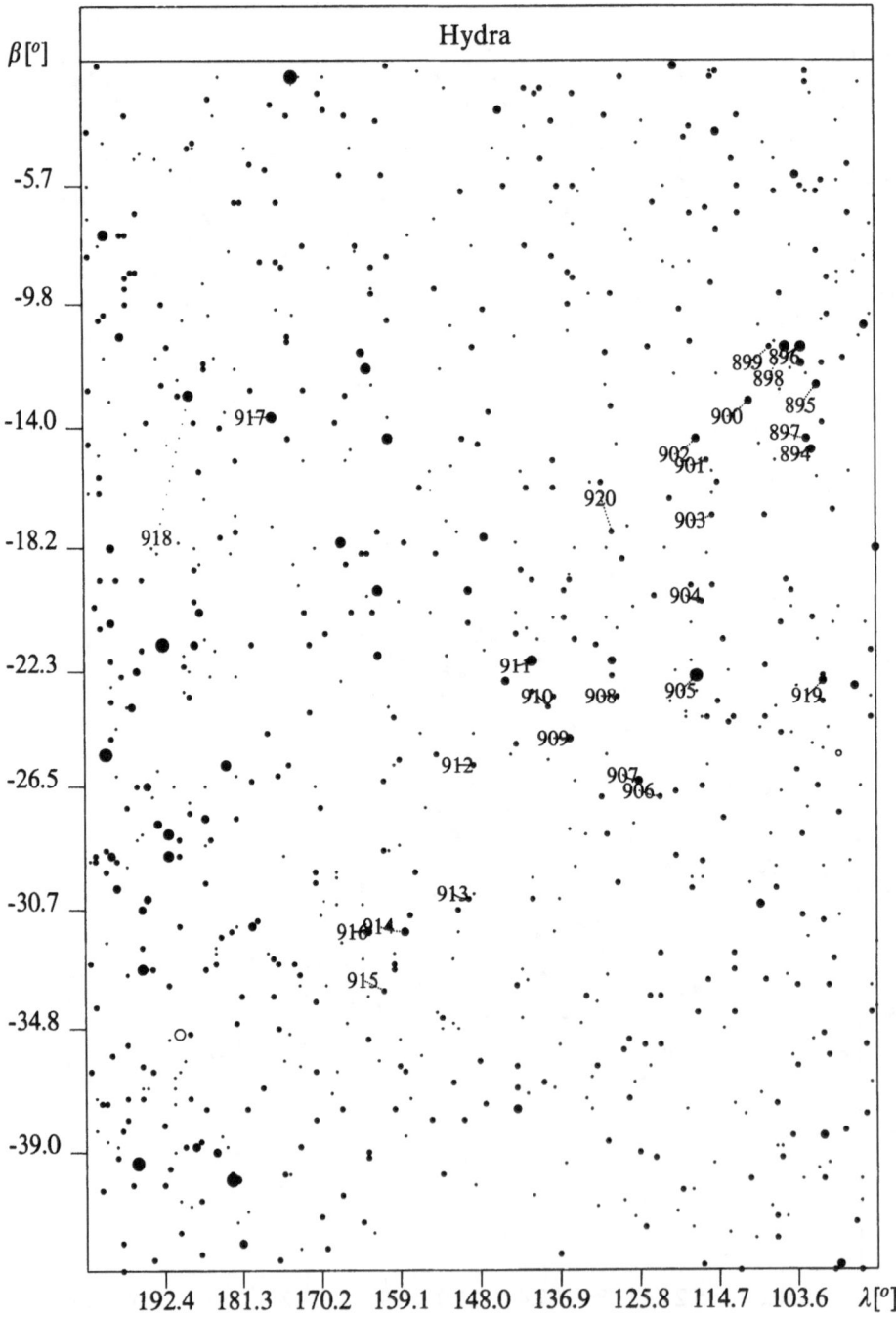

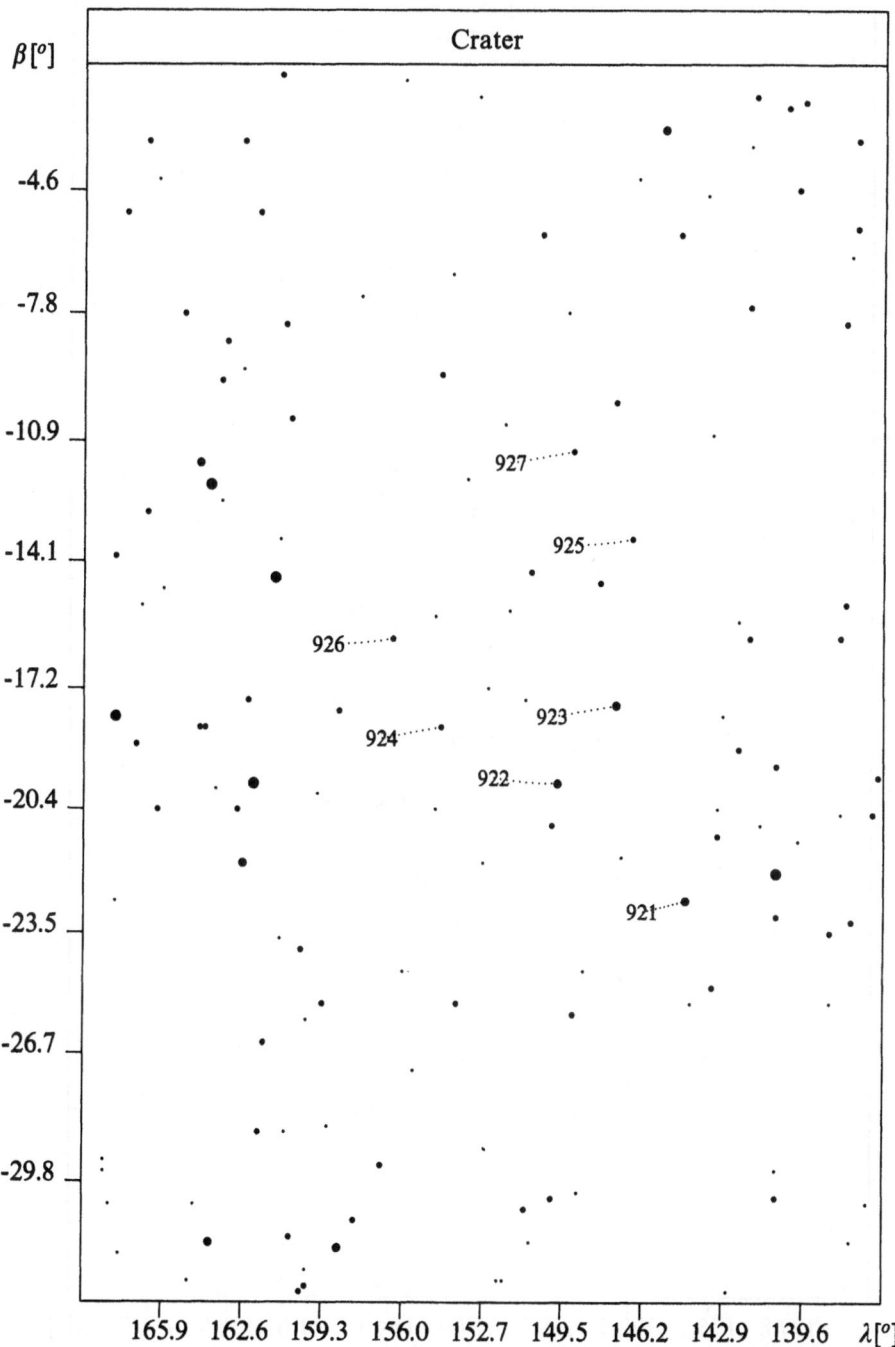

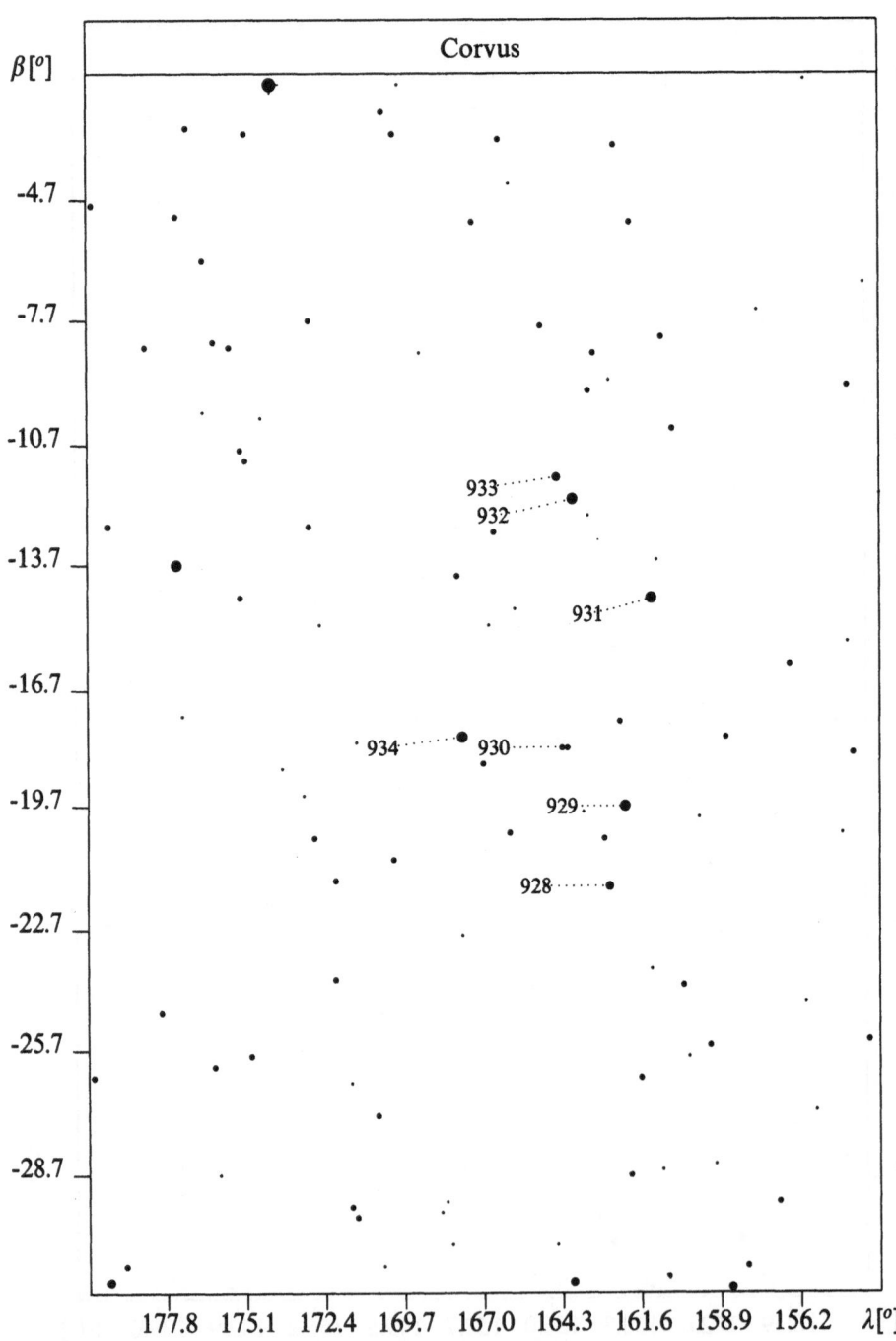

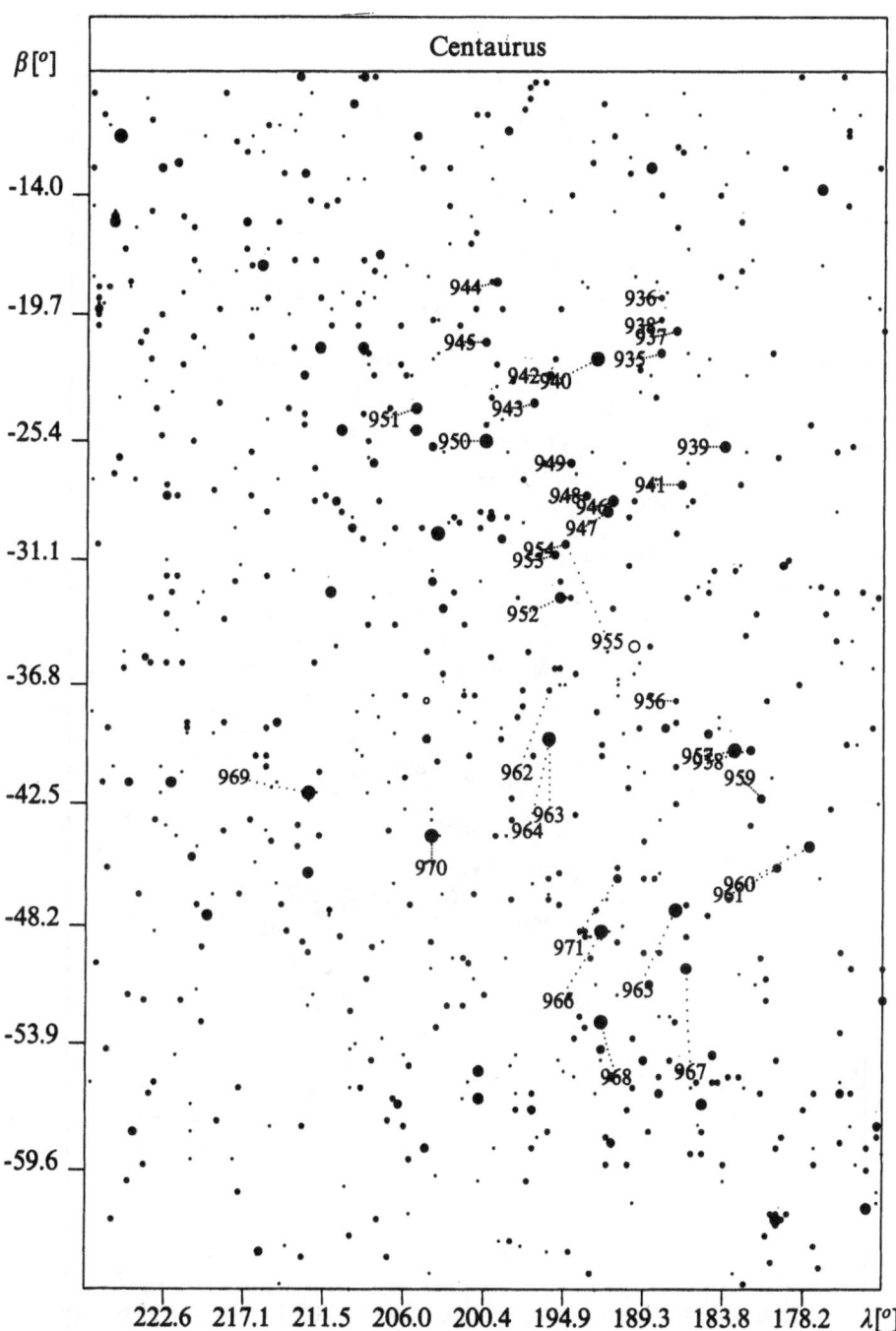

Centaurus

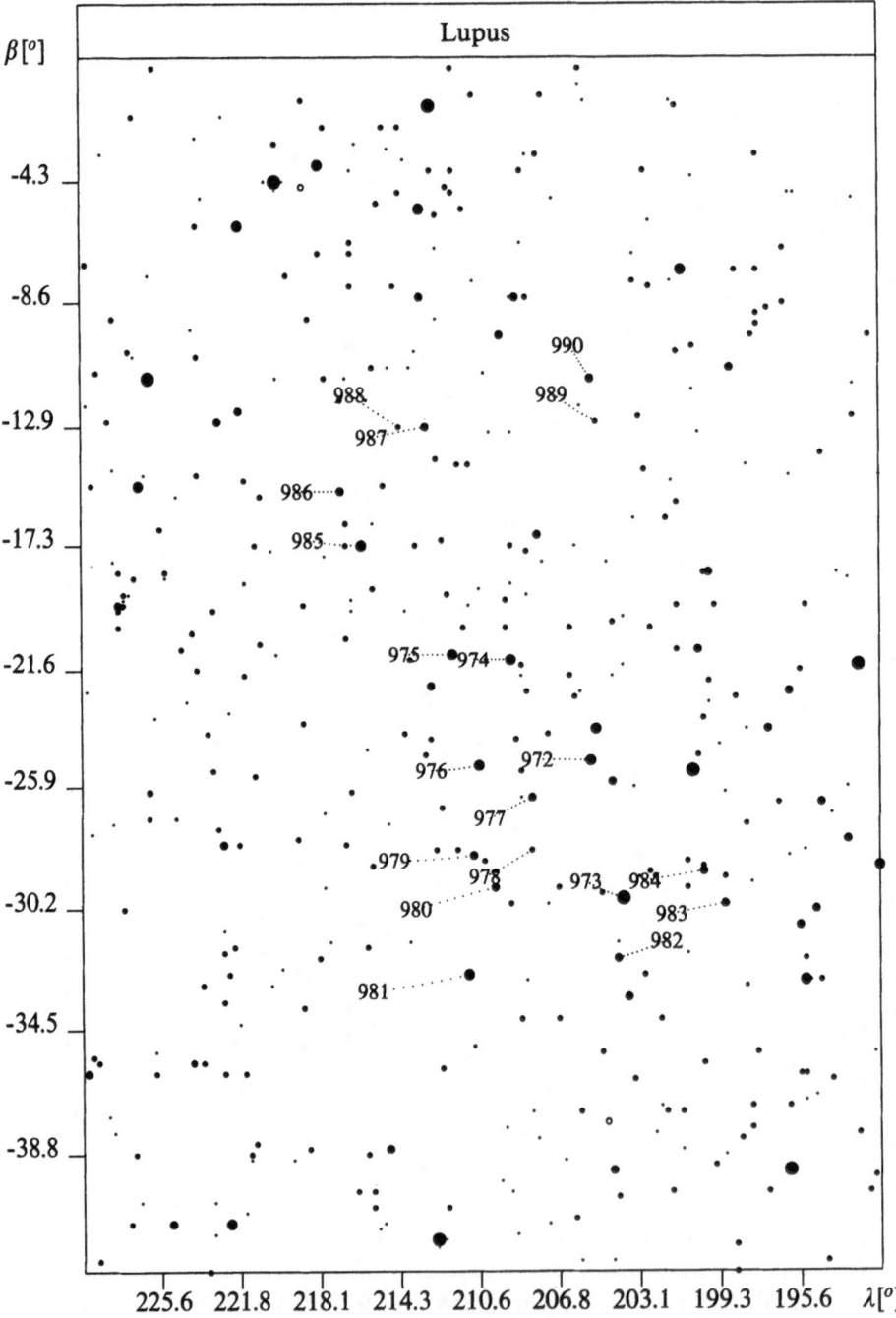

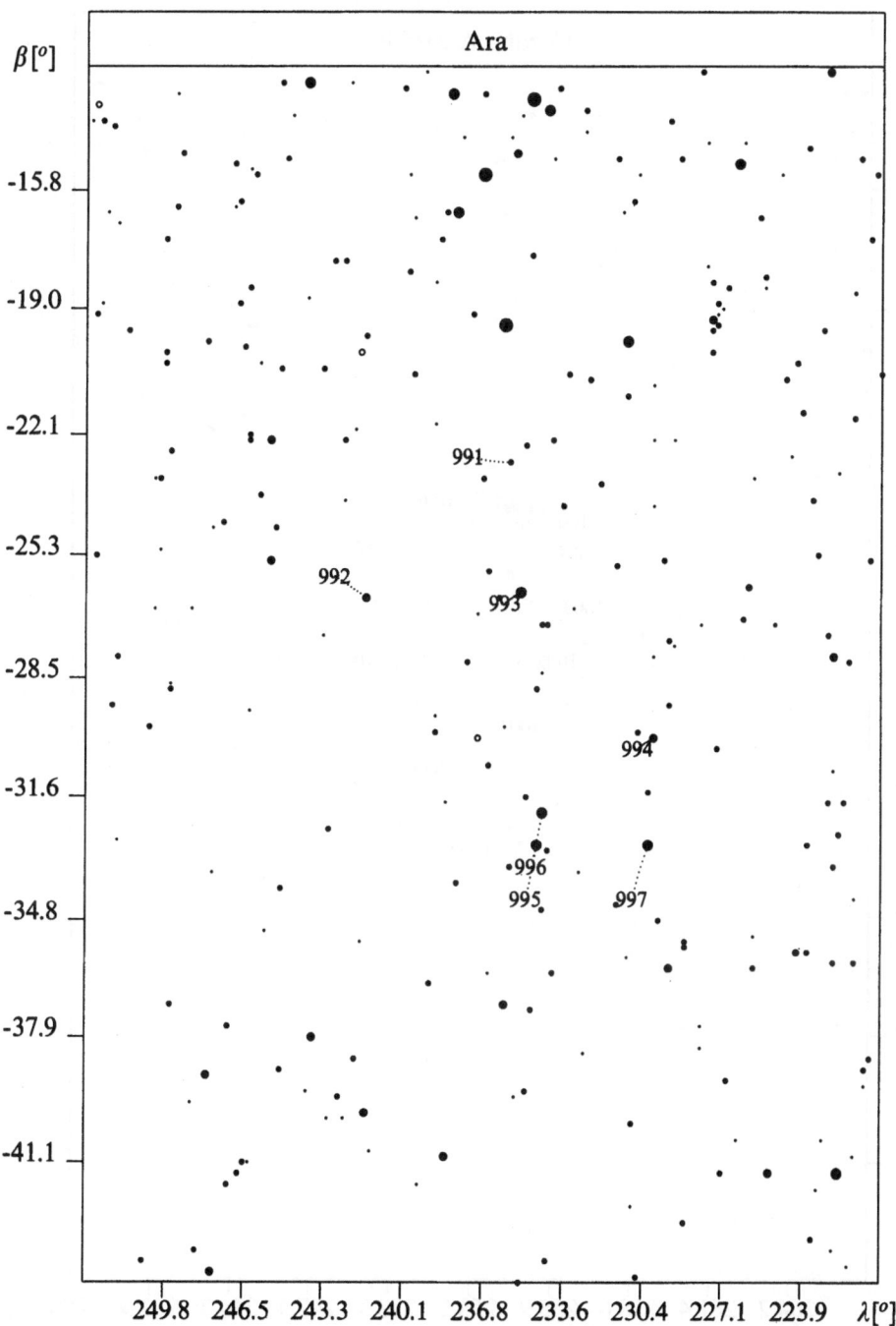

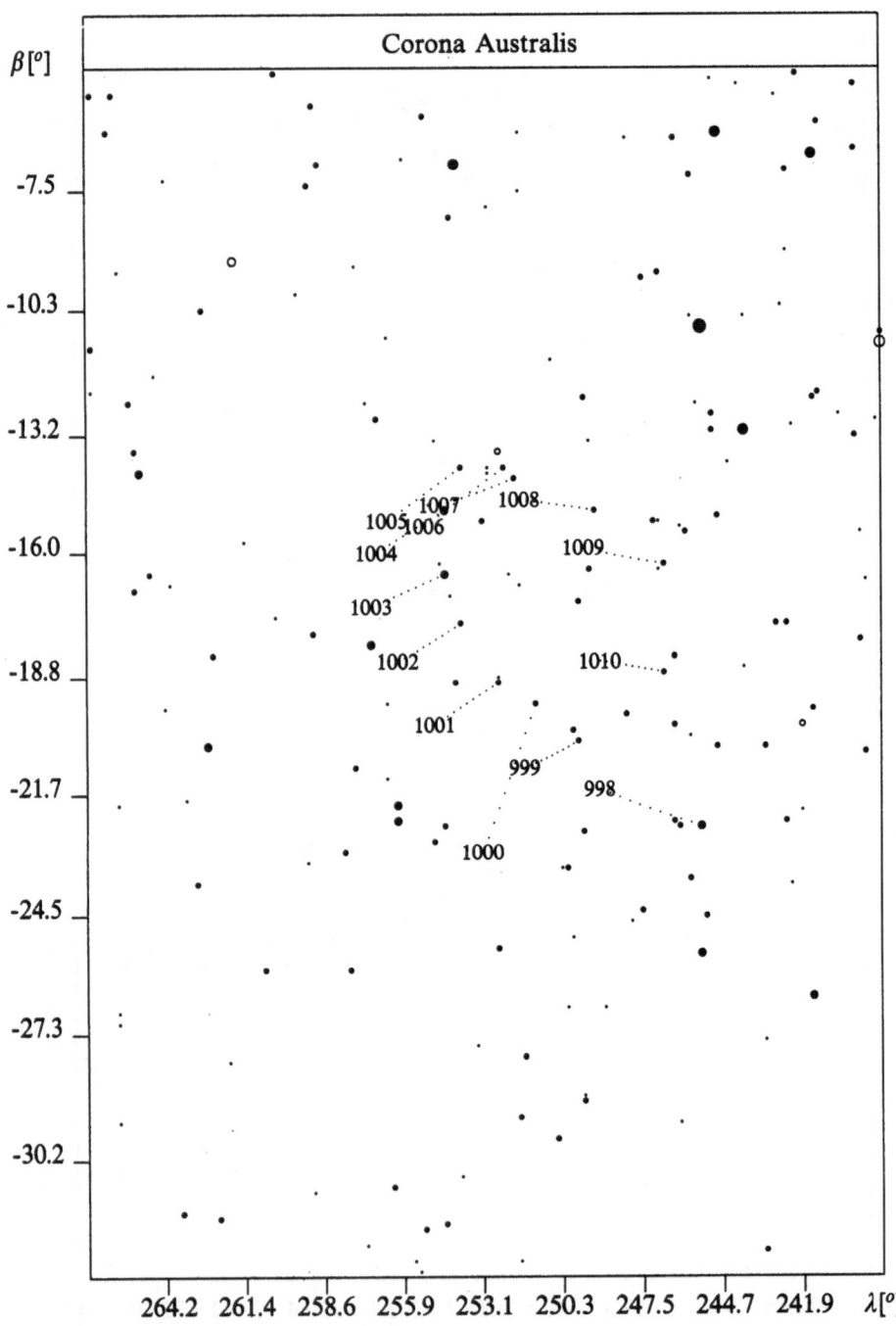

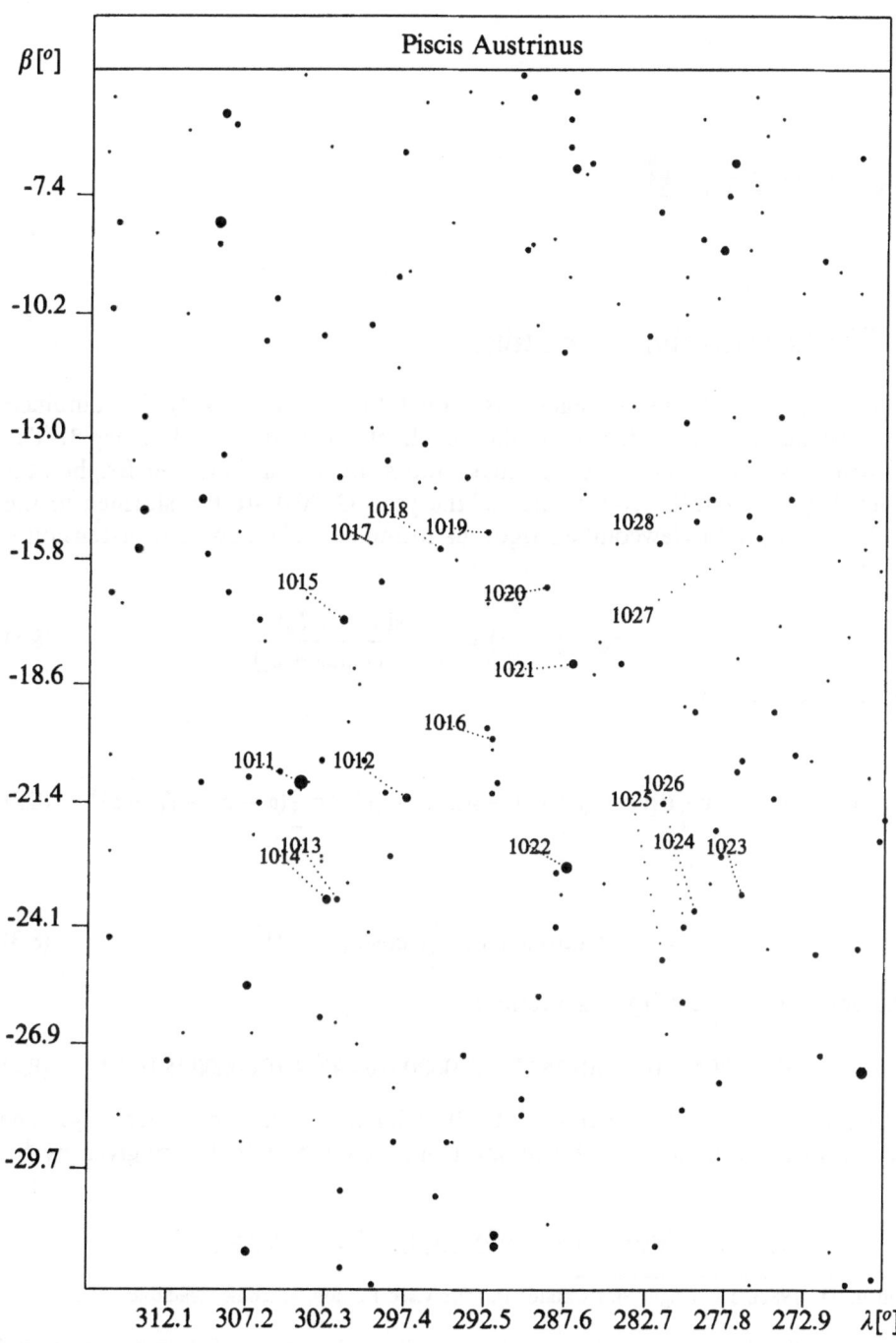

8. Appendix B

8.1 Transformation Formulae

The star catalogue of the Almagest, as edited by Toomer (1984), is combined with the calculated data reduced for the epoch of the year -128, January 1. The stellar positions, proper motions and magnitudes are taken from the Bright Star Catalogue.[1] The equatorial coordinates of the year J2000.0 are transformed to the year -128 according to Newcombe's rigorous formulae.[2] The new right ascension α is given by

$$\tan(\alpha - \alpha_0 - \zeta_0 - z) = \frac{q \sin(\alpha_0 + \zeta_0)}{1 - q \cos(\alpha_0 + \zeta_0)} \tag{8.1}$$

and the declination δ by

$$\tan \frac{1}{2}(\delta - \delta_0) = \tan \frac{1}{2}\theta \{\cos(\alpha_0 + \zeta_0) - \sin(\alpha_0 + \zeta_0) \tan \frac{1}{2}(\alpha - \alpha_0 - \zeta_0 - z)\} \tag{8.2}$$

with

$$q = \sin \theta \{\tan \delta_0 + \tan \frac{1}{2}\theta \cos(\alpha_0 + \zeta_0)\} \tag{8.3}$$

The obliquity of the ecliptic amounts to[3]

$$\epsilon = 23\overset{\circ}{.}452294 - 0\overset{\circ}{.}0130125\, T - 0\overset{\circ}{.}00000164\, T^2 + 0\overset{\circ}{.}000000503\, T^3 \tag{8.4}$$

with T as the interval $T = (t - J1900.0)$ in julian centuries of 36525 days. The coefficients of a coordinate transformation from the epoch J2000.0 are given by:[4]

$$\zeta_0 \quad = \quad 0\overset{\circ}{.}6406161\, T + 0\overset{\circ}{.}0000839\, T^2 + 0\overset{\circ}{.}0000050\, T^3$$

[1]Hoffleit, D. (1982), The Bright Star Catalogue, 4th revised Edition, Machine-Readable Version, New Haven.

[2]Explanatory Supplement to the Astronomical Ephemeris and the American Ephemeris and Nautical Almanac, London, Her Majesty's Stationary Office, 1961, p. 31.

[3]Explanatory Supplement, p. 98.

[4]Green, R.M. (1985), Spherical Astronomy, Cambridge, p. 219.

$$z = 0°\!.6406161\, T + 0°\!.0003041\, T^2 + 0°\!.0000051\, T^3$$
$$\theta = 0°\!.5567530\, T - 0°\!.0001185\, T^2 - 0°\!.0000116\, T^3 \tag{8.5}$$

with T as Julian centuries from the year J2000.0.

The proper motion of the stars and the perspective acceleration due to the radial velocity change the coordinates to[5]

$$\alpha = \alpha_0 + \{\mu_\alpha + \frac{1}{2}\frac{d\mu_\alpha}{dt}t\}t \tag{8.6}$$

$$\delta = \delta_0 + \{\mu_\delta + \frac{1}{2}\frac{d\mu_\delta}{dt}t\}t \tag{8.7}$$

with

$$\frac{d\mu_\alpha}{dt} = 2\mu_\alpha\mu_\delta \tan\delta - 2\mu_\alpha\frac{1}{r}\frac{dr}{dt} \tag{8.8}$$

$$\frac{d\mu_\delta}{dt} = -\mu_\alpha^2 \sin\delta \cos\delta - 2\mu_\delta\frac{1}{r}\frac{dr}{dt} \tag{8.9}$$

The perspective acceleration adds to the components

$$\frac{d\mu_i}{dt} = -0.00000205\mu_i \, \pi \, \frac{dr}{dt} \tag{8.10}$$

with μ and π in seconds of arc and the radial velocity dr/dt in km/s.

For a selection of bright stars the calculated positions can be compared with Hawkins, G. S., Rosenthal, S. K. (1967), 5000 and 10000-Year Star Catalogs, Smithsonian Contributions to Astrophysics, 10.2.

8.2 Column Headings

No: The stars in the catalogue of the Almagest are numbered according to Baily's and Peters-Knobel's numbering of stars. It is the same as in Toomer's edition with the exception of the pairs No. 611/612 and 665/666.

Name: The modern name of stars with a Greek letter and the abbreviated name of the constellation. The notation was introduced by Bayer and modified by Lacaille.[6] All entries are taken from the Bright Star Catalogue. Components of close multiple stars are indicated by a superscript number. Two superscript numbers indicate a double component closer than a minute of arc. Clusters are given by their Messier number. In all other cases only the abbreviation of the constellation name is noted.

HR: Number of the star in the Bright Star Catalogue.[7] Sometimes authors use the abbreviation "BSC" for the same reference number. In the case of clusters the NGC number is catalogued with an "N" preceding the number.

---

[5]Woolard, E., Clemence, G. (1966), Spherical Astronomy, New York, pp. 305–307.
[6]Compare: Werner, H., Schmeidler, F. (1986), Synopsis der Nomenklatur der Fixsterne, Stuttgart.
[7]This takes over the numbering of Pickering, E. C. (1908), Harvard Revised Photometry, Annals of Harvard College Observatory **50**.

C: Uncertain values in the Almagest are noted as "L" in longitude, "B" in latitude, and "M" in magnitude following Toomer, G. (1984). Additional variants on the basis of Kunitzsch's work in Ptolemy, C. (1986), p. 173ff, are referred to by small letters. A "D" indicates stars listed in two constellations.

λ_{Ptol}: Ecliptical longitude in the Almagest according to Toomer (1984). The numerical value of the following longitudes are changed.

No.	coordinate	corrected value
90	λ	155°40′
479	λ	132°10′
786	λ	14°50′
921	λ	26°20′

The following longitude is not changed but there is some evidence also for the value listed in the following table.

No.	coordinate	alternative value
458	λ	105°10′

β_{Ptol}: Ecliptical latitude in the Almagest according to Toomer· (1984). The latitudes of the following stars are changed.

No.	coordinate	corrected value
575	β	3°10′
748	β	-20°20′
814	β	-44°0′
864	β	-57°45′
1000	β	20°20′

The following latitudes are not changed but there is also some evidence in favour for the values listed in the following table.

No.	coordinate	alternative value
131	β	56°
233	β	16°
240	β	17°20′
277	β	10°30′
285	β	38°20′
399	β	north
502	β	1°30′
504	β	2°30′
518	β	7°10′
564	β	-15°20′
625	β	south
645	β	south
787	β	-23°50′
852	β	46°10′
908	β	26°15′
913	β	36°0′
989	β	11°30′

λ_{-128}: Ecliptical longitude of the reference star for the epoch of the year -128 according to Newcombe's formulae. It is rounded to the minute of arc.

β_{-128}: Ecliptical latitude of the reference star rounded to the minute of arc.

$\Delta\lambda$: Difference ($\lambda_{Ptol} - \lambda_{-128}$) of the ecliptical longitudes.

$\Delta\beta$: Difference ($\beta_{Ptol} - \beta_{-128}$) of the ecliptical latitudes.

$\alpha_{Ptol}, \beta_{Ptol}$: Transformation of Ptolemy's ecliptical coordinates into equatorial coordinates, defined by

$$\sin\delta = \sin\beta\cos\epsilon + \cos\beta\sin\epsilon\sin\lambda$$
$$\cos\delta\sin\alpha = -\sin\beta\sin\epsilon + \cos\beta\cos\epsilon\sin\lambda$$
$$\cos\delta\cos\alpha = \cos\beta\cos\lambda \tag{8.11}$$

with Ptolemy's obliquity of the ecliptic $\epsilon = 23;51°$.

m_{Pt}: Magnitudes of stars according to the Almagest. Ptolemy added for intermediate values to the magnitude classes 'greater' or 'less', which are represented by the signs '>' and '<'. Numerically they are interpreted by a constant difference of 0.3^m. The column lists the stars of the Almagest classified as 'faint' abbreviated as 'fnt' and 'nebulous' as 'neb'. In total the Almagest catalogues 12 faint stars and 5 instances of nebulous stars.

classification	Star No.
faint	40,41,42,43,219,311,312,313,314,494,495,496
nebulous	191,449,567,577,734

Changed are the following magnitudes:

Star No.	m	correction
130	5	> 4
132	3	5
133	4	3
854	3	4

m_{HR}: Magnitude of the correlated modern star according to the Bright Star Catalogue. In the case of multiple component stars with a distance of less than one minute of arc the combined magnitude is calculated.

Δm: The difference $(m_{Ptol} - m_{HR})$. Faint and nebulous stars of the Almagest count as stars of magnitude six.

No	Name	HR	C	λ_{Ptol}	β_{Ptol}	λ_{-128}	β_{-128}	$\Delta\lambda$	$\Delta\beta$
1	α UMi	424		60;10	66;00	58;58	65;50	1;12	0;10
2	δ UMi	6789		62;30	70;00	61;38	69;41	0;52	0;19
3	ϵ UMi	6322	L	70;10	74;20	69;26	73;38	0;44	0;42
4	ζ UMi	5903		89;40	75;40	87;21	74;51	2;19	0;49
5	η UMi	6116		93;40	77;40	90;30	77;42	3;10	-0;02
6	β UMi	5563	L	107;30	72;50	103;07	72;48	4;23	0;02
7	γ UMi	5735		116;10	74;50	111;13	75;04	4;57	-0;14
8	UMi	5430		103;00	71;10	98;19	71;13	4;41	-0;03
9	o UMa	3323		85;20	39;50	83;23	40;04	1;57	-0;14
10	UMa	3354		85;50	43;00	81;57	44;21	3;53	-1;21
11	π^2 UMa	3403		86;20	43;00	83;10	43;45	3;10	-0;45
12	ρ UMa	3576		86;10	47;10	84;17	47;41	1;53	-0;31
13	σ^2 UMa	3616		86;40	47;00	85;34	47;37	1;06	-0;37
14	UMa	3771		88;10	50;30	86;40	50;58	1;30	-0;28
15	τ UMa	3624		90;30	43;50	87;47	44;20	2;43	-0;30
16	UMa	3757		92;30	44;20	91;01	44;52	1;29	-0;32
17	υ UMa	3888		99;00	42;00	96;41	42;35	2;19	-0;35
18	φ UMa	3894	B	101;00	44;00	99;38	38;01	1;22	5;59
19	θ UMa	3775		100;40	35;00	98;02	35;11	2;38	-0;11
20	ι UMa	3569		95;30	29;20	93;21	29;33	2;09	-0;13
21	κ UMa	3594		96;20	28;20	94;16	28;46	2;04	-0;26
22	UMa	3662		95;40	36;00	93;35	35;49	2;05	0;11
23	UMa	3619		95;50	33;00	93;30	33;15	2;20	-0;15
24	α UMa	4301		107;40	49;00	105;25	49;33	2;15	-0;33
25	β UMa	4295		112;10	44;30	109;35	44;54	2;35	-0;24
26	δ UMa	4660		123;10	51;00	121;05	51;28	2;05	-0;28
27	γ UMa	4554		123;00	46;30	120;34	46;57	2;26	-0;27
28	λ UMa	4033		112;40	29;20	109;56	29;45	2;44	-0;25
29	μ UMa	4069		114;10	28;15	111;35	28;49	2;35	-0;34
30	ψ UMa	4335		121;40	35;15	119;05	35;25	2;35	-0;10
31	ν UMa	4377		129;50	25;50	126;58	26;02	2;52	-0;12
32	ξ UMa	4375	1	130;20	25;00	127;43	25;02	2;37	-0;02
33	ϵ UMa	4905	1	132;10	53;30	128;53	54;10	3;17	-0;40
34	ζ UMa	5054		138;00	55;40	135;35	56;16	2;25	-0;36
35	η UMa	5191		149;50	54;00	147;02	54;25	2;48	-0;25
36	$\alpha^{2,1}$ CVn	4915		147;50	39;45	144;55	40;08	2;55	-0;23
37	β CVn	4785		140;10	41;20	138;28	40;33	1;42	0;47
38	α Lyn	3705		105;00	17;15	102;20	17;47	2;40	-0;32
39	Lyn	3690		103;20	19;10	100;54	19;57	2;26	-0;47
40	LMi	3800	I	106;10	20;00	104;09	20;33	2;01	-0;33
41		3809	LBI	102;10	22;30	103;08	23;36	-0;58	-1;06
42		3612	BI	101;10	20;20	97;54	20;41	3;16	-0;21
43	Lyn	3275		90;00	22;15	87;55	22;54	2;05	-0;39
44	μ Dra	6370		206;40	76;30	204;43	76;26	1;57	0;04
45	ν^1 Dra	6554		221;50	78;30	219;53	78;22	1;57	0;08
46	β Dra	6536		223;10	75;40	222;04	75;33	1;06	0;07
47	ξ Dra	6688		237;20	80;20	234;41	80;31	2;39	-0;11
48	γ Dra	6705		239;40	75;30	238;21	75;13	1;19	0;17
49	Dra	6923		264;40	82;20	263;25	82;00	1;15	0;20
50	Dra	7049		272;20	78;15	270;57	78;07	1;23	0;08

No	Name	HR	α_{Ptol}	δ_{Ptol}	α_{-128}	δ_{-128}	m_{Pt}	m_{HR}	Δm
1	α UMi	424	347;01	78;01	347;44	77;32	3	2.02	0.98
2	δ UMi	6789	327;01	79;09	329;24	78;57	4	4.36	-0.36
3	ϵ UMi	6322	300;16	79;32	304;25	79;55	4	4.23	-0.23
4	ζ UMi	5903	270;30	80;29	274;38	81;23	4	4.32	-0.32
5	η UMi	6116	266;06	78;26	269;28	78;35	4	4.95	-0.95
6	β UMi	5563	235;27	81;00	240;53	82;04	2	2.08	-0.08
7	γ UMi	5735	236;41	77;53	241;05	78;53	2	3.05	-1.05
8	UMi	5430	232;37	83;08	242;23	84;14	4	4.25	-0.25
9	o UMa	3323	81;56	63;33	78;36	63;31	4	3.36	0.64
10	UMa	3354	82;16	66;44	74;45	67;39	5	5.47	-0.47
11	π^2 UMa	3403	83;12	66;46	77;12	67;10	5	4.60	0.40
12	ρ UMa	3576	82;01	70;55	78;00	71;10	5	4.76	0.24
13	σ^2 UMa	3616	83;05	70;46	80;42	71;11	5	4.80	0.20
14	UMa	3771	85;41	74;19	82;05	74;36	5	4.56	0.44
15	τ UMa	3624	90;57	67;41	85;46	68;01	4	4.67	-0.67
16	UMa	3757	94;48	68;08	91;59	68;35	4	3.67	0.33
17	v UMa	3888	106;11	65;20	102;09	66;01	4	3.80	0.20
18	φ UMa	3894	110;37	67;03	105;53	61;13	< 4	4.59	-0.29
19	θ UMa	3775	106;44	58;13	102;38	58;33	3	3.17	-0.17
20	ι UMa	3569	97;59	53;02	94;53	53;12	3	3.14	-0.14
21	κ UMa	3594	99;04	51;59	96;09	52;24	3	3.60	-0.60
22	UMa	3662	99;06	59;40	95;44	59;28	4	4.83	-0.83
23	UMa	3619	98;55	56;40	95;23	56;55	4	4.48	-0.48
24	α UMa	4301	126;45	70;34	122;55	71;29	2	1.79	0.21
25	β UMa	4295	130;00	65;15	125;59	66;10	2	2.37	-0.37
26	δ UMa	4660	154;03	67;29	151;38	68;34	3	3.31	-0.31
27	γ UMa	4554	147;57	63;45	144;53	64;54	2	2.44	-0.44
28	λ UMa	4033	122;00	50;39	118;23	51;31	3	3.45	-0.45
29	μ UMa	4069	123;33	49;17	120;17	50;16	3	3.05	-0.05
30	ψ UMa	4335	136;49	53;59	133;24	54;48	> 4	3.01	0.69
31	v UMa	4377	141;40	42;41	138;20	43;41	3	3.48	-0.48
32	ξ UMa	4375	141;50	41;45	138;45	42;31	3	3.66	-0.66
33	ϵ UMa	4905	168;55	65;59	166;05	67;45	2	1.77	0.23
34	ζ UMa	5054	178;28	65;13	176;55	66;36	2	1.52	0.48
35	η UMa	5191	186;24	59;15	184;22	60;41	2	1.86	0.14
36	$\alpha^{2,1}$ CVn	4915	169;55	48;37	167;07	50;05	3	2.15	0.85
37	β CVn	4785	163;19	52;59	160;39	52;56	5	4.26	0.74
38	α Lyn	3705	108;51	40;06	105;35	40;50	4	3.13	0.87
39	Lyn	3690	107;06	42;13	104;06	43;08	4	3.82	0.18
40	LMi	3800	110;51	42;40	108;21	43;22	fnt	4.55	1.45
41		3809	106;11	45;39	107;36	46;31	fnt	4.81	1.19
42		3612	104;32	43;37	100;19	44;07	fnt	4.56	1.44
43	Lyn	3275	90;00	46;06	87;12	46;36	fnt	4.25	1.75
44	μ Dra	6370	246;54	57;53	246;07	58;16	4	5.05	-1.05
45	v^1 Dra	6554	254;00	57;24	253;12	57;38	> 4	4.88	-1.18
46	β Dra	6536	251;43	54;51	251;09	55;01	3	2.79	0.21
47	ξ Dra	6688	260;15	57;37	259;37	58;06	4	3.75	0.25
48	γ Dra	6705	257;53	52;57	257;10	52;56	3	2.23	0.77
49	Dra	6923	268;38	58;31	268;16	58;20	4	4.98	-0.98
50	Dra	7049	270;49	54;24	270;20	54;23	4	5.04	-1.04

No	Name	HR	C	λ_{Ptol}	β_{Ptol}	λ_{-128}	β_{-128}	$\Delta\lambda$	$\Delta\beta$
51	Dra	6978		268;50	80;20	266;49	80;01	2;01	0;19
52	o Dra	7125	b	289;30	81;30	286;18	81;01	3;12	0;29
53	π Dra	7371		338;00	81;40	335;38	81;49	2;22	-0;09
54	δ Dra	7310		350;30	83;00	349;22	82;51	1;08	0;09
55	ϵ Dra	7582		7;40	78;50	4;20	79;23	3;20	-0;33
56	ρ Dra	7685		352;50	77;50	351;55	78;04	0;55	-0;14
57	σ Dra	7462		10;40	80;30	9;06	80;50	1;34	-0;20
58	υ Dra	7180	B	21;40	81;20	22;14	83;02	-0;34	-1;42
59	τ Dra	7352		26;10	80;15	25;49	80;26	0;21	-0;11
60	ψ^1 Dra	6636		73;20	84;30	73;15	83;45	0;05	0;45
61	χ Dra	6927	B	50;20	83;30	50;01	83;10	0;19	0;20
62	φ Dra	6920		41;50	84;50	42;24	84;37	-0;34	0;13
63	Dra	6566		118;40	87;30	113;53	86;45	4;47	0;45
64	ω Dra	6596		111;40	86;50	101;45	86;49	9;55	0;01
65	Dra	6223		159;00	81;15	152;47	81;39	6;13	-0;24
66	Dra	6315	b	159;20	83;00	152;43	83;12	6;37	-0;12
67	ζ Dra	6396		158;20	84;50	150;47	84;47	7;33	0;03
68	η Dra	6132		160;00	78;00	163;41	78;30	-3;41	-0;30
69	θ Dra	5986	1	163;00	74;40	167;06	74;31	-4;06	0;09
70	ι Dra	5744		162;40	70;00	154;33	71;07	8;07	-1;07
71	Dra	5226		127;20	64;40	124;46	65;15	2;34	-0;35
72	α Dra	5291		131;10	65;30	127;19	66;16	3;51	-0;46
73	κ Dra	4787		109;10	61;15	106;20	61;35	2;50	-0;20
74	λ Dra	4434		103;10	56;15	100;28	57;02	2;42	-0;47
75	κ Cep	7750		35;00	75;40	33;58	75;14	1;02	0;26
76	γ Cep	8974		33;00	64;15	30;39	64;21	2;21	-0;06
77	β Cep	8238		7;20	71;10	6;38	70;59	0;42	0;11
78	α Cep	8162		346;40	69;00	343;44	68;55	2;56	0;05
79	η Cep	7957		339;20	72;00	334;36	71;33	4;44	0;27
80	θ Cep	7850		340;00	74;00	336;15	73;56	3;45	0;04
81	ξ Cep	8417		358;30	65;30	354;58	65;43	3;32	-0;13
82	ι Cep	8694		7;30	62;30	4;18	62;28	3;12	0;02
83	ϵ Cep	8494	b	346;20	60;15	343;33	60;03	2;47	0;12
84	ζ Cep	8465	b	347;20	61;15	344;52	61;05	2;28	0;10
85	λ Cep	8469		349;00	61;20	346;55	61;49	2;05	-0;29
86	μ Cep	8316		343;40	64;00	340;43	64;09	2;57	-0;09
87	δ Cep	8571		351;20	59;30	348;29	59;28	2;51	0;02
88	κ^2 Boo	5329		152;20	58;40	149;50	58;52	2;30	-0;12
89	ι Boo	5350		154;10	58;20	151;14	58;51	2;56	-0;31
90	θ Boo	5404		155;40	60;10	152;25	60;23	3;15	-0;13
91	λ Boo	5351		159;40	54;40	157;13	54;40	2;27	0;00
92	γ Boo	5435		169;40	49;00	167;54	49;35	1;46	-0;35
93	β Boo	5602		176;40	53;50	174;19	54;18	2;21	-0;28
94	δ Boo	5681		185;40	48;40	183;10	49;10	2;30	-0;30
95	μ^1 Boo	5733		185;40	53;15	183;25	53;34	2;15	-0;19
96	v^1 Boo	5763		185;00	57;30	182;33	57;14	2;27	0;16
97	η CrB	5727	Ib	187;40	46;30	187;08	47;02	0;32	-0;32
98	o CrB	5709		188;30	45;30	186;51	46;09	1;39	-0;39
99	Boo	5634	I	188;10	41;20	185;20	40;41	2;50	0;39
100	ψ Boo	5616	I	186;40	41;40	183;49	42;22	2;51	-0;42

No	Name	HR	α_{Ptol}	δ_{Ptol}	α_{-128}	δ_{-128}	m_{Pt}	m_{HR}	Δm
51	Dra	6978	269;39	56;29	269;00	56;19	4	4.77	-0.77
52	o Dra	7125	275;21	58;01	274;41	57;34	4	4.66	-0.66
53	π Dra	7371	286;38	62;00	286;00	61;57	4	4.59	-0.59
54	δ Dra	7310	285;59	64;07	286;13	64;03	4	3.07	0.93
55	ε Dra	7582	297;14	65;12	295;39	64;53	4	3.83	0.17
56	ρ Dra	7685	296;30	62;04	295;59	62;08	4	4.51	-0.51
57	σ Dra	7462	293;37	66;07	292;48	66;04	5	4.68	0.32
58	υ Dra	7180	291;52	67;55	287;26	68;00	5	4.82	0.18
59	τ Dra	7352	294;43	68;41	294;22	68;44	5	4.45	0.55
60	ψ^1 Dra	6636	274;56	71;21	275;53	72;10	4	4.58	-0.58
61	χ Dra	6927	282;39	70;44	283;36	71;02	4	3.57	0.43
62	φ Dra	6920	280;55	69;15	281;25	69;32	4	4.22	-0.22
63	Dra	6566	266;45	68;19	266;18	69;13	6	5.05	0.95
64	ω Dra	6596	266;44	69;04	268;09	69;24	6	4.80	1.20
65	Dra	6223	247;54	67;49	249;02	68;50	5	4.83	0.17
66	Dra	6315	252;31	67;42	253;15	68;35	5	4.89	0.11
67	ζ Dra	6396	257;20	67;34	257;33	68;22	3	3.17	-0.17
68	η Dra	6132	239;25	67;25	240;50	66;53	3	2.74	0.26
69	θ Dra	5986	231;37	65;58	232;01	65;00	> 4	4.01	-0.31
70	ι Dra	5744	221;18	64;15	220;55	67;15	3	3.29	-0.29
71	Dra	5226	191;49	74;38	191;54	75;52	4	4.65	-0.65
72	α Dra	5291	196;48	73;26	197;09	75;12	3	3.65	-0.65
73	κ Dra	4787	158;52	80;15	154;21	81;28	3	3.87	-0.87
74	λ Dra	4434	128;36	78;18	123;01	79;32	3	3.84	-0.84
75	κ Cep	7750	307;45	70;39	309;14	70;29	4	4.39	-0.39
76	γ Cep	8974	337;55	66;51	336;40	66;05	4	3.21	0.79
77	β Cep	8238	312;52	61;55	313;05	61;44	4	3.23	0.77
78	α Cep	8162	307;35	55;08	306;27	54;27	3	2.44	0.56
79	η Cep	7957	300;50	55;40	299;28	54;28	4	3.43	0.57
80	θ Cep	7850	298;36	57;15	297;24	56;37	4	4.22	-0.22
81	ξ Cep	8417	317;39	55;53	315;42	55;05	5	4.29	0.71
82	ι Cep	8694	326;27	56;41	324;49	55;40	> 4	3.52	0.18
83	ε Cep	8494	316;27	48;18	315;03	47;25	5	4.19	0.81
84	ζ Cep	8465	316;08	49;24	314;57	48;39	4	3.35	0.65
85	λ Cep	8469	317;02	49;57	315;28	49;49	5	5.04	-0.04
86	μ Cep	8316	311;28	50;33	309;48	50;00	5	4.08	0.92
87	δ Cep	8571	320;11	49;13	318;34	48;23	4	3.75	0.25
88	κ^2 Boo	5329	195;08	61;30	193;24	62;39	5	4.54	0.46
89	ι Boo	5350	195;56	60;34	194;24	62;06	5	4.75	0.25
90	θ Boo	5404	199;35	61;19	197;46	62;37	5	4.05	0.95
91	λ Boo	5351	195;04	55;50	193;00	56;49	5	4.18	0.82
92	γ Boo	5435	197;01	47;33	196;01	48;44	3	3.03	-0.03
93	β Boo	5602	206;36	48;47	205;14	50;03	> 4	3.50	0.20
94	δ Boo	5681	208;56	41;20	207;20	42;42	> 4	3.47	0.23
95	μ^1 Boo	5733	212;25	45;09	211;00	46;15	4	4.31	-0.31
96	v^1 Boo	5763	215;39	48;48	213;41	49;28	4	5.02	-1.02
97	η CrB	5727	208;57	38;47	208;48	39;30	> 4	5.58	-1.88
98	o CrB	5709	208;56	37;37	208;00	38;50	5	5.51	-0.51
99	Boo	5634	206;08	34;07	203;24	34;39	5	4.93	0.07
100	ψ Boo	5616	205;08	34;57	203;13	36;40	5	4.54	0.46

No	Name	HR	C	λ_{Ptol}	β_{Ptol}	λ_{-128}	β_{-128}	$\Delta\lambda$	$\Delta\beta$
101	Boo	5638	I	187;00	42;30	185;07	42;04	1;53	0;26
102	ω Boo	5600	I	187;20	40;20	184;00	40;22	3;20	-0;02
103	ϵ Boo	5506		180;00	40;15	178;20	40;46	1;40	-0;31
104	σ Boo	5447		175;40	41;40	173;59	42;08	1;41	-0;28
105	ρ Boo	5429		175;00	42;10	173;04	42;31	1;56	-0;21
106	ζ Boo	5478		185;20	28;00	183;17	28;02	2;03	-0;02
107	η Boo	5235		171;20	28;00	169;33	28;23	1;47	-0;23
108	τ Boo	5185		170;30	26;30	168;30	26;42	2;00	-0;12
109	υ Boo	5200		171;20	25;00	169;33	25;17	1;47	-0;17
110	α Boo	5340		177;00	31;30	174;39	32;11	2;21	-0;41
111	α CrB	5793		194;40	44;30	192;22	44;33	2;18	-0;03
112	β CrB	5747	b	191;40	46;30	189;27	46;14	2;13	0;16
113	θ CrB	5778		191;50	48;00	189;37	48;45	2;13	-0;45
114	π CrB	5855		193;40	50;30	192;22	50;40	1;18	-0;10
115	γ CrB	5849		197;10	44;45	195;10	44;42	2;00	0;03
116	δ CrB	5889		199;10	44;50	197;17	45;02	1;53	-0;12
117	ϵ CrB	5947		201;20	46;10	199;22	46;20	1;58	-0;10
118	ι CrB	5971		201;40	49;20	199;13	49;23	2;27	-0;03
119	$\alpha^{1,2}$ Her	6406		227;40	37;30	226;31	37;33	1;09	-0;03
120	β Her	6148		213;40	43;00	211;27	42;59	2;13	0;01
121	γ Her	6095		211;40	40;10	209;32	40;14	2;08	-0;04
122	κ Her	6008		208;00	37;10	206;01	37;28	1;59	-0;18
123	δ Her	6410		226;40	48;00	225;05	48;04	1;35	-0;04
124	λ Her	6526		232;00	49;30	230;14	49;34	1;46	-0;04
125	μ Her	6623		237;40	52;00	235;52	51;50	1;48	0;10
126	o Her	6779		245;30	52;50	243;07	52;28	2;23	0;22
127	ν Her	6707		241;40	54;00	239;52	53;55	1;48	0;05
128	ξ Her	6703		241;30	53;00	239;31	52;59	1;59	0;01
129	ζ Her	6212	B	213;50	53;10	212;13	53;12	1;37	-0;02
130	ϵ Her	6324	M	220;10	53;30	218;38	53;30	1;32	-0;00
131	Her	6332	b	220;00	56;30	218;15	56;10	1;45	0;20
132		6377	b	221;10	58;30	219;54	58;44	1;16	-0;14
133	π Her	6418		224;00	59;50	222;21	59;50	1;39	0;00
134	Her	6436	B	225;20	60;20	223;16	60;21	2;04	-0;01
135	ρ Her	6484		226;20	61;15	225;42	60;24	0;38	0;51
136	θ Her	6695	l	240;50	61;00	238;52	60;58	1;58	0;02
137	ι Her	6588		232;10	69;20	230;10	69;33	2;00	-0;13
138	Her	6464	b	225;20	70;15	220;48	69;16	4;32	0;59
139	Her	6509		226;50	71;15	222;46	71;29	4;04	-0;14
140	Her	6574		229;40	72;15	227;44	72;01	1;56	0;14
141	η Her	6220	B	210;40	60;15	208;52	60;35	1;48	-0;20
142	σ Her	6168	B	205;20	63;00	203;20	63;22	2;00	-0;22
143	τ Her	6092		195;40	65;30	194;22	66;01	1;18	-0;31
144	φ Her	6023		193;40	63;40	191;39	63;57	2;01	-0;17
145	υ Her	5982		190;10	64;15	188;13	64;31	1;57	-0;16
146	χ Her	5914		191;10	60;00	188;06	60;02	3;04	-0;02
147	ν^1 Boo	5763	D	185;00	57;30	182;33	57;14	2;27	0;16
148	ω Her	6117	I	212;40	38;10	211;51	35;27	0;49	2;43
149	α Lyr	7001		257;20	62;00	255;34	61;52	1;46	0;08
150	ϵ^1 Lyr	7051		260;20	62;40	259;11	62;38	1;09	0;02

No	Name	HR	α_{Ptol}	δ_{Ptol}	α_{-128}	δ_{-128}	m_{Pt}	m_{HR}	Δm
101	Boo	5638	205;54	35;34	204;04	35;55	5	5.67	-0.67
102	ω Boo	5600	204;53	33;33	202;08	34;51	5	4.81	0.19
103	ϵ Boo	5506	198;54	36;13	197;46	37;20	3	1.95	1.05
104	σ Boo	5447	196;15	39;07	195;05	40;13	4	4.46	-0.46
105	ρ Boo	5429	196;03	39;48	194;35	40;53	> 4	3.58	0.12
106	ζ Boo	5478	196;46	23;20	194;57	24;12	3	3.68	-0.68
107	η Boo	5235	184;28	28;54	182;59	29;58	3	2.68	0.32
108	τ Boo	5185	182;56	27;54	181;10	28;52	4	4.50	-0.50
109	υ Boo	5200	182;56	26;13	181;24	27;12	4	4.07	-0.07
110	α Boo	5340	191;19	29;44	189;34	31;18	1	0.04	0.96
111	α CrB	5793	213;02	34;37	211;13	35;30	> 2	2.23	-0.53
112	β CrB	5747	211;58	37;23	210;02	37;59	> 4	3.68	0.02
113	θ CrB	5778	213;03	38;37	211;49	40;05	5	4.14	0.86
114	π CrB	5855	216;01	40;10	215;07	40;49	6	5.56	0.44
115	γ CrB	5849	215;04	34;00	213;27	34;42	4	3.84	0.16
116	δ CrB	5889	216;38	33;25	215;15	34;17	4	4.63	-0.63
117	ϵ CrB	5947	218;59	33;55	217;33	34;46	4	4.15	-0.15
118	ι CrB	5971	221;01	36;37	219;13	37;30	4	4.99	-0.99
119	$\alpha^{1,2}$ Her	6406	235;41	18;38	234;45	19;03	3	2.73	0.27
120	β Her	6148	226;44	27;23	224;58	28;04	3	2.77	0.23
121	γ Her	6095	223;59	25;19	222;18	26;06	3	3.75	-0.75
122	κ Her	6008	219;49	23;39	218;19	24;38	4	5.00	-1.00
123	δ Her	6410	238;22	28;52	237;13	29;23	3	3.14	-0.14
124	λ Her	6526	242;44	29;15	241;27	29;45	> 4	4.41	-0.71
125	μ Her	6623	247;29	30;41	246;09	30;56	> 4	3.42	0.28
126	o Her	6779	253;06	30;26	251;22	30;30	> 4	3.83	-0.13
127	ν Her	6707	250;47	32;03	249;31	32;20	> 4	4.41	-0.71
128	ξ Her	6703	250;24	31;06	249;01	31;29	4	3.70	0.30
129	ζ Her	6212	231;38	36;40	230;29	37;12	3	2.81	0.19
130	ϵ Her	6324	236;05	35;27	235;00	35;55	> 4	3.92	-0.17
131	Her	6332	237;25	38;16	236;03	38;27	5	5.25	-0.25
132		6377	239;11	39;51	238;28	40;27	5	5.39	-0.39
133	π Her	6418	241;37	40;31	240;33	40;56	3	3.16	0.16
134	Her	6436	242;40	40;43	241;22	41;15	4	4.65	-0.65
135	ρ Her	6484	243;44	41;23	242;53	40;50	> 4	4.72	-1.02
136	θ Her	6695	252;19	38;58	251;05	39;18	4	3.86	0.14
137	ι Her	6588	251;08	47;59	250;14	48;35	4	3.80	0.20
138	Her	6464	248;25	49;47	245;30	49;44	6	5.59	0.41
139	Her	6509	249;47	50;28	248;03	51;23	6	5.80	0.20
140	Her	6574	251;44	51;00	250;40	51;09	6	5.37	0.63
141	η Her	6220	233;46	43;46	232;50	44;35	> 4	3.53	0.17
142	σ Her	6168	232;40	47;25	231;45	48;20	4	4.20	-0.20
143	τ Her	6092	229;40	51;54	229;26	52;44	> 4	3.89	-0.19
144	φ Her	6023	226;45	51;01	225;50	51;54	4	4.26	-0.26
145	υ Her	5982	225;27	52;27	224;33	53;17	4	4.76	-0.76
146	χ Her	5914	221;49	48;51	219;51	49;53	4	4.62	-0.62
147	ν^1 Boo	5763	215;39	48;48	213;41	49;28	4	5.02	-1.02
148	ω Her	6117	223;57	23;11	222;10	20;59	5	4.57	0.43
149	α Lyr	7001	262;27	38;29	261;21	38;35	1	0.03	0.97
150	ϵ^1 Lyr	7051	264;18	39;01	263;37	39;10	> 4	5.06	-1.36

No	Name	HR	C	λ_{Ptol}	β_{Ptol}	λ_{-128}	β_{-128}	$\Delta\lambda$	$\Delta\beta$
151	ζ^1 Lyr	7056		260;20	61;00	258;38	60;37	1;42	0;23
152	δ^2 Lyr	7139	I	263;40	60;00	262;16	59;35	1;24	0;25
153	η Lyr	7298		272;00	61;20	270;43	60;56	1;17	0;24
154	θ Lyr	7314		271;40	60;20	271;11	59;49	0;29	0;31
155	β Lyr	7106		261;00	56;10	259;26	56;16	1;34	-0;06
156	v^2 Lyr	7102	I	260;50	55;00	259;09	55;28	1;41	-0;28
157	γ Lyr	7178		264;10	55;20	262;29	55;17	1;41	0;03
158	λ Lyr	7192	L	264;10	54;45	262;41	54;42	1;29	0;03
159	$\beta^{1,2}$ Cyg	7417	B	274;30	49;00	271;50	49;13	2;40	-0;13
160	φ Cyg	7478		279;00	50;30	275;32	50;50	3;28	-0;20
161	η Cyg	7615		286;20	54;30	283;39	54;30	2;41	0;00
162	γ Cyg	7796		298;30	57;20	295;37	57;18	2;53	0;02
163	α Cyg	7924		309;10	60;00	306;11	60;02	2;59	-0;02
164	δ Cyg	7528	L	289;20	64;40	287;00	64;36	2;20	0;04
165	θ Cyg	7469		292;30	69;40	289;28	69;41	3;02	-0;01
166	ι^2 Cyg	7420		291;10	71;30	288;52	71;35	2;18	-0;05
167	κ Cyg	7328	l	286;40	74;00	285;48	73;58	0;52	0;02
168	ϵ Cyg	7949		300;50	49;30	298;02	49;29	2;48	0;01
169	λ Cyg	7963		303;50	52;10	300;28	51;46	3;22	0;24
170	ζ Cyg	8115		306;40	44;00	303;43	43;52	2;57	0;08
171	v Cyg	8028		310;00	55;10	306;56	55;03	3;04	0;07
172	ξ Cyg	8079		314;30	57;00	311;37	56;41	2;53	0;19
173	Cyg	7735		301;10	64;00	298;58	63;45	2;12	0;15
174	Cyg	7751		302;40	64;30	300;43	64;26	1;57	0;04
175	ω^2 Cyg	7851		312;10	64;45	307;43	64;17	4;27	0;28
176	τ Cyg	8130		310;40	49;40	309;02	50;29	1;38	-0;49
177	σ Cyg	8143		313;50	51;40	311;06	51;36	2;44	0;04
178	ζ Cas	153		7;50	45;20	5;42	44;34	2;08	0;46
179	α Cas	168		10;50	46;45	8;25	46;29	2;25	0;16
180	η Cas	219		13;00	47;50	10;23	47;27	2;37	0;23
181	γ Cas	264		16;40	49;00	14;34	48;38	2;06	0;22
182	δ Cas	403		20;40	45;30	18;21	46;19	2;19	-0;49
183	ϵ Cas	542		27;00	47;45	25;21	47;20	1;39	0;25
184	ι Cas	707	L	31;40	47;20	32;48	48;43	-1;08	-1;23
185	θ Cas	343	I	14;40	44;20	12;16	43;01	2;24	1;19
186	φ Cas	382	I	17;40	45;00	16;08	44;54	1;32	0;06
187	σ Cas	9071		2;20	50;00	0;50	49;16	1;30	0;44
188	κ Cas	130		15;00	52;40	13;17	52;05	1;43	0;35
189	β Cas	21		7;50	51;40	5;33	51;20	2;17	0;20
190	ρ Cas	9045		3;40	51;40	1;46	51;01	1;54	0;39
191	h	N869	I	26;40	40;30	24;46	40;08	1;54	0;22
192	η Per	834		31;10	37;30	29;13	37;15	1;57	0;15
193	γ Per	915		32;40	34;30	30;32	34;17	2;08	0;13
194	θ Per	799		27;30	32;20	24;59	31;30	2;31	0;50
195	τ Per	854		30;40	34;30	28;26	34;08	2;14	0;22
196	ι Per	937		31;30	31;10	28;58	30;40	2;32	0;30
197	α Per	1017		34;50	30;00	32;34	29;53	2;16	0;07
198	σ Per	1052		35;20	27;50	33;05	27;47	2;15	0;03
199	ψ Per	1087		37;00	27;40	34;13	27;44	2;47	-0;04
200	δ Per	1122		37;40	27;20	35;16	27;04	2;24	0;16

No	Name	HR	α_{Ptol}	δ_{Ptol}	α_{-128}	δ_{-128}	m_{Pt}	m_{HR}	Δm
151	ζ^1 Lyr	7056	264;07	37;21	263;02	37;10	> 4	4.36	-0.66
152	δ^2 Lyr	7139	266;05	36;14	265;10	36;00	4	4.30	-0.30
153	η Lyr	7298	271;13	37;30	270;26	37;12	4	4.39	-0.39
154	θ Lyr	7314	271;02	36;29	270;44	36;06	4	4.36	-0.36
155	β Lyr	7106	264;04	32;30	263;02	32;48	3	3.45	-0.45
156	v^2 Lyr	7102	263;51	31;21	262;46	32;02	< 4	5.25	-0.95
157	γ Lyr	7178	266;07	31;34	264;59	31;42	3	3.24	-0.24
158	λ Lyr	7192	266;05	30;59	265;04	31;06	< 4	4.93	-0.63
159	$\beta^{1,2}$ Cyg	7417	273;16	25;12	271;20	25;30	3	2.33	0.67
160	φ Cyg	7478	276;24	26;51	273;56	27;12	5	4.69	0.31
161	η Cyg	7615	281;01	31;17	279;13	31;13	> 4	3.89	-0.19
162	γ Cyg	7796	288;24	35;19	286;35	35;03	3	2.20	0.80
163	α Cyg	7924	294;08	39;27	292;21	39;07	2	1.25	0.75
164	δ Cyg	7528	280;55	41;34	279;38	41;27	3	2.87	0.13
165	θ Cyg	7469	281;11	46;42	279;42	46;37	4	4.48	-0.48
166	ι^2 Cyg	7420	279;56	48;23	278;52	48;27	> 4	3.79	-0.09
167	κ Cyg	7328	277;09	50;34	276;49	50;37	> 4	3.77	-0.07
168	ϵ Cyg	7949	292;09	28;02	290;10	27;44	3	2.46	0.54
169	λ Cyg	7963	293;30	31;05	291;19	30;18	> 4	4.53	-0.83
170	ζ Cyg	8115	297;59	23;42	295;48	23;09	3	3.20	-0.20
171	v Cyg	8028	296;38	35;01	294;41	34;29	> 4	3.94	-0.24
172	ξ Cyg	8079	298;48	37;35	297;08	36;51	> 4	3.72	-0.02
173	Cyg	7735	287;48	42;06	286;41	41;43	4	3.79	0.21
174	Cyg	7751	288;27	42;46	287;25	42;35	4	3.98	0.02
175	ω^2 Cyg	7851	293;37	44;23	291;25	43;23	5	5.44	-0.44
176	τ Cyg	8130	299;07	29;55	297;43	30;29	> 4	3.72	-0.02
177	σ Cyg	8143	300;36	32;27	298;46	31;57	> 4	4.23	-0.53
178	ζ Cas	153	343;59	43;34	342;57	42;08	> 4	3.66	0.04
179	α Cas	168	345;17	45;55	343;41	44;47	3	2.23	0.77
180	η Cas	219	346;07	47;39	344;28	46;21	4	3.44	0.56
181	γ Cas	264	348;03	50;02	346;50	48;56	> 3	2.47	0.23
182	δ Cas	403	354;35	48;48	352;02	48;33	3	2.68	0.32
183	ϵ Cas	542	358;05	53;10	357;11	52;10	4	3.38	0.62
184	ι Cas	707	2;47	54;43	2;35	56;17	4	4.52	-0.52
185	θ Cas	343	350;25	45;26	349;31	43;24	4	4.33	-0.33
186	φ Cas	382	352;25	47;11	351;20	46;30	5	4.98	0.02
187	σ Cas	9071	336;01	45;20	335;35	44;14	6	4.88	1.12
188	κ Cas	130	343;06	52;15	342;33	51;11	< 4	4.16	0.14
189	β Cas	21	338;41	48;44	337;25	47;39	3	2.27	0.73
190	ρ Cas	9045	335;35	47;11	334;52	46;01	6	4.54	1.46
191	h	N869	4;10	47;03	2;48	45;58	neb	4.40	1.60
192	η Per	834	10;47	46;17	9;11	45;16	4	3.76	0.24
193	γ Per	915	14;23	44;16	12;30	43;12	< 3	2.93	0.37
194	θ Per	799	10;37	40;19	8;48	38;33	4	4.12	-0.12
195	τ Per	854	12;22	43;28	10;31	42;14	4	3.95	0.05
196	ι Per	937	15;18	40;51	13;10	39;23	4	4.05	-0.05
197	α Per	1017	19;24	41;06	17;14	40;05	2	1.79	0.21
198	σ Per	1052	21;08	39;20	18;58	38;23	4	4.36	-0.36
199	ψ Per	1087	22;58	39;48	20;09	38;46	4	4.23	-0.23
200	δ Per	1122	23;51	39;45	21;35	38;34	3	3.01	-0.01

No	Name	HR	C	λ_{Ptol}	β_{Ptol}	λ_{-128}	β_{-128}	$\Delta\lambda$	$\Delta\beta$
201	κ Per	941		30;30	27;00	28;06	25;57	2;24	1;03
202	β Per	936		29;40	23;00	26;39	22;11	3;01	0;49
203	ω Per	947		29;10	21;00	26;51	20;43	2;19	0;17
204	ρ Per	921		27;40	21;00	25;20	20;25	2;20	0;35
205	π Per	879		26;50	22;15	24;22	21;31	2;28	0;44
206		1324	B	44;50	28;00	42;15	28;13	2;35	-0;13
207	λ Per	1261		43;00	28;10	40;14	28;38	2;46	-0;28
208	Per	1273		42;20	25;00	39;57	25;59	2;23	-0;59
209	μ Per	1303		44;00	26;15	41;15	26;27	2;45	-0;12
210	Per	1350		44;10	24;30	42;04	24;22	2;06	0;08
211	Per	1454	M	46;20	18;45	44;02	18;44	2;18	0;01
212	ν Per	1135		36;50	21;50	34;18	21;54	2;32	-0;04
213	ε Per	1220		38;40	19;15	36;08	18;52	2;32	0;23
214	ξ Per	1228		38;20	14;45	35;25	14;41	2;55	0;04
215	o Per	1131		34;10	12;00	31;35	11;56	2;35	0;04
216	ζ Per	1203		36;20	11;00	33;34	11;05	2;46	-0;05
217	Per	1306		41;50	18;00	39;35	18;40	2;15	-0;40
218		1314		45;00	31;00	42;47	31;28	2;13	-0;28
219	Per	840		24;40	20;40	22;13	20;49	2;27	-0;09
220	δ Aur	2077		62;30	30;00	60;17	30;38	2;13	-0;38
221	ξ Aur	2029	b	62;20	31;50	59;34	31;58	2;46	-0;08
222	α Aur	1708		55;00	22;30	52;16	22;50	2;44	-0;20
223	β Aur	2088		62;50	20;00	60;22	21;13	2;28	-1;13
224	ν Aur	2012		61;10	15;15	58;42	15;26	2;28	-0;11
225	θ Aur	2095		62;50	13;20	60;20	13;32	2;30	-0;12
226	ε Aur	1605		52;00	20;40	49;17	20;40	2;43	0;00
227	η Aur	1641		52;10	18;00	49;52	18;02	2;18	-0;02
228	ζ Aur	1612		52;00	18;00	49;04	17;56	2;56	0;04
229	ι Aur	1577	b	49;50	10;10	47;04	10;11	2;46	-0;01
230	β Tau	1791	D	55;40	5;00	52;59	5;12	2;41	-0;12
231	χ Aur	1843		56;00	8;30	54;35	8;36	1;25	-0;06
232	φ Aur	1805		56;20	12;10	53;38	10;58	2;42	1;12
233	Aur	1706	ILB	50;40	10;20	50;56	9;19	-0;16	1;01
234	α Oph	6556		234;50	36;00	232;44	36;15	2;06	-0;15
235	β Oph	6603		238;00	27;15	235;46	28;08	2;14	-0;53
236	γ Oph	6629		239;00	26;30	237;03	26;27	1;57	0;03
237	ι Oph	6281		223;20	33;00	221;01	32;48	2;19	0;12
238	κ Oph	6299		224;40	31;50	222;23	32;08	2;17	-0;18
239	λ Oph	6149	B	218;20	23;45	215;58	23;52	2;22	-0;07
240	δ Oph	6056		215;00	17;00	212;41	17;35	2;19	-0;35
241	ε Oph	6075		216;00	16;30	213;50	16;40	2;10	-0;10
242	μ Oph	6567		236;40	15;00	234;44	15;31	1;56	-0;31
243	ν Oph	6698		242;20	13;40	240;10	14;02	2;10	-0;22
244	τ Oph	6733		243;20	14;20	241;11	15;34	2;09	-1;14
245	η Oph	6378		231;10	7;30	228;21	7;25	2;49	0;05
246	ξ Oph	6445	IL	233;40	2;15	231;10	2;25	2;30	-0;10
247	Oph	6401	IB	233;00	-2;15	230;39	-2;34	2;21	0;19
248	θ Oph	6453	IB	234;20	-1;30	231;49	-1;33	2;31	0;03
249	Oph	6486	IB	235;00	-0;20	232;45	-0;36	2;15	0;16
250	Oph	6519	IB	235;50	-0;15	233;53	-0;23	1;57	0;08

No	Name	HR	α_{Ptol}	δ_{Ptol}	α_{-128}	δ_{-128}	m_{Pt}	m_{HR}	Δm
201	κ Per	941	16;41	36;44	14;56	34;50	4	3.80	0.20
202	β Per	936	17;55	32;48	15;25	30;51	2	2.12	-0.12
203	ω Per	947	18;24	30;47	16;19	29;36	4	4.63	-0.63
204	ρ Per	921	16;55	30;12	14;59	28;44	4	3.39	0.61
205	π Per	879	15;30	31;01	13;32	29;22	4	4.70	-0.70
206		1324	31;13	42;56	28;22	42;09	4	4.61	-0.61
207	λ Per	1261	29;07	42;26	25;58	41;49	4	4.29	-0.29
208	Per	1273	30;02	39;18	27;04	39;18	4	4.04	-0.04
209	μ Per	1303	31;13	41;02	28;14	40;11	4	4.14	-0.14
210	Per	1350	32;16	39;28	30;09	38;33	5	4.85	0.15
211	Per	1454	37;13	34;49	34;49	33;59	5	4.25	0.75
212	ν Per	1135	25;46	34;25	23;12	33;29	> 4	3.77	-0.07
213	ε Per	1220	28;51	32;41	26;28	31;22	3	2.89	0.11
214	ξ Per	1228	30;26	28;23	27;33	27;14	4	4.04	-0.04
215	o Per	1131	27;20	24;21	24;51	23;18	< 3	3.83	-0.53
216	ζ Per	1203	29;54	24;11	27;09	23;13	> 3	2.85	-0.15
217	Per	1306	32;43	32;37	30;09	32;25	5	4.71	0.29
218		1314	29;44	45;44	27;07	45;19	5	5.19	-0.19
219	Per	840	14;09	28;44	11;48	27;52	fnt	4.23	1.77
220	δ Aur	2077	51;22	50;10	48;19	50;05	4	3.72	0.28
221	ξ Aur	2029	50;17	51;52	46;45	51;10	4	4.99	-0.99
222	α Aur	1708	45;24	41;00	42;11	40;25	1	0.08	0.92
223	β Aur	2088	55;35	40;36	52;17	41;06	2	1.90	0.10
224	ν Aur	2012	55;05	35;37	52;16	35;06	4	3.97	0.03
225	θ Aur	2095	57;33	34;07	54;41	33;39	> 4	2.62	1.08
226	ε Aur	1605	42;42	38;23	39;44	37;28	> 4	2.99	0.71
227	η Aur	1641	43;56	35;54	41;26	35;10	> 4	3.17	0.53
228	ζ Aur	1612	43;45	35;51	40;35	34;49	4	3.75	0.25
229	ι Aur	1577	44;10	27;45	41;17	26;51	< 3	2.69	0.61
230	β Tau	1791	51;55	24;21	49;04	23;45	> 3	1.65	1.05
231	χ Aur	1843	51;18	27;49	49;47	27;27	5	4.76	0.24
232	φ Aur	1805	50;34	31;27	48;02	29;29	5	5.07	-0.07
233	Aur	1706	45;00	28;09	45;39	27;10	6	5.02	0.98
234	α Oph	6556	241;03	15;40	239;24	16;27	> 3	2.08	0.62
235	β Oph	6603	241;42	6;33	239;56	7;57	> 4	2.77	0.93
236	γ Oph	6629	242;25	5;37	240;41	6;03	4	3.75	0.25
237	ι Oph	6281	230;45	15;24	228;45	15;54	4	4.38	-0.38
238	κ Oph	6299	231;30	13;56	229;39	14;55	4	3.20	0.80
239	λ Oph	6149	223;32	7;59	221;29	8;53	4	3.82	0.18
240	δ Oph	6056	218;21	2;37	216;27	3;59	3	2.74	0.26
241	ε Oph	6075	219;06	1;50	217;11	2;45	3	3.24	-0.24
242	μ Oph	6567	237;48	-5;08	236;05	-4;06	4	4.62	-0.62
243	ν Oph	6698	242;56	-7;35	240;56	-6;42	< 4	3.34	0.96
244	τ Oph	6733	244;01	-7;06	242;12	-5;23	4	5.19	-1.19
245	η Oph	6378	230;41	-11;07	227;57	-10;22	3	2.43	0.57
246	ξ Oph	6445	231;47	-16;50	229;21	-15;56	> 4	4.39	-0.69
247	Oph	6401	229;54	-21;01	227;25	-20;36	4	5.33	-1.33
248	θ Oph	6453	231;29	-20;38	228;54	-19;55	> 4	3.27	0.43
249	Oph	6486	232;29	-19;40	230;07	-19;15	4	4.17	-0.17
250	Oph	6519	233;21	-19;47	231;21	-19;20	5	4.81	0.19

No	Name	HR	C	λ_{Ptol}	β_{Ptol}	λ_{-128}	β_{-128}	$\Delta\lambda$	$\Delta\beta$
251	Oph	6595	I	237;10	1;00	236;37	2;01	0;33	-1;01
252	ζ Oph	6175		222;10	11;50	219;37	11;39	2;33	0;11
253	φ Oph	6147		221;40	5;20	219;06	5;30	2;34	-0;10
254	χ Oph	6118		220;40	3;10	218;23	3;31	2;17	-0;21
255	ψ Oph	6104	B	219;50	1;40	217;58	1;51	1;52	-0;11
256	ω Oph	6153		222;20	0;40	220;02	0;41	2;18	-0;01
257	ρ Oph	6112		220;40	-0;45	218;51	-1;29	1;49	0;44
258	Oph	6712		242;00	28;10	240;29	28;06	1;31	0;04
259	Oph	6714		242;40	26;20	240;35	26;40	2;05	-0;20
260	Oph	6723	I	243;00	25;00	240;53	25;03	2;07	-0;03
261	Oph	6752		243;40	27;00	241;46	26;52	1;54	0;08
262	Oph	6771		244;40	33;00	242;37	33;14	2;03	-0;14
263	ι Ser	5842		198;50	38;00	197;28	38;21	1;22	-0;21
264	ρ Ser	5899		201;40	40;00	199;48	40;13	1;52	-0;13
265	γ Ser	5933	L	204;20	36;00	202;37	36;07	1;43	-0;07
266	β Ser	5867		202;00	34;15	200;10	34;34	1;50	-0;19
267	κ Ser	5879		201;20	37;15	200;04	37;23	1;16	-0;08
268	π Ser	5972		203;10	42;30	202;22	42;40	0;48	-0;10
269	δ Ser	5788		201;40	29;15	198;42	29;07	2;58	0;08
270	λ Ser	5868		204;50	26;30	202;51	26;50	1;59	-0;20
271	α Ser	5854		204;20	25;20	202;19	25;41	2;01	-0;21
272	ε Ser	5892		206;20	24;00	204;36	24;11	1;44	-0;11
273	μ Ser	5881		208;50	16;30	206;21	16;30	2;29	-0;00
274	υ Oph	6129	B	218;10	13;15	216;59	13;29	1;11	-0;14
275	ν Ser	6446		233;40	10;30	230;40	10;33	3;00	-0;03
276	ξ Ser	6561		237;00	8;30	234;59	8;16	2;01	0;14
277	o Ser	6581		237;50	10;50	235;51	10;49	1;59	0;01
278	ζ Ser	6710		243;40	20;00	240;27	20;04	3;13	-0;04
279	η Ser	6869		248;40	21;10	246;28	21;07	2;12	0;03
280	$\theta^{1,2}$ Ser	7141		258;20	27;00	256;10	27;09	2;10	-0;09
281	γ Sge	7635		280;10	39;20	277;33	39;25	2;37	-0;05
282	ζ Sge	7546		276;40	39;10	274;34	39;39	2;06	-0;29
283	δ Sge	7536		275;50	39;30	273;55	39;09	1;55	0;21
284	α Sge	7479		274;40	39;00	271;36	39;03	3;04	-0;03
285	β Sge	7488	B	273;20	38;40	271;44	38;29	1;36	0;11
286	τ Aql	7669		277;10	26;50	275;31	27;15	1;39	-0;25
287	β Aql	7602		274;50	27;10	272;57	27;11	1;53	-0;01
288	α Aql	7557		273;50	29;10	271;52	29;24	1;58	-0;14
289	o Aql	7560	I	274;40	30;00	272;35	31;10	2;05	-1;10
290	γ Aql	7525		273;10	31;30	271;26	31;30	1;44	0;00
291	φ Aql	7610		276;00	31;30	274;26	31;45	1;34	-0;15
292	μ Aql	7429		269;40	28;40	267;10	29;02	2;30	-0;22
293	σ Aql	7474		271;10	26;40	268;17	26;44	2;53	-0;04
294	ζ Aql	7235		262;10	36;40	260;17	36;31	1;53	0;09
295	η Aql	7570		273;40	21;40	270;54	21;47	2;46	-0;07
296	θ Aql	7710		278;50	19;10	275;21	18;58	3;29	0;12
297	δ Aql	7377		266;00	25;00	263;56	25;04	2;04	-0;04
298	ι Aql	7447		268;30	20;00	266;18	20;17	2;12	-0;17
299	κ Aql	7446	I	269;40	15;30	265;18	14;37	4;22	0;53
300	λ Aql	7236		261;10	18;10	257;47	17;54	3;23	0;16

No	Name	HR	α_{Ptol}	δ_{Ptol}	α_{-128}	δ_{-128}	m_{Pt}	m_{HR}	Δm
251	Oph	6595	235;03	-18;53	234;45	-17;40	5	4.87	0.13
252	ζ Oph	6175	223;19	-4;29	220;52	-3;48	3	2.56	0.44
253	φ Oph	6147	220;51	-10;31	218;26	-9;29	> 5	4.28	0.42
254	χ Oph	6118	219;11	-12;16	217;07	-11;09	5	4.42	0.58
255	ψ Oph	6104	217;54	-13;26	216;10	-12;35	> 5	4.50	0.20
256	ω Oph	6153	220;01	-15;10	217;47	-14;21	5	4.45	0.55
257	ρ Oph	6112	217;55	-15;59	215;54	-16;01	4	4.27	-0.27
258	Oph	6712	245;22	6;43	244;02	7;02	4	4.64	-0.64
259	Oph	6714	245;37	4;48	243;50	5;37	4	3.97	0.03
260	Oph	6723	245;39	3;26	243;47	3;58	4	4.45	-0.45
261	Oph	6752	246;37	5;18	244;55	5;36	4	4.03	-0.03
262	Oph	6771	248;33	11;03	246;52	11;43	4	3.73	0.27
263	ι Ser	5842	212;51	27;24	211;52	28;15	4	4.52	-0.52
264	ρ Ser	5899	216;04	28;16	214;39	29;09	4	4.76	-0.76
265	γ Ser	5933	216;21	23;45	214;58	24;30	3	3.85	-0.85
266	β Ser	5867	213;41	22;56	212;17	23;54	3	3.67	-0.67
267	κ Ser	5879	214;30	25;53	213;29	26;30	4	4.09	-0.09
268	π Ser	5972	218;28	30;02	217;53	30;31	4	4.83	-0.83
269	δ Ser	5788	211;15	18;28	208;39	19;26	3	3.05	-0.05
270	λ Ser	5868	212;50	14;50	211;15	15;54	4	4.43	-0.43
271	α Ser	5854	211;57	13;56.	210;20	15;01	3	2.65	0.35
272	ε Ser	5892	213;10	12;01	211;42	12;51	3	3.71	-0.71
273	μ Ser	5881	212;38	4;10	210;24	5;06	4	3.53	0.47
274	υ Oph	6129	220;02	-1;55	219;01	-1;15	5	4.63	0.37
275	ν Ser	6446	233;52	-8;50	231;01	-7;57	4	4.33	-0.33
276	ξ Ser	6561	236;39	-11;33	234;37	-11;13	> 4	3.54	0.16
277	o Ser	6581	237;59	-9;27	236;04	-8;56	4	4.26	-0.26
278	ζ Ser	6710	245;21	-1;35	242;24	-0;50	4	4.62	-0.62
279	η Ser	6869	250;10	-1;12	248;08	-0;49	> 4	3.26	0.44
280	$\theta^{1,2}$ Ser	7141	259;36	3;35	257;41	4;01	4	3.87	0.13
281	γ Sge	7635	278;09	15;47	276;03	15;52	4	3.47	0.53
282	ζ Sge	7546	275;21	15;27	273;39	15;59	6	5.00	1.00
283	δ Sge	7536	274;40	15;45	273;09	15;29	5	3.82	1.18
284	α Sge	7479	273;45	15;13	271;17	15;21	5	4.37	0.63
285	β Sge	7488	272;42	14;51	271;24	14;47	5	4.37	0.63
286	τ Aql	7669	276;24	3;09	274;55	3;38	4	5.52	-1.52
287	β Aql	7602	274;18	3;23	272;38	3;30	3	3.71	-0.71
288	α Aql	7557	273;22	5;22	271;38	5;41	> 2	0.77	0.93
289	o Aql	7560	274;04	6;13	272;14	7;28	< 3	5.11	-1.81
290	γ Aql	7525	272;43	7;41	271;14	7;47	3	2.72	0.28
291	φ Aql	7610	275;10	7;46	273;48	8;05	5	5.28	-0.28
292	μ Aql	7429	269;42	4;49	267;31	5;20	5	4.45	0.55
293	σ Aql	7474	271;03	2;49	268;28	3;01	> 5	5.17	-0.47
294	ζ Aql	7235	263;33	13;00	262;00	13;04	3	2.99	0.01
295	η Aql	7570	273;25	-2;08	270;50	-1;56	3	3.90	-0.90
296	θ Aql	7710	278;22	-4;25	275;05	-4;39	3	3.23	-0.23
297	δ Aql	7377	266;22	1;12	264;30	1;27	> 4	3.36	0.34
298	ι Aql	7447	268;35	-3;51	266;31	-3;24	3	4.36	-1.36
299	κ Aql	7446	269;41	-8;21	265;24	-9;02	5	4.95	0.05
300	λ Aql	7236	261;34	-5;25	258;20	-5;20	3	3.44	-0.44

No	Name	HR	C	λ_{Ptol}	β_{Ptol}	λ_{-128}	β_{-128}	$\Delta\lambda$	$\Delta\beta$
301	ϵ Del	7852		287;40	29;10	284;35	29;18	3;05	-0;08
302	ι Del	7883		288;40	29;00	285;49	29;02	2;51	-0;02
303	κ Del	7896		288;40	27;45	285;34	27;45	3;06	0;00
304	β Del	7882		288;30	32;00	286;49	32;10	1;41	-0;10
305	α Del	7906	LB	290;10	33;50	287;53	33;14	2;17	0;36
306	δ Del	7928		291;20	32;00	288;41	32;10	2;39	-0;10
307	$\gamma^{2,1}$ Del	7948		293;10	33;10	289;59	33;00	3;11	0;10
308	η Del	7858	B	287;30	30;15	285;18	30;53	2;12	-0;38
309	ζ Del	7871	1	287;30	31;50	286;16	32;22	1;14	-0;32
310	θ Del	7892		289;00	31;30	286;45	30;49	2;15	0;41
311	α Equ	8131		296;20	20;30	293;36	20;21	2;44	0;09
312	β Equ	8178		298;00	20;40	295;54	21;12	2;06	-0;32
313	γ Equ	8097		296;20	25;30	293;57	25;27	2;23	0;03
314	δ Equ	8123		297;40	25;00	295;00	25;05	2;40	-0;05
315	α And	15		347;50	26;00	344;49	25;44	3;01	0;16
316	γ Peg	39		342;10	12;30	339;38	12;34	2;32	-0;04
317	β Peg	8775		332;10	31;00	329;48	31;07	2;22	-0;07
318	α Peg	8781		326;40	19;40	323;58	19;29	2;42	0;11
319	τ Peg	8880		334;30	25;30	331;35	25;35	2;55	-0;05
320	υ Peg	8905		335;00	25;00	332;24	24;49	2;36	0;11
321	η Peg	8650		329;00	35;00	326;20	35;09	2;40	-0;09
322	o Peg	8641		328;30	34;30	325;32	34;28	2;58	0;02
323	λ Peg	8667		326;10	29;00	323;36	28;51	2;34	0;09
324	μ Peg	8684		327;00	29;30	324;53	29;29	2;07	0;01
325	ζ Peg	8634		318;50	18;00	316;37	17;47	2;13	0;13
326	ξ Peg	8665		320;30	19;00	318;27	18;48	2;03	0;12
327	ρ Peg	8717		321;20	15;00	319;00	14;34	2;20	0;26
328	σ Peg	8697		320;30	16;00	318;31	15;52	1;59	0;08
329	θ Peg	8450	L	309;20	16;30	307;10	16;30	2;10	-0;00
330	ν Peg	8413		308;00	16;00	305;43	15;47	2;17	0;13
331	ϵ Peg	8308		305;20	22;30	302;23	22;15	2;57	0;15
332	π^2 Peg	8454	b	323;20	41;10	320;14	41;03	3;06	0;07
333	ι Peg	8430		317;20	34;15	314;49	34;24	2;31	-0;09
334	κ Peg	8315		312;20	36;50	309;31	36;45	2;49	0;05
335	δ And	165		355;20	24;30	352;18	24;20	3;02	0;10
336	π And	154		356;20	27;00	353;13	27;03	3;07	-0;03
337	ϵ And	163		354;20	23;00	351;39	23;00	2;41	0;00
338	σ And	68		353;40	32;00	351;02	31;30	2;38	0;30
339	θ And	63		354;40	33;30	351;50	33;17	2;50	0;13
340	ρ And	82		355;00	32;20	352;11	32;19	2;49	0;01
341	ι And	8965		349;40	41;00	346;43	40;58	2;57	0;02
342	κ And	8976		350;40	42;00	347;55	41;40	2;45	0;20
343	λ And	8961		352;10	44;00	349;03	43;57	3;07	0;03
344	ζ And	215		354;10	17;30	351;09	17;32	3;01	-0;02
345	η And	271		355;40	15;50	352;54	15;50	2;46	-0;00
346	β And	337		3;50	26;20	0;52	25;54	2;58	0;26
347	μ And	269		1;50	30;00	359;38	29;33	2;12	0;27
348	ν And	226		2;00	32;30	359;43	32;26	2;17	0;04
349	$\gamma^{1,2}$ And	603		16;50	28;00	14;44	27;39	2;06	0;21
350	φ Per	496		17;10	37;20	15;09	36;40	2;01	0;40

No	Name	HR	α_{Ptol}	δ_{Ptol}	α_{-128}	δ_{-128}	m_{Pt}	m_{HR}	Δm
301	ϵ Del	7852	285;28	6;17	282;46	6;14	< 3	4.03	-0.73
302	ι Del	7883	286;21	6;13	283;52	6;05	< 4	5.43	-1.13
303	κ Del	7896	286;31	4;59	283;47	4;46	4	5.05	-1.05
304	β Del	7882	285;49	9;11	284;22	9;17	< 3	3.63	-0.33
305	α Del	7906	286;58	11;11	285;09	10;28	< 3	3.77	-0.47
306	δ Del	7928	288;14	9;31	285;58	9;29	< 3	4.43	-1.13
307	$\gamma^{2,1}$ Del	7948	289;36	10;54	286;57	10;28	< 3	3.52	-0.22
308	η Del	7858	285;11	7;20	283;13	7;52	6	5.38	0.62
309	ζ Del	7871	284;59	8;54	283;53	9;26	6	4.68	1.32
310	θ Del	7892	286;19	8;44	284;29	7;57	6	5.72	0.28
311	α Equ	8131	294;33	-1;06	292;03	-1;33	fnt	3.92	2.08
312	β Equ	8178	296;04	-0;39	294;02	-0;22	fnt	5.16	0.84
313	γ Equ	8097	293;39	3;49	291;32	3;32	fnt	4.69	1.31
314	δ Equ	8123	294;56	3;33	292;33	3;19	fnt	4.49	1.51
315	α And	15	338;15	18;56	335;48	17;37	< 2	2.06	0.24
316	γ Peg	39	338;46	4;25	336;28	3;35	< 2	2.83	-0.53
317	β Peg	8775	322;51	18;01	320;52	17;27	< 2	2.42	-0.12
318	α Peg	8781	322;15	5;39	319;55	4;43	< 2	2.49	-0.19
319	τ Peg	8880	326;59	13;41	324;28	12;52	4	4.60	-0.60
320	υ Peg	8905	327;36	13;24	325;27	12;25	4	4.40	-0.40
321	η Peg	8650	318;39	20;44	316;28	20;10	3	2.94	0.06
322	o Peg	8641	318;27	20;07	316;04	19;17	5	4.79	0.21
323	λ Peg	8667	318;34	14;16	316;27	13;27	4	3.95	0.05
324	μ Peg	8684	319;05	15;00	317;20	14;25	4	3.48	0.52
325	ζ Peg	8634	315;45	1;41	313;48	0;56	3	3.40	-0.40
326	ξ Peg	8665	316;57	3;08	315;10	2;25	4	4.19	-0.19
327	ρ Peg	8717	318;57	-0;25	316;57	-1;27	5	4.90	0.10
328	σ Peg	8697	317;53	0;17	316;06	-0;21	5	5.16	-0.16
329	θ Peg	8450	307;28	-2;18	305;26	-2;43	3	3.53	-0.53
330	ν Peg	8413	306;21	-3;06	304;15	-3;45	4	4.84	-0.84
331	ϵ Peg	8308	302;20	2;36	299;44	1;50	> 3	2.39	0.31
332	π^2 Peg	8454	311;43	24;51	309;23	24;01	> 4	4.29	-0.59
333	ι Peg	8430	309;24	16;45	307;18	16;22	> 4	3.76	-0.06
334	κ Peg	8315	304;31	18;00	302;18	17;24	> 4	4.13	-0.43
335	δ And	165	345;27	20;27	342;55	19;09	3	3.27	-0.27
336	π And	154	345;09	23;05	342;29	21;58	4	4.36	-0.36
337	ϵ And	163	345;15	18;42	342;56	17;42	4	4.37	-0.37
338	σ And	68	340;25	26;32	338;29	25;09	4	4.52	-0.52
339	θ And	63	340;30	28;16	338;17	27;02	4	4.61	-0.61
340	ρ And	82	341;23	27;21	339;04	26;19	5	5.18	-0.18
341	ι And	8965	332;21	33;03	330;06	32;02	4	4.29	-0.29
342	κ And	8976	332;33	34;17	330;39	33;04	4	4.14	-0.14
343	λ And	8961	332;32	36;34	330;13	35;28	4	3.82	0.18
344	ζ And	215	347;30	13;38	344;50	12;32	4	4.06	-0.06
345	η And	271	349;33	12;43	347;06	11;39	4	4.42	-0.42
346	β And	337	352;04	25;28	349;42	23;55	3	2.06	0.94
347	μ And	269	348;27	27;56	346;50	26;41	4	3.87	0.13
348	ν And	226	347;17	30;13	345;25	29;18	4	4.53	-0.53
349	$\gamma^{1,2}$ And	603	2;59	32;12	1;19	31;01	3	1.51	1.49
350	φ Per	496	357;42	40;30	356;26	39;07	< 4	4.07	0.23

No	Name	HR	C	λ_{Ptol}	β_{Ptol}	λ_{-128}	β_{-128}	$\Delta\lambda$	$\Delta\beta$
351	And	464		15;10	35;40	13;00	35;18	2;10	0;22
352	υ And	458		12;20	29;00	9;18	28;58	3;02	0;02
353	τ And	477		12;00	28;00	9;26	27;46	2;34	0;14
354	φ And	335		10;10	35;30	7;00	36;12	3;10	-0;42
355	And	430	I	12;40	34;30	10;41	34;23	1;59	0;07
356	χ And	469	I	14;10	32;30	11;03	31;18	3;07	1;12
357	o And	8762		341;40	44;00	338;27	43;43	3;13	0;17
358	α Tri	544		11;00	16;30	7;23	16;45	3;37	-0;15
359	β Tri	622		16;00	20;40	12;46	20;26	3;14	0;14
360	δ Tri	660		16;20	19;40	13;28	19;30	2;52	0;10
361	γ Tri	664		16;50	19;00	13;59	18;47	2;51	0;13
362	$\gamma^{2,1}$ Ari	546		6;40	7;20	3;36	7;05	3;04	0;15
363	β Ari	553		7;40	8;20	4;23	8;25	3;17	-0;05
364	η Ari	646		11;00	7;40	8;29	7;15	2;31	0;25
365	θ Ari	669		11;30	6;00	9;19	5;34	2;11	0;26
366	ι Ari	563	L	6;30	5;30	3;56	5;19	2;34	0;11
367	ν Ari	773		17;40	6;00	14;35	5;59	3;05	0;01
368	ε Ari	888		21;20	4;50	18;56	3;57	2;24	0;53
369	δ Ari	951		23;50	1;40	21;11	1;38	2;39	0;02
370	ζ Ari	972		25;20	2;30	22;23	2;42	2;57	-0;12
371	τ^2 Ari	1015		27;00	1;50	24;05	1;54	2;55	-0;04
372	ρ^3 Ari	869	B	19;40	1;30	17;13	1;08	2;27	0;22
373	σ Ari	847		18;00	-1;30	15;20	-1;29	2;40	-0;01
374	μ Cet	813	I	15;00	-5;15	12;11	-5;41	2;49	0;26
375	α Ari	617	B	10;40	10;00	8;02	9;55	2;38	0;05
376	Ari	838		21;40	10;10	18;38	10;19	3;02	-0;09
377	Ari	824		21;20	12;40	18;46	12;22	2;34	0;18
378	Ari	801		19;40	11;10	17;23	11;07	2;17	0;03
379	Ari	782		19;10	10;40	16;33	10;43	2;37	-0;03
380	Tau	1066		26;20	-4;00	23;58	-6;09	2;22	0;09
381	Tau	1061		26;00	-7;15	23;29	-7;40	2;31	0;25
382	ξ Tau	1038	L	24;20	-8;30	22;16	-9;00	2;04	0;30
383	o Tau	1030		24;20	-9;15	21;36	-9;32	2;44	0;17
384	Tau	1174	b	29;40	-9;30	27;43	-8;52	1;57	-0;38
385	λ Tau	1239		33;40	-6;00	31;02	-8;12	2;38	0;12
386	μ Tau	1320		36;40	-12;40	33;57	-12;26	2;43	-0;14
387	ν Tau	1251	l	33;00	-14;50	30;17	-14;42	2;43	-0;08
388	Tau	1473		42;10	-8;00	40;06	-9;46	2;04	-0;14
389	Tau	1458	Lb	43;00	-11;00	39;09	-11;59	3;51	-1;01
390	γ Tau	1346		39;00	-5;45	36;08	-5;58	2;52	0;13
391	δ^1 Tau	1373	b	40;20	-4;15	37;13	-4;12	3;07	-0;03
392	θ^1 Tau	1411		40;50	-5;50	38;18	-5;59	2;32	0;09
393	α Tau	1457		42;40	-5;10	40;10	-5;37	2;30	0;27
394	ε Tau	1409	L	41;50	-1;00	38;49	-2;48	3;01	-0;12
395	Tau	1547	L	47;10	-2;00	44;07	-3;53	3;03	-0;07
396	Tau	1656		50;20	-3;00	47;37	-4;30	2;43	-0;30
397	Tau	1658		50;00	-3;30	48;13	-2;44	1;47	-0;46
398	ζ Tau	1910		57;40	-2;30	55;12	-2;28	2;28	-0;02
399	τ Tau	1497	B	45;40	-0;15	42;34	0;27	3;06	-0;42
400	β Tau	1791	b	55;40	5;00	52;59	5;12	2;41	-0;12

No	Name	HR	α_{Ptol}	δ_{Ptol}	α_{-128}	δ_{-128}	m_{Pt}	m_{HR}	Δm
351	And	464	356;59	38;16	355;22	37;04	> 4	3.57	0.13
352	υ And	458	358;19	31;16	355;40	30;01	4	4.09	-0.09
353	τ And	477	358;33	30;14	356;25	29;00	4	4.94	-0.94
354	φ And	335	352;39	36;06	349;34	35;29	5	4.25	0.75
355	And	430	355;28	36;14	353;52	35;21	5	5.27	-0.27
356	χ And	469	358;00	35;05	355;58	32;47	5	4.98	0.02
357	o And	8762	324;27	32;57	322;13	31;44	3	3.62	-0.62
358	α Tri	544	3;12	19;30	359;48	18;16	3	3.41	-0.41
359	β Tri	622	5;55	25;17	3;05	23;46	3	3.00	0.00
360	δ Tri	660	6;42	24;31	4;10	23;12	4	4.87	-0.87
361	γ Tri	664	7;29	24;07	4;58	22;46	3	4.01	-1.01
362	$\gamma^{2,1}$ Ari	546	3;07	9;24	0;26	7;56	< 3	4.00	-0.70
363	β Ari	553	3;38	10;43	0;36	9;27	3	2.64	0.36
364	η Ari	646	6;58	11;27	4;50	10;03	5	5.27	-0.27
365	θ Ari	669	8;07	10;07	6;18	8;51	5	5.62	-0.62
366	ι Ari	563	3;43	7;39	1;27	6;27	5	5.10	-0.10
367	ν Ari	773	13;51	12;34	11;01	11;18	6	5.30	0.70
368	ϵ Ari	888	17;46	12;55	15;53	11;09	5	3.88	1.12
369	δ Ari	951	21;22	10;57	18;55	9;52	4	4.35	-0.35
370	ζ Ari	972	22;28	12;17	19;38	11;19	4	4.89	-0.89
371	τ^2 Ari	1015	24;18	12;17	21;32	11;13	4	5.09	-1.09
372	ρ^3 Ari	869	17;31	9;12	15;24	7;53	5	5.63	-0.63
373	σ Ari	847	17;08	5;48	14;40	4;45	5	5.49	-0.49
374	μ Cet	813	15;50	1;11	13;25	-0;22	> 4	4.27	-0.57
375	α Ari	617	5;42	13;27	3;20	12;18	> 3	2.00	0.70
376	Ari	838	.15;55	17;58	13;02	16;53	4	3.63	0.37
377	Ari	824	14;33	20;08	12;17	18;48	5	4.51	0.49
378	Ari	801	13;37	18;06	11;31	17;07	5	4.66	0.34
379	Ari	782	13;21	17;27	10;55	16;25	5	5.30	-0.30
380	Tau	1066	26;34	4;45	24;27	3;42	4	4.11	-0.11
381	Tau	1061	26;43	3;28	24;33	2;07	4	5.14	-1.14
382	ξ Tau	1038	25;38	1;42	23;56	0;25	4	3.74	0.26
383	o Tau	1030	25;55	1;00	23;31	-0;19	4	3.60	0.40
384	Tau	1174	30;55	2;40	28;54	2;30	5	5.07	-0.07
385	λ Tau	1239	34;07	5;26	31;44	4;17	3	3.47	-0.47
386	μ Tau	1320	38;27	2;00	35;53	1;17	4	4.29	-0.29
387	ν Tau	1251	35;49	-1;13	33;18	-2;04	4	3.91	0.09
388	Tau	1473	42;45	6;14	40;44	5;45	4	4.27	-0.27
389	Tau	1458	44;26	3;37	40;33	3;21	4	4.25	-0.25
390	γ Tau	1346	38;25	9;18	35;47	8;05	< 3	3.65	-0.35
391	δ^1 Tau	1373	39;13	11;08	36;13	10;06	< 3	3.76	-0.46
392	θ^1 Tau	1411	40;12	9;47	37;50	8;46	< 3	3.84	-0.54
393	α Tau	1457	41;45	10;59	39;31	9;42	1	0.85	0.15
394	ϵ Tau	1409	40;16	12;48	37;18	11;57	< 3	3.53	-0.23
395	Tau	1547	45;48	13;25	42;48	12;33	4	5.10	-1.10
396	Tau	1656	49;12	13;19	46;24	12;59	5	5.00	0.00
397	Tau	1658	48;27	14;40	46;29	14;50	5	5.29	-0.29
398	ζ Tau	1910	55;55	17;33	53;25	16;53	3	3.00	0.00
399	τ Tau	1497	43;11	16;34	39;55	16;13	4	4.28	-0.28
400	β Tau	1791	51;55	24;21	49;04	23;45	3	1.65	1.35

No	Name	HR	C	λ_{Ptol}	β_{Ptol}	λ_{-128}	β_{-128}	$\Delta\lambda$	$\Delta\beta$
401	υ Tau	1392		42;00	0;30	38;51	0;52	3;09	-0;22
402	κ^1 Tau	1387		41;40	0;15	38;34	0;24	3;06	-0;09
403	Tau	1256		37;00	0;40	33;49	1;03	3;11	-0;23
404	ω^2 Tau	1329	B	39;00	-1;00	36;30	-0;59	2;30	-0;01
405	Tau	1287	ILB	38;00	5;00	36;06	5;04	1;54	-0;04
406	ψ Tau	1269	I	38;30	7;20	35;46	7;41	2;44	-0;21
407	χ Tau	1369		42;00	3;00	38;32	3;46	3;28	-0;46
408	φ Tau	1348		41;40	5;00	38;21	5;35	3;19	-0;35
409	Tau	1145	I	32;10	4;30	29;59	4;18	2;11	0;12
410	Tau	1156	I	32;30	3;40	30;07	3;44	2;23	-0;04
411	Tau	1178	I	33;40	3;20	30;46	3;42	2;54	-0;22
412		1188	M	33;40	5;00	31;21	5;10	2;19	-0;10
413	Tau	1101	1	25;00	-17;30	22;31	-18;25	2;29	0;55
414	ι Tau	1620		50;00	-0;00	47;10	-1;27	2;50	-0;33
415	Tau	1739	L	54;00	-1;45	50;59	-1;15	3;01	-0;30
416	Tau	1810		56;00	-0;00	52;54	-1;35	3;06	-0;25
417	Tau	1946	I	59;00	-6;20	55;53	-7;07	3;07	0;47
418	Tau	1985	I	59;00	-7;40	57;12	-7;52	1;48	0;12
419	Tau	1875	I	57;00	0;40	54;48	0;26	2;12	0;14
420	Tau	1928	I	59;00	1;00	55;50	2;16	3;10	-1;16
421	Tau	2002	I	61;00	1;20	57;55	0;53	3;05	0;27
422	Tau	2034	I	62;20	3;20	58;56	3;54	3;24	-0;34
423	Tau	2084	I	63;20	1;15	59;58	2;14	3;22	-0;59
424	α Gem	2891	B	83;20	9;30	80;44	9;54	2;36	-0;24
425	β Gem	2990		86;40	6;15	83;59	6;31	2;41	-0;16
426	θ Gem	2540		76;40	10;00	71;31	10;46	5;09	-0;46
427	τ Gem	2697		78;40	7;20	75;52	7;30	2;48	-0;10
428	ι Gem	2821		82;00	5;30	79;25	5;33	2;35	-0;03
429	υ Gem	2905		84;00	4;50	81;45	5;01	2;15	-0;11
430	κ Gem	2985		86;40	2;40	84;05	2;51	2;35	-0;11
431	Gem	2808		81;40	2;40	79;18	2;43	2;22	-0;03
432	Gem	2810	ILB	83;10	0;20	79;34	0;38	3;36	-0;18
433	ϵ Gem	2473		73;00	1;30	70;21	1;47	2;39	-0;17
434	ζ Gem	2650	L	78;10	-2;30	75;25	-2;19	2;45	-0;11
435	δ Gem	2777		81;40	-0;30	78;57	-0;27	2;43	-0;03
436	λ Gem	2763	LB	81;40	-4;00	79;14	-5;53	2;26	-0;07
437	η Gem	2216		66;30	-1;30	63;53	-1;10	2;37	-0;20
438	μ Gem	2286	L	68;30	-1;15	65;41	-1;03	2;49	-0;12
439	ν Gem	2343	b	70;10	-3;30	67;13	-3;20	2;57	-0;10
440	γ Gem	2421		72;00	-7;30	69;30	-7;00	2;30	-0;30
441	ξ Gem	2484		74;40	-10;30	71;41	-10;16	2;59	-0;14
442	Gem	2134		64;10	-0;40	61;22	-0;24	2;48	-0;16
443	κ Aur	2219		66;30	5;50	63;49	5;58	2;41	-0;08
444	Gem	2529		75;10	-2;15	72;22	-1;25	2;48	-0;50
445	Gem	3086	I	88;20	-1;20	87;28	-1;07	0;52	-0;13
446	Gem	3003	I	86;20	-3;20	85;33	-2;52	0;47	-0;28
447	Gem	2938	I	86;00	-4;30	84;05	-4;01	1;55	-0;29
448	ζ^1 Cnc	3208	IL	90;40	-2;40	91;42	-2;27	-1;02	-0;13
449	M44	N2632		100;20	0;20	91;50	21;16	8;30	-20;56
450	η Cnc	3366		97;40	1;15	95;50	1;22	1;50	-0;07

No	Name	HR	α_{Ptol}	δ_{Ptol}	α_{-128}	δ_{-128}	m_{Pt}	m_{HR}	Δm
401	υ Tau	1392	39;19	16;10	36;07	15;27	5	4.28	0.72
402	κ^1 Tau	1387	39;04	15;50	36;00	14;54	5	4.22	0.78
403	Tau	1256	34;21	14;43	31;09	13;55	5	4.36	0.64
404	ω^2 Tau	1329	36;52	13;48	34;27	12;55	6	4.94	1.06
405	Tau	1287	33;49	19;08	31;57	18;29	5	5.41	-0.41
406	ψ Tau	1269	33;28	21;30	30;40	20;49	5	5.23	-0.23
407	χ Tau	1369	38;29	18;33	34;48	18;04	5	5.37	-0.37
408	φ Tau	1348	37;28	20;20	34;00	19;43	5	4.95	0.05
409	Tau	1145	28;16	16;38	26;15	15;37	5	4.30	0.70
410	Tau	1156	28;54	15;59	26;36	15;08	5	4.18	0.82
411	Tau	1178	30;09	16;05	27;15	15;20	5	3.63	1.37
412		1188	29;32	17;38	27;16	16;55	4	5.26	-1.26
413	Tau	1101	29;34	-6;26	27;41	-8;14	4	4.28	-0.28
414	ι Tau	1620	48;02	16;07	45;04	15;46	5	4.64	0.36
415	Tau	1739	52;00	17;24	48;50	17;01	5	4.94	0.06
416	Tau	1810	54;06	17;39	50;52	17;12	5	4.88	0.12
417	Tau	1946	58;09	14;06	55;15	12;33	5	4.86	0.14
418	Tau	1985	58;26	12;48	56;43	12;06	5	6.00	-1.00
419	Tau	1875	54;27	20;28	52;17	19;37	5	5.38	-0.38
420	Tau	1928	56;28	21;15	52;52	21;38	5	5.18	-0.18
421	Tau	2002	58;29	22;01	55;23	20;47	5	4.86	0.14
422	Tau	2034	59;27	24;15	55;42	23;57	5	4.58	0.42
423	Tau	2084	60;58	22;24	57;12	22;33	5	4.82	0.18
424	α Gem	2891	82;08	33;10	79;03	33;16	2	0.84	1.16
425	β Gem	2990	86;10	30;03	83;05	30;05	2	1.14	0.86
426	θ Gem	2540	74;16	33;07	68;11	33;05	4	3.60	0.40
427	τ Gem	2697	76;54	30;40	73;41	30;25	4	4.41	-0.41
428	ι Gem	2821	80;53	29;06	77;58	28;49	4	3.79	0.21
429	υ Gem	2905	83;11	28;32	80;39	28;28	4	4.06	-0.06
430	κ Gem	2985	86;17	26;28	83;24	26;26	4	3.57	0.43
431	Gem	2808	80;43	26;15	78;06	25;59	5	5.03	-0.03
432	Gem	2810	82;31	24;00	78;35	23;56	5	6.02	-1.02
433	ϵ Gem	2473	71;18	24;14	68;24	24;02	3	2.98	0.02
434	ζ Gem	2650	77;20	20;49	74;24	20;36	3	3.79	-0.79
435	δ Gem	2777	80;56	23;05	78;00	22;49	3	3.53	-0.53
436	λ Gem	2763	81;18	17;36	78;46	17;25	3	3.58	-0.58
437	η Gem	2216	64;51	20;17	62;04	20;02	> 4	3.28	0.42
438	μ Gem	2286	66;55	20;52	63;55	20;29	> 4	2.88	0.82
439	ν Gem	2343	69;02	18;54	65;57	18;29	> 4	4.15	-0.45
440	γ Gem	2421	71;29	15;11	68;53	15;12	3	1.93	1.07
441	ξ Gem	2484	74;33	12;31	71;33	12;16	4	3.36	0.64
442	Gem	2134	62;14	20;41	59;16	20;17	4	4.16	-0.16
443	κ Aur	2219	63;26	27;30	60;29	27;01	> 4	4.35	-0.65
444	Gem	2529	74;07	20;46	71;04	21;09	5	5.27	-0.27
445	Gem	3086	88;12	22;30	87;16	22;35	5	5.35	-0.35
446	Gem	3003	86;06	20;28	85;14	20;47	5	4.88	0.12
447	Gem	2938	85;46	19;17	83;45	19;35	5	5.05	-0.05
448	ζ^1 Cnc	3208	90;43	21;11	91;50	21;16	4	5.44	-1.44
449	M44	N2632	101;18	23;46	92;24	44;58	neb	5.44	0.56
450	η Cnc	3366	98;27	24;52	96;26	24;57	< 4	5.33	-1.03

No	Name	HR	C	λ_{Ptol}	β_{Ptol}	λ_{-128}	β_{-128}	$\Delta\lambda$	$\Delta\beta$
451	θ Cnc	3357		98;00	-1;10	96;10	-0;58	1;50	-0;12
452	γ Cnc	3449		100;20	2;40	98;00	3;00	2;20	-0;20
453	δ Cnc	3461		101;20	-0;10	99;06	-0;01	2;14	-0;09
454	α Cnc	3572		106;30	-5;30	104;03	-5;17	2;27	-0;13
455	ι Cnc	3474		98;20	11;50	96;43	10;13	1;37	1;37
456	μ^2 Cnc	3176		92;40	1;00	89;52	1;08	2;48	-0;08
457	β Cnc	3249		97;10	-7;30	94;43	-10;30	2;27	3;00
458	π^2 Cnc	3669	l	109;40	-2;20	107;05	-1;08	2;35	-1;12
459	κ Cnc	3623	l	111;10	-5;40	106;37	-5;46	4;33	0;06
460	ν Cnc	3595	I	104;00	7;15	101;26	7;04	2;34	0;11
461	ξ Cnc	3627	Ib	107;00	4;50	103;36	5;12	3;24	-0;22
462	κ Leo	3731		108;20	10;00	105;41	10;15	2;39	-0;15
463	λ Leo	3773		111;10	7;30	108;16	7;43	2;54	-0;13
464	μ Leo	3905		114;20	12;00	111;54	12;14	2;26	-0;14
465	ϵ Leo	3873		114;10	9;30	111;06	9;33	3;04	-0;03
466	ζ Leo	4031	l	120;10	11;00	117;55	11;42	2;15	-0;42
467	$\gamma^{1,2}$ Leo	4057		122;10	8;30	119;47	8;41	2;23	-0;11
468	η Leo	3975		120;40	4;30	118;18	4;43	2;22	-0;13
469	α Leo	3982		122;30	0;10	120;23	0;22	2;07	-0;12
470	Leo	3980		123;30	-1;50	120;52	-1;31	2;38	-0;19
471	ν Leo	3937		120;00	-0;15	117;46	-0;06	2;14	-0;09
472	ψ Leo	3866		117;20	0;00	113;54	0;11	3;26	-0;11
473	ξ Leo	3782	M	114;10	-3;40	112;06	-3;16	2;04	-0;24
474	o Leo	3852		117;20	-4;10	114;45	-3;53	2;35	-0;17
475	π Leo	3950		122;30	-4;15	119;45	-4;03	2;45	-0;12
476	ρ Leo	4133		129;10	-0;10	126;48	0;02	2;22	-0;12
477	Leo	4127		127;00	4;00	124;53	4;27	2;07	-0;27
478	Leo	4209	l	130;20	5;20	128;05	5;53	2;15	-0;33
479	Leo	4227	l	132;10	2;20	130;05	2;44	2;05	-0;24
480	Leo	4300	M	131;20	12;15	129;13	12;48	2;07	-0;33
481	δ Leo	4357		134;10	13;40	131;33	14;16	2;37	-0;36
482	Leo	4408	ILb	134;20	11;10	135;59	11;40	-1;39	-0;30
483	θ Leo	4359		136;20	9;40	133;48	9;39	2;32	0;01
484	ι Leo	4399		140;20	5;50	137;50	6;02	2;30	-0;12
485	σ Leo	4386		141;40	1;15	139;09	1;40	2;31	-0;25
486	τ Leo	4418		144;40	-0;50	141;55	-0;36	2;45	-0;14
487	υ Leo	4471	B	147;30	-3;10	145;28	-3;06	2;02	-0;04
488	β Leo	4534		144;30	11;50	142;13	12;24	2;17	-0;34
489	LMi	4192	l	126;00	13;20	123;54	13;51	2;06	-0;31
490	Leo	4259		128;10	15;30	125;53	16;24	2;17	-0;54
491	χ Leo	4310		137;30	1;10	135;05	1;22	2;25	-0;12
492	Leo	4294		137;10	-0;30	134;26	-0;16	2;44	-0;14
493	Leo	4291		138;00	-2;40	135;20	-2;35	2;40	-0;05
494	γ Com	4737	IM	144;50	30;00	144;10	28;26	0;40	1;34
495	Com	4667	I	144;20	25;00	143;56	23;27	0;24	1;33
496	Com	4789	I	148;30	25;30	148;46	24;07	-0;16	1;23
497	ν Vir	4517	L	146;20	4;15	144;31	4;39	1;49	-0;24
498	ξ Vir	4515	L	147;00	5;40	143;41	6;04	3;19	-0;24
499	o Vir	4608		150;40	8;00	148;12	8;32	2;28	-0;32
500	π Vir	4589		150;30	5;30	147;56	6;09	2;34	-0;39

No	Name	HR	α_{Ptol}	δ_{Ptol}	α_{-128}	δ_{-128}	m_{Pt}	m_{HR}	Δm
451	θ Cnc	3357	98;40	22;26	96;41	22;37	< 4	5.35	-1.05
452	γ Cnc	3449	101;31	26;06	98;55	26;28	> 4	4.66	-0.96
453	δ Cnc	3461	102;21	23;11	99;56	23;23	> 4	3.94	-0.24
454	α Cnc	3572	107;14	17;21	104;42	17;43	4	4.25	-0.25
455	ι Cnc	3474	100;01	35;23	97;58	33;45	4	5.82	-1.82
456	μ^2 Cnc	3176	92;56	24;49	89;51	24;51	5	5.30	-0.30
457	β Cnc	3249	97;24	16;10	94;46	13;09	> 4	3.52	0.18
458	π^2 Cnc	3669	110;59	20;04	108;24	21;29	< 4	5.34	-1.04
459	κ Cnc	3623	112;01	16;33	107;18	16;57	< 4	5.24	-0.94
460	ν Cnc	3595	106;08	30;18	103;10	30;16	5	5.45	-0.45
461	ξ Cnc	3627	109;11	27;32	105;25	28;12	5	5.14	-0.14
462	κ Leo	3731	111;32	32;28	108;29	32;58	4	4.46	-0.46
463	λ Leo	3773	114;18	29;33	111;02	30;06	4	4.31	-0.31
464	μ Leo	3905	118;52	33;24	116;05	33;59	3	3.88	-0.88
465	ϵ Leo	3873	118;06	30;59	114;36	31;28	> 3	2.98	-0.28
466	ζ Leo	4031	125;12	31;11	122;50	32;16	3	3.44	-0.44
467	$\gamma^{1,2}$ Leo	4057	126;43	28;17	124;07	28;54	2	1.86	0.14
468	η Leo	3975	124;03	24;44	121;31	25;21	3	3.52	-0.52
469	α Leo	3982	124;54	20;06	122;43	20;40	1	1.35	-0.35
470	Leo	3980	125;26	17;55	122;47	18;43	4	4.37	-0.37
471	ν Leo	3937	122;12	20;15	119;52	20;46	5	5.26	-0.26
472	ψ Leo	3866	119;28	21;03	115;51	21;45	5	5.35	-0.35
473	ξ Leo	3782	115;27	18;02	113;22	18;39	5	4.97	0.03
474	o Leo	3852	118;36	16;58	115;59	17;37	4	3.52	0.48
475	π Leo	3950	123;50	15;48	121;05	16;29	4	4.70	-0.70
476	ρ Leo	4133	131;39	18;07	129;16	18;49	4	3.85	0.15
477	Leo	4127	130;36	22;42	128;28	23;35	6	5.46	0.54
478	Leo	4209	134;28	23;04	132;15	24;08	6	5.48	0.52
479	Leo	4227	135;36	19;38	133;23	20;33	6	5.25	0.75
480	Leo	4300	137;48	29;23	135;39	30;26	6	4.42	1.58
481	δ Leo	4357	141;19	29;52	138;41	31;10	< 2	2.56	-0.26
482	Leo	4408	140;35	27;27	142;27	27;19	5	5.57	-0.57
483	θ Leo	4359	142;07	25;23	139;26	26;05	3	3.34	-0.34
484	ι Leo	4399	144;50	20;28	142;21	21;24	3	3.94	-0.94
485	σ Leo	4386	144;33	15;42	142;11	16;50	4	4.05	-0.05
486	τ Leo	4418	146;45	12;44	144;08	13;48	4	4.95	-0.95
487	υ Leo	4471	148;39	9;35	146;43	10;16	5	4.30	0.70
488	β Leo	4534	151;16	24;40	149;07	25;56	< 1	2.14	-0.84
489	LMi	4192	132;23	31;57	130;10	32;55	5	5.08	-0.08
490	Leo	4259	135;31	33;25	133;14	34;50	5	4.50	0.50
491	χ Leo	4310	140;25	16;58	138;02	17;48	< 4	4.63	-0.33
492	Leo	4294	139;33	15;29	136;53	16;26	5	4.99	0.01
493	Leo	4291	139;40	13;10	137;04	13;57	5	4.84	0.16
494	γ Com	4737	160;16	41;13	158;35	40;01	fnt	4.36	1.64
495	Com	4667	157;00	36;53	155;44	35;34	fnt	4.95	1.05
496	Com	4789	161;31	35;46	160;59	34;22	fnt	4.81	1.19
497	ν Vir	4517	150;11	16;56	148;31	17;53	5	4.03	0.97
498	ξ Vir	4515	151;22	18;02	148;12	19;30	5	4.85	0.15
499	o Vir	4608	155;50	18;52	153;35	20;13	5	4.12	0.88
500	π Vir	4589	154;23	16;44	152;25	18;05	5	4.66	0.34

No	Name	HR	C	λ_{Ptol}	β_{Ptol}	λ_{-128}	β_{-128}	$\Delta\lambda$	$\Delta\beta$
501	β Vir	4540	B	149;00	0;10	147;06	0;39	1;54	-0;29
502	η Vir	4689	B	158;15	1;10	155;16	1;25	2;59	-0;15
503	γ Vir	4825		163;10	2;50	160;51	2;58	2;19	-0;08
504	Vir	4925	IB	167;30	2;50	165;41	2;55	1;49	-0;05
505	θ Vir	4963		171;00	1;40	168;39	1;51	2;21	-0;11
506	δ Vir	4910	l	164;20	8;30	162;04	8;48	2;16	-0;18
507	ρ Vir	4828	B	158;10	13;30	155;47	13;36	2;23	-0;06
508	Vir	4847		160;10	11;40	157;51	11;37	2;19	0;03
509	ε Vir	4932	B	162;10	15;10	160;26	16;18	1;44	-1;08
510	α Vir	5056		176;40	-0;00	174;17	-1;55	2;23	-0;05
511	ζ Vir	5107		174;50	8;40	172;40	8;47	2;10	-0;07
512	Vir	5095	I	176;20	3;20	174;02	3;17	2;18	0;03
513	Vir	5100	LB	177;15	0;10	175;40	-0;17	1;35	0;27
514	Vir	5150	I	180;00	1;30	177;10	1;50	2;50	-0;20
515	Vir	5064	Ib	178;00	-1;00	175;15	-3;11	2;45	0;11
516	Vir	5173	b	181;40	-1;30	179;28	-1;15	2;12	-0;15
517	Vir	5232	I	178;00	8;30	177;36	9;46	0;24	-1;16
518	ι Vir	5338	IL	186;40	7;30	184;06	7;36	2;34	-0;06
519	κ Vir	5315		187;20	2;40	184;55	3;00	2;25	-0;20
520	φ Vir	5409		188;20	11;40	185;54	11;57	2;26	-0;17
521	λ Vir	5359		190;00	0;30	187;23	0;39	2;37	-0;09
522	μ Vir	5487		192;40	9;50	190;23	10;01	2;17	-0;11
523	χ Vir	4813		164;40	-3;30	162;37	-3;23	2;03	-0;07
524	ψ Vir	4902		169;00	-3;30	166;38	-3;20	2;22	-0;10
525	Vir	4955	Lb	172;15	-3;20	170;10	-3;10	2;05	-0;10
526	Vir	4981	B	177;10	-7;10	173;06	-7;41	4;04	0;31
527	Vir	5019	I	178;10	-8;20	175;49	-8;21	2;21	0;01
528	Vir	5196		185;00	-7;50	182;27	-6;11	2;33	-1;39
529	α² Lib	5531		198;00	0;40	195;33	0;36	2;27	0;04
530	μ Lib	5523		197;00	2;30	194;36	2;15	2;24	0;15
531	β Lib	5685		202;10	8;50	199;48	8;45	2;22	0;05
532	δ Lib	5586		197;40	8;30	195;42	8;28	1;58	0;02
533	ι¹ Lib	5652		204;00	-1;40	201;26	-1;36	2;34	-0;04
534	ν Lib	5622		201;20	1;15	199;12	1;26	2;08	-0;11
535	γ Lib	5787		207;50	4;45	205;30	4;37	2;20	0;08
536	θ Lib	5908	b	213;00	3;30	210;14	3;38	2;46	-0;08
537	Lib	5777		206;10	9;00	203;52	9;13	2;18	-0;13
538	Lib	5941		213;40	6;40	210;48	6;21	2;52	0;19
539	ξ Sco	5978		214;20	9;15	211;44	9;31	2;36	-0;16
540	λ Lib	5902		213;30	0;30	210;53	0;21	2;37	0;09
541	κ Lib	5838	IB	210;40	0;20	208;11	0;17	2;29	0;03
542		5810		211;10	-1;30	207;36	-1;15	3;34	-0;15
543	σ Lib	5603		203;00	-7;30	201;09	-7;23	1;51	-0;07
544	υ Lib	5794	B	211;10	-8;30	209;03	-8;16	2;07	-0;14
545	τ Lib	5812		212;00	-9;40	209;48	-9;45	2;12	0;05
546	β¹,² Sco	5984		216;20	1;20	213;36	1;17	2;44	0;03
547	δ Sco	5953		215;40	-1;40	212;59	-1;43	2;41	0;03
548	π Sco	5944		215;40	-3;00	213;22	-5;12	2;18	0;12
549	ρ Sco	5928		216;00	-7;50	213;35	-8;20	2;25	0;30
550	ν Sco	6027		217;00	1;40	215;03	1;55	1;57	-0;15

No	Name	HR	α_{Ptol}	δ_{Ptol}	α_{-128}	δ_{-128}	m_{Pt}	m_{HR}	Δm
501	β Vir	4540	151;16	12;11	149;36	13;14	3	3.61	-0.61
502	η Vir	4689	160;24	9;42	157;40	11;00	3	3.89	-0.89
503	γ Vir	4825	165;39	9;20	163;31	10;19	3	2.90	0.10
504	Vir	4925	169;40	7;37	168;00	8;23	5	5.99	-0.99
505	θ Vir	4963	172;26	5;09	170;20	6;15	4	4.38	-0.38
506	δ Vir	4910	169;02	14;04	167;01	15;13	3	3.38	-0.38
507	ρ Vir	4828	165;20	21;05	163;02	22;04	5	4.88	0.12
508	Vir	4847	166;27	18;37	164;10	19;27	6	5.22	0.78
509	ϵ Vir	4932	169;51	21;02	168;39	22;44	> 3	2.83	-0.13
510	α Vir	5056	176;09	-0;29	174;00	0;33	1	0.98	0.02
511	ζ Vir	5107	178;48	10;01	176;51	10;59	3	3.37	-0.37
512	Vir	5095	178;00	4;32	175;51	5;24	5	4.69	0.31
513	Vir	5100	177;33	1;16	175;56	1;29	6	5.21	0.79
514	Vir	5150	180;36	1;22	178;09	2;49	< 4	5.01	-0.71
515	Vir	5064	176;57	-1;56	174;23	-1;01	5	5.25	-0.25
516	Vir	5173	180;55	-2;03	179;00	-0;55	5	5.51	-0.51
517	Vir	5232	181;38	8;35	181;46	9;54	5	5.15	-0.15
518	ι Vir	5338	189;07	4;11	186;48	5;19	4	4.08	-0.08
519	κ Vir	5315	187;47	-0;31	185;42	0;46	4	4.19	-0.19
520	φ Vir	5409	192;19	7;20	190;13	8;35	4	4.81	-0.81
521	λ Vir	5359	189;22	-3;34	187;01	-2;22	4	4.52	-0.52
522	μ Vir	5487	195;30	3;57	193;30	5;02	4	3.88	0.12
523	χ Vir	4813	164;33	2;55	162;42	3;47	5	4.66	0.34
524	ψ Vir	4902	168;32	1;13	166;25	2;16	5	4.79	0.21
525	Vir	4955	171;34	0;04	169;43	1;02	5	5.19	-0.19
526	Vir	4981	174;30	-5;25	170;37	-4;16	6	5.04	0.96
527	Vir	5019	174;56	-6;53	172;48	-5;58	5	4.74	0.26
528	Vir	5196	181;23	-9;11	179;44	-6;39	6	4.97	1.03
529	α^2 Lib	5531	196;49	-6;34	194;31	-5;38	2	2.75	-0.75
530	μ Lib	5523	196;36	-4;29	194;18	-3;45	5	5.31	-0.31
531	β Lib	5685	203;46	-0;35	201;35	0;15	2	2.61	-0.61
532	δ Lib	5586	199;32	0;48	197;43	1;33	5	4.92	0.08
533	ι^1 Lib	5652	201;31	-11;01	199;10	-9;56	4	4.54	-0.54
534	ν Lib	5622	200;08	-7;18	198;14	-6;17	4	5.20	-1.20
535	γ Lib	5787	207;31	-6;27	205;18	-5;41	4	3.91	0.09
536	θ Lib	5908	211;56	-9;26	209;23	-8;17	< 4	4.15	0.15
537	Lib	5777	207;30	-1;54	205;28	-0;48	5	4.62	0.38
538	Lib	5941	213;40	-6;41	210;53	-5;56	< 4	4.88	-0.58
539	ξ Sco	5978	215;10	-4;29	212;50	-3;17	< 4	4.02	0.28
540	λ Lib	5902	211;22	-12;26	208;50	-11;35	6	5.03	0.97
541	κ Lib	5838	208;17	-11;28	206;14	-10;41	5	4.74	0.26
542		5810	208;24	-13;29	205;06	-11;54	4	5.84	-1.84
543	σ Lib	5603	198;17	-16;01	196;37	-15;10	3	3.29	-0.29
544	υ Lib	5794	205;46	-20;00	203;50	-18;57	4	3.58	0.42
545	τ Lib	5812	206;07	-21;24	203;58	-20;37	4	3.66	0.34
546	$\beta^{1,2}$ Sco	5984	214;23	-12;36	211;46	-11;40	3	1.87	1.13
547	δ Sco	5953	212;42	-15;12	210;07	-14;16	3	2.32	0.68
548	π Sco	5944	211;30	-18;20	209;12	-17;40	3	2.89	0.11
549	ρ Sco	5928	210;47	-21;06	208;14	-20;39	3	3.88	-0.88
550	ν Sco	6027	215;09	-12;31	213;23	-11;34	4	4.01	-0.01

No	Name	HR	C	λ_{Ptol}	β_{Ptol}	λ_{-128}	β_{-128}	$\Delta\lambda$	$\Delta\beta$
551	ω^1 Sco	5993	I	216;20	0;30	214;05	0;30	2;15	0;00
552	σ Sco	6084	b	220;40	-3;45	218;13	-3;45	2;27	0;00
553	α Sco	6134		222;40	-2;00	220;11	-4;17	2;29	0;17
554	τ Sco	6165		224;30	-5;30	221;53	-5;50	2;37	0;20
555	Sco	6028		219;20	-6;30	216;41	-6;24	2;39	-0;06
556		6070		220;40	-6;40	218;07	-6;48	2;33	0;08
557	ϵ Sco	6241		228;30	-9;00	226;07	-11;16	2;23	0;16
558	μ^1 Sco	6247		228;50	-13;00	226;36	-15;07	2;14	0;07
559	ζ^1 Sco	6262	ILB	230;00	-18;40	227;33	-19;22	2;27	0;42
560	ζ^2 Sco	6271	ILB	230;10	-16;00	227;44	-19;13	2;26	1;13
561	η Sco	6380		233;10	-19;30	231;09	-19;44	2;01	0;14
562	θ Sco	6553		238;10	-18;50	236;01	-19;21	2;09	0;31
563	ι^1 Sco	6615		240;30	-16;40	237;56	-16;25	2;34	-0;15
564	κ Sco	6580	b	239;00	-15;10	236;53	-15;20	2;07	0;10
565	λ Sco	6527		237;30	-13;20	235;00	-13;29	2;30	0;09
566	υ Sco	6508		237;00	-13;30	234;26	-13;42	2;34	0;12
567	M7	N6475		241;10	-13;15	239;10	-11;05	2;00	-2;10
568	Oph	6492		235;30	-6;10	233;17	-6;16	2;13	0;06
569	Sgr	6616	LB	239;30	-4;10	237;39	-4;08	1;51	-0;02
570	γ^2 Sgr	6746		244;30	-6;30	241;42	-6;35	2;48	0;05
571	δ Sgr	6859		247;40	-6;30	244;58	-6;10	2;42	-0;20
572	ϵ Sgr	6879		248;00	-10;50	245;31	-10;41	2;29	-0;09
573	λ Sgr	6913		249;00	-1;30	246;46	-1;44	2;14	0;14
574	μ Sgr	6812		246;40	2;50	243;37	2;38	3;03	0;12
575	σ Sgr	7121		255;20	-3;10	252;47	-3;08	2;33	-0;02
576	φ Sgr	7039	B	253;00	-3;30	250;34	-3;40	2;26	0;10
577	v^1 Sgr	7116		255;10	0;45	252;53	0;24	2;17	0;21
578	ξ^2 Sgr	7150		255;40	2;10	253;51	1;57	1;49	0;13
579	o Sgr	7217		257;40	1;30	255;22	1;11	2;18	0;19
580	π Sgr	7264		259;10	2;00	256;40	1;44	2;30	0;16
581	Sgr	7304		261;20	2;50	258;46	3;33	2;34	-0;43
582	ρ^1 Sgr	7340	l	262;20	4;30	259;53	4;29	2;27	0;01
583	υ Sgr	7342		262;50	6;30	260;09	6;23	2;41	0;07
584	Sgr	7489	L	265;40	5;30	265;03	5;26	0;37	0;04
585	Sgr	7614		269;30	5;50	268;52	5;24	0;38	0;26
586	Sgr	7561	I	267;40	2;00	266;50	2;09	0;50	-0;09
587	χ^1 Sgr	7362	IL	262;40	-1;50	259;43	-2;11	2;57	0;21
588	Sgr	7431	I	264;50	-2;50	262;07	-2;47	2;43	-0;03
589	ψ Sgr	7292		260;00	-2;30	257;26	-2;38	2;34	0;08
590	τ Sgr	7234		257;40	-4;30	255;17	-4;40	2;23	0;10
591	ζ Sgr	7194		256;20	-6;45	254;03	-6;54	2;17	0;09
592	β^1 Sgr	7337	Im	257;40	-21;00	256;09	-21;51	1;31	-1;09
593	α Sgr	7348	m	257;00	-16;00	257;01	-18;02	-0;01	0;02
594	η Sgr	6832		246;40	-11;00	244;07	-13;00	2;33	-0;00
595	θ^1 Sgr	7623	I	267;20	-13;30	265;15	-14;06	2;05	0;36
596	ι Sgr	7581	LB	266;50	-20;10	262;55	-20;25	3;55	0;15
597	ω Sgr	7597	L	267;40	-4;50	266;07	-5;11	1;33	0;21
598	Sgr	7618		268;50	-4;50	266;56	-5;13	1;54	0;23
599	Sgr	7604		268;50	-5;50	266;19	-6;03	2;31	0;13
600	Sgr	7650		269;40	-6;30	267;26	-6;51	2;14	0;21

No	Name	HR	α_{Ptol}	δ_{Ptol}	α_{-128}	δ_{-128}	m_{Pt}	m_{HR}	Δm
551	ω^1 Sco	5993	214;06	-13;23	211;57	-12;34	4	3.96	0.04
552	σ Sco	6084	216;54	-18;50	214;30	-17;58	3	2.89	0.11
553	α Sco	6134	218;49	-19;42	216;16	-19;06	2	0.96	1.04
554	τ Sco	6165	220;10	-21;42	217;27	-21;07	3	2.82	0.18
555	Sco	6028	214;36	-20;59	212;02	-19;56	5	4.59	0.41
556		6070	215;53	-21;35	213;19	-20;47	5	4.78	0.22
557	ϵ Sco	6241	222;28	-28;09	219;54	-27;36	3	2.29	0.71
558	μ^1 Sco	6247	221;24	-32;03	219;00	-31;24	3	3.08	-0.08
559	ζ^1 Sco	6262	221;16	-35;53	218;22	-35;43	4	4.73	-0.73
560	ζ^2 Sco	6271	221;43	-35;18	218;37	-35;37	4	3.62	0.38
561	η Sco	6380	224;29	-37;37	222;11	-37;09	3	3.33	-0.33
562	θ Sco	6553	230;28	-38;21	227;51	-38;11	3	1.87	1.13
563	ι^1 Sco	6615	233;53	-36;50	231;05	-35;52	3	3.03	-0.03
564	κ Sco	6580	232;38	-35;01	230;14	-34;33	3	2.41	0.59
565	λ Sco	6527	231;30	-32;52	228;44	-32;16	3	1.63	1.37
566	υ Sco	6508	230;54	-32;54	228;02	-32;20	4	2.69	1.31
567	M7	N6475	235;40	-33;40	234;05	-31;00	neb	3.20	2.80
568	Oph	6492	231;25	-25;26	229;05	-24;51	> 5	4.29	0.41
569	Sgr	6616	236;13	-24;27	234;18	-23;53	5	4.54	0.46
570	γ^2 Sgr	6746	241;05	-27;47	238;02	-27;11	3	2.99	0.01
571	δ Sgr	6859	244;35	-28;22	241;42	-27;26	3	2.70	0.30
572	ϵ Sgr	6879	244;05	-32;42	241;18	-31;58	3	1.85	1.15
573	λ Sgr	6913	246;59	-23;40	244;33	-23;24	3	2.81	0.19
574	μ Sgr	6812	245;16	-19;00	242;05	-18;32	4	3.86	0.14
575	σ Sgr	7121	253;38	-26;10	250;52	-25;42	3	2.02	0.98
576	φ Sgr	7039	251;01	-26;13	248;20	-25;55	4	3.17	0.83
577	υ^1 Sgr	7116	253;57	-22;16	251;28	-22;13	neb	4.83	1.17
578	ξ^2 Sgr	7150	254;39	-20;55	252;42	-20;47	4	3.51	0.49
579	o Sgr	7217	256;42	-21;46	254;13	-21;44	4	3.77	0.23
580	π Sgr	7264	258;22	-21;24	255;41	-21;19	4	2.89	1.11
581	Sgr	7304	260;44	-20;44	258;05	-19;43	5	4.96	0.04
582	ρ^1 Sgr	7340	261;54	-19;08	259;20	-18;52	4	3.93	0.07
583	υ Sgr	7342	262;33	-17;10	259;46	-17;00	4	4.61	-0.61
584	Sgr	7489	265;27	-18;17	264;48	-18;12	6	5.06	0.94
585	Sgr	7614	269;29	-18;01	268;49	-18;19	5	5.02	-0.02
586	Sgr	7561	267;29	-21;50	266;36	-21;32	6	5.92	0.08
587	χ^1 Sgr	7362	261;53	-25;28	258;36	-25;29	5	5.03	-0.03
588	Sgr	7431	264;14	-26;35	261;13	-26;16	4	5.65	-1.65
589	ψ Sgr	7292	258;53	-25;57	256;02	-25;44	5	4.85	0.15
590	τ Sgr	7234	256;05	-27;45	253;25	-27;32	> 4	3.32	0.38
591	ζ Sgr	7194	254;18	-29;51	251;43	-29;36	3	2.60	0.40
592	β^1 Sgr	7337	253;31	-46;08	251;48	-44;42	2	4.01	-2.01
593	α Sgr	7348	253;30	-41;06	253;33	-41;00	< 2	3.97	-1.67
594	η Sgr	6832	242;03	-34;35	239;09	-33;57	3	3.11	-0.11
595	θ^1 Sgr	7623	266;44	-37;19	264;10	-37;44	3	4.37	-1.37
596	ι Sgr	7581	265;52	-43;58	260;46	-43;55	3	4.13	-1.13
597	ω Sgr	7597	267;21	-28;40	265;35	-28;51	5	4.70	0.30
598	Sgr	7618	268;40	-28;41	266;31	-28;54	5	4.83	0.17
599	Sgr	7604	268;40	-29;41	265;47	-29;43	5	4.52	0.48
600	Sgr	7650	269;37	-30;21	267;03	-30;33	5	4.58	0.42

No	Name	HR	C	λ_{Ptol}	β_{Ptol}	λ_{-128}	β_{-128}	$\Delta\lambda$	$\Delta\beta$
601	α^1 Cap	7747		277;20	7;20	274;11	7;14	3;09	0;06
602	v Cap	7773		277;40	6;40	274;51	6;50	2;49	-0;10
603	β Cap	7776		277;20	5;00	274;27	4;50	2;53	0;10
604	ξ^2 Cap	7715	L	275;00	8;00	272;50	7;33	2;10	0;27
605	o Cap	7830		279;00	0;45	275;38	0;41	3;22	0;04
606	π Cap	7814		278;40	1;45	275;07	1;09	3;33	0;36
607	ρ Cap	7822		278;50	1;30	275;36	1;27	3;14	0;03
608	σ Cap	7761		276;10	0;40	273;05	0;42	3;05	-0;02
609	τ Cap	7889		281;40	3;50	278;43	3;36	2;57	0;14
610	v Cap	7900	B	281;50	0;50	278;05	0;28	3;45	0;22
611	ω Cap	7980		281;40	-8;40	278;21	-8;44	3;19	0;04
612	ψ Cap	7936		280;50	-6;30	277;36	-6;43	3;14	0;13
613	Cap	8080		286;40	-7;40	282;15	-7;52	4;25	0;12
614	ζ Cap	8204		290;10	-6;50	287;19	-6;48	2;51	-0;02
615	Cap	8213	1	290;20	-4;00	287;54	-6;20	2;26	0;20
616	φ Cap	8127		288;30	-4;15	285;25	-4;19	3;05	0;04
617	χ Cap	8087		286;40	-2;00	283;41	-4;18	2;59	0;18
618	η Cap	8060		286;40	-2;50	283;10	-2;46	3;30	-0;04
619	θ Cap	8075		286;40	0;00	284;13	-0;20	2;27	0;20
620	ι Cap	8167		291;00	-0;50	288;04	-1;10	2;56	0;20
621	ϵ Cap	8260		293;20	-4;45	290;35	-4;47	2;45	0;02
622	κ Cap	8288		295;00	-4;30	291;57	-4;37	3;03	0;07
623	γ Cap	8278		294;50	-2;10	292;06	-2;19	2;44	0;09
624	δ Cap	8322		296;20	-0;00	293;51	-2;13	2;29	0;13
625	Cap	8283		296;50	0;20	293;31	0;07	3;19	0;13
626	μ Cap	8351		298;40	0;00	296;04	-0;28	2;36	0;28
627	λ Cap	8319		297;40	2;50	295;25	2;07	2;15	0;43
628	Cap	8311		298;40	4;20	295;50	4;23	2;50	-0;03
629	Aqr	8277		300;20	15;45	298;28	15;32	1;52	0;13
630	α Aqr	8414		306;20	11;00	303;48	10;48	2;32	0;12
631	o Aqr	8402		305;10	9;40	302;33	9;19	2;37	0;21
632	β Aqr	8232	b	296;30	8;50	293;50	8;48	2;40	0;02
633	ξ Aqr	8264		297;20	6;15	294;30	6;10	2;50	0;05
634	v Aqr	8093		287;40	5;30	286;47	5;00	0;53	0;30
635	μ Aqr	7990		286;10	8;00	283;29	8;29	2;41	-0;29
636	ϵ Aqr	7950		284;40	8;40	282;09	8;19	2;31	0;21
637	γ Aqr	8518		309;30	8;45	307;05	8;23	2;25	0;22
638	π Aqr	8539		311;40	10;45	309;03	10;35	2;37	0;10
639	ζ^2 Aqr	8559		312;00	9;00	309;14	8;59	2;46	0;01
640	η Aqr	8597		313;20	8;30	310;49	8;18	2;31	0;12
641	θ Aqr	8499		306;10	3;00	303;38	2;53	2;32	0;07
642	ρ Aqr	8512	B	307;00	3;10	304;27	2;30	2;33	0;40
643	σ Aqr	8573		308;40	-0;50	305;48	-1;05	2;52	0;15
644	ι Aqr	8418		301;40	-1;40	299;07	-1;53	2;33	0;13
645	Aqr	8452	B	303;10	0;15	300;53	-0;08	2;17	0;23
646	δ Aqr	8709	b	311;40	-7;30	309;16	-8;05	2;24	0;35
647	τ^2 Aqr	8679		311;20	-3;00	309;00	-5;32	2;20	0;32
648	Aqr	8544		304;40	-5;40	302;28	-6;18	2;12	0;38
649	Aqr	8670		308;20	-8;00	306;16	-10;50	2;04	0;50
650	Aqr	8649		307;50	-7;00	305;37	-9;49	2;13	0;49

No	Name	HR	α_{Ptol}	δ_{Ptol}	α_{-128}	δ_{-128}	m_{Pt}	m_{HR}	Δm
601	α^1 Cap	7747	277;35	-16;19	274;20	-16;25	3	4.24	-1.24
602	ν Cap	7773	277;58	-16;58	275;02	-16;48	6	4.76	1.24
603	β Cap	7776	277;43	-18;39	274;41	-18;49	3	3.08	-0.08
604	ξ^2 Cap	7715	275;09	-15;46	272;56	-16;08	6	5.85	0.15
605	o Cap	7830	279;46	-22;47	276;07	-22;55	6	5.94	0.06
606	π Cap	7814	279;20	-21;49	275;33	-22;28	6	5.25	0.75
607	ρ Cap	7822	279;32	-22;03	276;02	-22;09	6	4.78	1.22
608	σ Cap	7761	276;42	-23;02	273;21	-22;59	5	5.28	-0.28
609	τ Cap	7889	282;22	-19;31	279;15	-19;51	6	5.22	0.78
610	υ Cap	7900	282;49	-22;29	278;48	-23;00	5	5.10	-0.10
611	ω Cap	7980	283;38	-31;57	279;46	-32;10	4	4.11	-0.11
612	ψ Cap	7936	282;26	-29;53	278;45	-30;12	4	4.14	-0.14
613	Cap	8080	289;14	-30;24	284;11	-30;59	4	4.50	-0.50
614	ζ Cap	8204	293;03	-29;03	289;49	-29;20	4	3.74	0.26
615	Cap	8213	293;05	-28;12	290;24	-28;47	5	4.51	0.49
616	φ Cap	8127	290;45	-26;45	287;19	-27;06	5	5.24	-0.24
617	χ Cap	8087	288;41	-26;45	285;23	-27;17	5	6.02	-1.02
618	η Cap	8060	288;31	-25;36	284;38	-25;48	5	4.84	0.16
619	θ Cap	8075	288;07	-22;47	285;30	-23;17	4	4.07	-0.07
620	ι Cap	8167	292;55	-23;00	289;47	-23;38	4	4.28	-0.28
621	ϵ Cap	8260	296;10	-26;28	293;07	-26;51	4	4.68	-0.68
622	κ Cap	8288	297;56	-25;55	294;36	-26;28	4	4.73	-0.73
623	γ Cap	8278	297;16	-23;39	294;20	-24;11	3	3.68	-0.68
624	δ Cap	8322	298;50	-23;13	296;12	-23;46	3	2.87	0.13
625	Cap	8283	298;53	-20;49	295;24	-21;31	4	5.18	-1.18
626	μ Cap	8351	300;52	-20;47	298;13	-21;38	5	5.08	-0.08
627	λ Cap	8319	299;13	-18;12	297;01	-19;14	5	5.58	-0.58
628	Cap	8311	299;56	-16;32	297;01	-16;55	5	5.09	-0.09
629	Aqr	8277	299;12	-5;02	297;28	-5;29	5	5.10	-0.10
630	α Aqr	8414	306;00	-8;21	303;36	-9;01	3	2.96	0.04
631	o Aqr	8402	305;12	-9;55	302;43	-10;45	5	4.69	0.31
632	β Aqr	8232	296;51	-12;32	294;11	-12;55	3	2.91	0.09
633	ξ Aqr	8264	298;11	-14;55	295;19	-15;24	5	4.69	0.31
634	ν Aqr	8093	288;26	-17;12	287;34	-17;42	3	4.51	-1.51
635	μ Aqr	7990	286;35	-14;55	283;47	-14;36	4	4.73	-0.73
636	ϵ Aqr	7950	284;59	-14;25	282;27	-14;52	3	3.77	-0.77
637	γ Aqr	8518	309;38	-9;45	307;22	-10;37	3	3.84	-0.84
638	π Aqr	8539	311;11	-7;15	308;42	-7;59	3	4.66	-1.66
639	ζ^2 Aqr	8559	311;59	-8;51	309;18	-9;30	3	4.42	-1.42
640	η Aqr	8597	313;24	-8;57	311;01	-9;44	3	4.02	-1.02
641	θ Aqr	8499	307;51	-16;09	305;17	-16;46	4	4.16	-0.16
642	ρ Aqr	8512	308;38	-15;47	306;12	-16;57	5	5.37	-0.37
643	σ Aqr	8573	311;25	-19;12	308;31	-20;06	4	4.82	-0.82
644	ι Aqr	8418	304;24	-21;45	301;44	-22;25	4	4.27	-0.27
645	Aqr	8452	305;29	-19;32	303;11	-20;20	6	5.46	0.54
646	δ Aqr	8709	316;32	-24;46	314;10	-25;55	3	3.27	-0.27
647	τ^2 Aqr	8679	315;24	-22;28	313;06	-23;33	4	4.01	-0.01
648	Aqr	8544	308;37	-24;55	306;25	-25;58	5	6.57	-1.57
649	Aqr	8670	313;50	-28;07	311;49	-29;23	5	5.26	-0.26
650	Aqr	8649	312;59	-27;17	310;49	-28;35	5	4.69	0.31

No	Name	HR	C	λ_{Ptol}	β_{Ptol}	λ_{-128}	β_{-128}	$\Delta\lambda$	$\Delta\beta$
651	κ Aqr	8610	Il	315;00	2;00	309;55	4;16	5;05	-2;16
652	λ Aqr	8698		314;50	0;10	311;58	-0;19	2;52	0;29
653	Aqr	8782		317;40	-1;10	314;44	-1;35	2;56	0;25
654	φ Aqr	8834		320;00	-0;30	317;34	-0;52	2;26	0;22
655	χ Aqr	8850	l	320;30	-1;40	317;28	-2;46	3;02	1;06
656	ψ^1 Aqr	8841		319;00	-3;30	316;30	-3;50	2;30	0;20
657	ψ^3 Aqr	8865	I	319;50	-4;10	317;10	-4;42	2;40	0;32
658	Aqr	8866	I	317;50	-8;15	315;33	-8;08	2;17	-0;07
659	ω^1 Aqr	8968	Lb	322;40	-9;00	320;00	-10;57	2;40	-0;03
660	ω^2 Aqr	8988		323;10	-10;50	320;31	-11;30	2;39	0;40
661	Aqr	8982		321;40	-12;00	318;57	-14;27	2;43	0;27
662	Aqr	8998		322;10	-14;45	319;17	-15;06	2;53	0;21
663	Aqr	9031	I	323;10	-15;40	320;37	-16;23	2;33	0;43
664	Aqr	8892		317;00	-14;10	313;53	-14;41	3;07	0;31
665	Aqr	8939		318;20	-15;45	315;44	-16;27	2;36	0;42
666	Aqr	8906		317;30	-13;00	314;17	-15;28	3;13	0;28
667	Aqr	8789	Bm	311;50	-14;45	308;38	-16;28	3;12	1;43
668	Aqr	8817	m	312;20	-15;20	309;55	-15;36	2;25	0;16
669	Aqr	8812	m	313;10	-12;00	310;19	-14;24	2;51	0;24
670	α PsA	8728		307;00	-20;20	304;01	-20;50	2;59	0;30
671	Cet	9098		326;40	-15;30	324;05	-16;11	2;35	0;41
672	Cet	33		329;40	-14;40	326;43	-15;08	2;57	0;28
673	Cet	48		329;00	-18;15	325;53	-18;43	3;07	0;28
674	β Psc	8773	M	321;40	9;15	319;02	9;07	2;38	0;08
675	γ Psc	8852		324;10	7;30	321;29	7;28	2;41	0;02
676	Psc	8878		326;00	9;20	323;27	8;58	2;33	0;22
677	θ Psc	8916		328;10	9;30	325;43	9;03	2;27	0;27
678	ι Psc	8969		330;40	7;30	328;00	7;30	2;40	0;00
679	κ Psc	8911		326;00	4;30	323;19	4;33	2;41	-0;03
680	λ Psc	8984		329;40	3;30	327;07	3;29	2;33	0;01
681	ω Psc	9072		336;00	6;20	332;58	6;27	3;02	-0;07
682	Psc	80		341;00	5;45	338;25	5;26	2;35	0;19
683	Psc	132		343;00	3;45	340;35	3;08	2;25	0;37
684	δ Psc	224		347;10	2;15	344;32	2;09	2;38	0;06
685	ϵ Psc	294		350;30	1;10	347;59	0;59	2;31	0;11
686	ζ Psc	361	b	353;00	-0;10	350;13	-0;15	2;47	0;05
687	Psc	330		352;20	-0;00	348;30	-1;32	3;50	-0;28
688	Psc	378		353;00	-3;00	349;45	-4;21	3;15	-0;39
689	μ Psc	434		356;30	-2;20	353;23	-3;05	3;07	0;45
690	ν Psc	489		358;40	-4;40	355;55	-4;49	2;45	0;09
691	ξ Psc	549		0;40	-7;45	357;53	-8;04	2;47	0;19
692	α Psc	595		2;30	-8;30	359;44	-9;12	2;46	0;42
693	o Psc	510		0;30	-1;40	358;06	-1;46	2;24	0;06
694	π Psc	463		0;10	1;50	357;22	1;44	2;48	0;06
695	η Psc	437	L	0;40	5;20	357;14	5;15	3;26	0;05
696	ρ Psc	413		0;30	9;00	357;33	9;15	2;57	-0;15
697	Psc	349		2;00	21;45	359;19	21;52	2;41	-0;07
698	τ Psc	352		1;40	21;40	358;48	20;38	2;52	1;02
699	Psc	274		358;40	20;00	355;25	20;51	3;15	-0;51
700	Psc	262		357;40	19;50	354;14	19;23	3;26	0;27

No	Name	HR	α_{Ptol}	δ_{Ptol}	α_{-128}	δ_{-128}	m_{Pt}	m_{HR}	Δm
651	κ Aqr	8610	316;56	-14;42	311;14	-13;52	4	5.03	-1.03
652	λ Aqr	8698	317;20	-16;30	314;35	-17;42	4	3.74	0.26
653	Aqr	8782	320;35	-16;55	317;45	-18;07	4	5.43	-1.43
654	φ Aqr	8834	322;40	-15;32	320;21	-16;35	4	4.22	-0.22
655	χ Aqr	8850	323;33	-16;29	320;53	-18;25	4	5.06	-1.06
656	ψ^1 Aqr	8841	322;41	-18;42	320;15	-19;44	4	4.21	-0.21
657	ψ^3 Aqr	8865	323;45	-19;04	321;13	-20;20	4	4.98	-0.98
658	Aqr	8866	323;09	-23;34	320;43	-24;06	5	5.08	-0.08
659	ω^1 Aqr	8968	329;05	-24;32	326;19	-25;19	5	5.00	0.00
660	ω^2 Aqr	8988	329;32	-24;12	327;03	-25;41	5	4.49	0.51
661	Aqr	8982	329;16	-27;41	326;35	-28;58	5	4.82	0.18
662	Aqr	8998	330;05	-28;13	327;11	-29;28	5	5.24	-0.24
663	Aqr	9031	331;29	-28;43	329;06	-30;13	5	5.18	-0.18
664	Aqr	8892	324;30	-29;25	321;21	-30;50	4	3.97	0.03
665	Aqr	8939	326;31	-30;28	324;00	-31;55	4	4.71	-0.71
666	Aqr	8906	325;21	-30;02	322;05	-31;27	4	4.39	-0.39
667	Aqr	8789	319;14	-31;37	316;19	-34;07	4	4.47	-0.47
668	Aqr	8817	320;21	-31;55	317;24	-32;55	4	4.69	-0.69
669	Aqr	8812	320;23	-30;30	317;25	-31;39	4	3.66	0.34
670	α PsA	8728	316;02	-38;22	312;44	-39;35	1	1.16	-0.16
671	Cet	9098	334;57	-27;18	332;34	-28;49	> 4	4.55	-0.85
672	Cet	33	337;35	-25;25	334;48	-26;53	> 4	4.89	-1.19
673	Cet	48	338;30	-28;58	335;30	-30;29	> 4	4.44	-0.74
674	β Psc	8773	321;06	-5;46	318;39	-6;37	4	4.53	-0.53
675	γ Psc	8852	324;01	-6;37	321;29	-7;26	4	3.69	0.31
676	Psc	8878	325;07	-4;17	322;51	-5;24	4	5.05	-1.05
677	θ Psc	8916	327;05	-3;24	324;57	-4;34	4	4.28	-0.28
678	ι Psc	8969	330;06	-4;25	327;36	-5;17	4	4.13	-0.13
679	κ Psc	8911	326;46	-8;50	324;10	-9;37	4	4.94	-0.94
680	λ Psc	8984	330;35	-8;31	328;10	-9;20	4	4.50	-0.50
681	ω Psc	9072	335;28	-3;35	332;37	-4;31	4	4.01	-0.01
682	Psc	80	340;18	-2;15	338;02	-3;29	6	5.37	0.63
683	Psc	132	342;55	-3;20	340;55	-4;47	6	5.67	0.33
684	δ Psc	224	347;20	-3;05	344;57	-4;10	4	4.43	-0.43
685	ϵ Psc	294	350;50	-2;45	348;35	-3;54	4	4.28	-0.28
686	ζ Psc	361	353;40	-2;59	351;08	-4;09	4	5.24	-1.24
687	Psc	330	353;47	-4;55	350;04	-6;01	6	5.52	0.48
688	Psc	378	355;37	-7;24	352;20	-8;06	6	5.16	0.84
689	μ Psc	434	357;45	-3;33	355;10	-5;29	4	4.84	-0.84
690	ν Psc	489	0;40	-4;48	358;12	-6;04	4	4.44	-0.44
691	ξ Psc	549	3;45	-6;49	1;20	-8;14	4	4.62	-0.62
692	α Psc	595	5;44	-6;46	3;29	-8;31	3	4.48	-1.48
693	o Psc	510	1;08	-1;19	358;58	-2;23	4	4.26	-0.26
694	π Psc	463	359;25	1;45	356;53	0;31	5	5.57	-0.57
695	η Psc	437	358;27	5;09	355;21	3;42	3	3.62	-0.62
696	ρ Psc	413	356;47	8;26	354;01	7;29	4	5.38	-1.38
697	Psc	349	352;37	20;37	350;13	19;40	5	5.16	-0.16
698	τ Psc	352	352;22	20;24	350;18	18;21	5	4.51	0.49
699	Psc	274	350;26	17;42	347;12	17;12	6	5.42	0.58
700	Psc	262	349;37	17;09	346;47	15;25	6	6.00	0.00

No	Name	HR	C	λ_{Ptol}	β_{Ptol}	λ_{-128}	β_{-128}	$\Delta\lambda$	$\Delta\beta$
701	Psc	230	B	357;00	20;20	353;06	20;26	3;54	-0;06
702	$\psi^{1,1}$ Psc	310		355;40	14;20	353;53	13;17	1;47	1;03
703	ψ^2 Psc	328	LB	356;20	13;15	354;05	12;27	2;15	0;48
704	χ Psc	351	I	357;40	12;00	354;59	12;20	2;41	-0;20
705	υ Psc	383		2;10	17;00	359;16	17;20	2;54	-0;20
706	φ Psc	360		359;50	15;20	356;56	15;24	2;54	-0;04
707	ψ^3 Psc	339	I	0;00	11;45	354;06	11;12	5;54	0;33
708	Psc	9067		331;10	-2;40	328;43	-3;05	2;27	0;25
709	Psc	9087		332;15	-2;30	329;36	-2;57	2;39	0;27
710	Psc	9089		330;40	-5;30	328;25	-5;40	2;15	0;10
711	Psc	3		332;20	-5;30	329;19	-5;49	3;01	0;19
712	λ Cet	896		17;40	-7;45	15;29	-7;59	2;11	0;14
713	α Cet	911		17;40	-12;20	14;42	-12;45	2;58	0;25
714	γ Cet	804		12;40	-11;30	9;54	-12;07	2;46	0;37
715	δ Cet	779		10;30	-12;00	7;55	-14;38	2;35	0;38
716	ν Cet	754	IL	10;10	-8;10	8;47	-9;21	1;23	1;11
717	ξ^2 Cet	718	I	12;40	-6;20	7;50	-6;01	4;50	-0;19
718	ξ^1 Cet	649	L	7;40	-4;10	4;27	-4;26	3;13	0;16
719	ρ Cet	708		3;00	-24;30	0;01	-25;23	2;59	0;53
720	σ Cet	740		3;20	-26;00	0;27	-28;37	2;53	0;37
721	ϵ Cet	781	b	6;40	-25;10	3;36	-26;00	3;04	0;50
722	π Cet	811		7;00	-27;30	4;03	-28;24	2;57	0;54
723	τ Cet	509		352;00	-25;20	348;54	-25;46	3;06	0;26
724	υ Cet	585		353;00	-30;50	349;36	-31;05	3;24	0;15
725	ζ Cet	539		355;00	-18;00	352;15	-20;25	2;45	0;25
726	θ Cet	402	B	349;40	-15;40	346;39	-15;45	3;01	0;05
727	η Cet	334		345;00	-15;40	342;01	-16;03	2;59	0;23
728	φ^2 Cet	235	I	341;00	-13;40	337;47	-14;42	3;13	1;02
729		227	Ib	340;40	-14;40	336;09	-17;17	4;31	2;37
730	φ^1 Cet	194	Ib	339;20	-11;00	336;16	-14;05	3;04	1;05
731		190	I	339;00	-12;00	335;36	-15;16	3;24	1;16
732	ι Cet	74		334;20	-9;40	331;18	-10;01	3;02	0;21
733	β Cet	188		335;40	-20;20	332;45	-20;45	2;55	0;25
734	λ Ori	1879	IB	57;00	-13;30	54;07	-13;39	2;53	0;09
735	α Ori	2061		62;00	-15;00	59;09	-16;19	2;51	-0;41
736	γ Ori	1790		54;00	-17;30	51;21	-17;06	2;39	-0;24
737	Ori	1839		55;00	-16;00	52;47	-17;34	2;13	-0;26
738	μ Ori	2124		64;20	-14;30	61;01	-14;04	3;19	-0;26
739	Ori	2241		66;20	-11;50	64;28	-11;31	1;52	-0;19
740	ξ Ori	2199	I	66;30	-8;00	63;21	-9;29	3;09	-0;31
741	ν Ori	2159		66;00	-9;45	62;16	-8;56	3;44	-0;49
742	Ori	2223		67;20	-8;15	64;09	-7;32	3;11	-0;43
743	Ori	2198		66;40	-8;15	63;20	-7;34	3;20	-0;41
744	χ^1 Ori	2047		61;40	-3;45	59;12	-3;24	2;28	-0;21
745	χ^2 Ori	2135	L	64;40	-4;15	61;20	-3;35	3;20	-0;40
746	ω Ori	1934		57;50	-19;40	54;54	-19;31	2;56	-0;09
747	Ori	1872		56;20	-18;00	53;37	-19;48	2;43	-0;12
748	Ori	1842		55;20	-20;20	52;46	-20;15	2;34	-0;05
749	ψ^2 Ori	1811		54;10	-20;40	51;35	-20;23	2;35	-0;17
750	Ori	1676		50;30	-6;00	48;12	-7;35	2;18	-0;25

No	Name	HR	α_{Ptol}	δ_{Ptol}	α_{-128}	δ_{-128}	m_{Pt}	m_{HR}	Δm
701	Psc	230	348;48	17;20	345;20	15;56	6	7.00	-1.00
702	$\psi^{1,1}$ Psc	310	350;11	11;21	349;03	9;42	4	4.59	-0.59
703	ψ^2 Psc	328	351;15	10;38	349;34	9;01	4	5.55	-1.55
704	χ Psc	351	352;58	10;01	350;25	9;17	4	4.66	-0.66
705	υ Psc	383	354;54	16;23	352;11	15;32	4	4.76	-0.76
706	φ Psc	360	353;31	13;56	350;55	12;51	4	4.65	-0.65
707	ψ^3 Psc	339	355;12	10;44	350;05	7;53	4	5.55	-1.55
708	Psc	9067	334;16	-13;44	332;02	-14;57	4	4.86	-0.86
709	Psc	9087	335;14	-13;11	332;50	-14;30	4	5.10	-1.10
710	Psc	9089	334;52	-16;33	332;42	-17;28	4	4.41	-0.41
711	Psc	3	336;28	-15;56	333;38	-17;16	4	4.61	-0.61
712	λ Cet	896	19;15	-0;06	17;20	-1;12	4	4.70	-0.70
713	α Cet	911	21;01	-4;20	18;29	-5;53	3	2.53	0.47
714	γ Cet	804	16;10	-5;29	13;54	-7;09	3	3.47	-0.47
715	δ Cet	779	15;13	-8;37	13;08	-10;14	3	4.07	-1.07
716	ν Cet	754	12;34	-3;24	11;47	-5;03	4	4.86	-0.86
717	ξ^2 Cet	718	14;07	-0;44	9;35	-2;23	4	4.28	-0.28
718	ξ^1 Cet	649	8;41	-0;44	5;51	-2;16	4	4.37	-0.37
719	ρ Cet	708	13;05	-21;06	10;49	-23;06	4	4.89	-0.89
720	σ Cet	740	15;02	-24;07	12;47	-25;50	4	4.75	-0.75
721	ϵ Cet	781	16;36	-20;16	14;16	-22;15	4	4.84	-0.84
722	π Cet	811	17;58	-22;15	15;48	-24;14	3	4.25	-1.25
723	τ Cet	509	3;42	-26;15	1;03	-27;53	3	3.50	-0.50
724	υ Cet	585	7;27	-30;44	4;29	-32;20	4	4.00	0.00
725	ζ Cet	539	3;52	-20;14	1;31	-21;43	3	3.73	-0.73
726	θ Cet	402	357;03	-18;28	354;15	-19;44	3	3.60	-0.60
727	η Cet	334	352;43	-20;21	350;02	-21;52	3	3.45	-0.45
728	φ^2 Cet	235	348;05	-20;07	345;25	-22;18	5	5.19	-0.19
729		227	348;13	-21;10	345;00	-25;18	5	5.59	-0.59
730	φ^1 Cet	194	346;13	-20;10	343;42	-22;19	> 5	4.76	-0.06
731		190	346;20	-21;13	343;35	-23;40	> 5	6.02	-1.32
732	ι Cet	74	340;03	-19;02	337;12	-20;27	< 3	3.56	-0.26
733	β Cet	188	346;01	-28;18	343;17	-29;47	3	2.04	0.96
734	λ Ori	1879	57;47	6;40	55;04	5;46	neb	2.91	3.09
735	α Ori	2061	63;15	4;15	60;26	4;15	< 1	0.50	0.80
736	γ Ori	1790	55;53	2;07	53;20	1;47	> 2	1.64	0.06
737	Ori	1839	56;55	1;51	54;46	1;40	< 4	4.20	0.10
738	μ Ori	2124	65;00	7;07	61;44	6;50	4	4.12	-0.12
739	Ori	2241	66;29	10;04	64;36	9;58	6	5.04	0.96
740	ξ Ori	2199	66;20	11;55	63;08	11;46	4	4.48	-0.48
741	ν Ori	2159	65;48	12;04	61;57	12;06	4	4.42	-0.42
742	Ori	2223	66;53	13;46	63;34	13;49	6	5.30	0.70
743	Ori	2198	66;13	13;40	62;45	13;39	6	4.95	1.05
744	χ^1 Ori	2047	60;17	17;11	57;43	16;54	5	4.41	0.59
745	χ^2 Ori	2135	63;28	17;15	59;56	17;10	5	4.63	0.37
746	ω Ori	1934	59;55	0;50	57;11	0;16	4	4.57	-0.57
747	Ori	1872	58;36	0;12	56;04	-0;19	6	5.36	0.64
748	Ori	1842	57;44	-0;11	55;24	-0;56	6	5.46	0.54
749	ψ^2 Ori	1811	56;47	-0;55	54;21	-1;20	5	4.59	0.41
750	Ori	1676	50;10	10;28	47;50	10;10	4	4.82	-0.82

No	Name	HR	C	λ_{Ptol}	β_{Ptol}	λ_{-128}	β_{-128}	$\Delta\lambda$	$\Delta\beta$
751	Ori	1638		49;20	-8;10	46;56	-7;39	2;24	-0;31
752	o^2 Ori	1580		48;00	-10;15	44;47	-9;19	3;13	-0;56
753	π^1 Ori	1570		46;20	-12;50	43;58	-12;34	2;22	-0;16
754	π^2 Ori	1544		45;10	-14;15	42;46	-13;45	2;24	-0;30
755	π^3 Ori	1543		44;50	-15;50	42;01	-15;38	2;49	-0;12
756	π^4 Ori	1552		44;50	-17;10	42;29	-17;03	2;21	-0;07
757	π^5 Ori	1567		45;20	-20;20	42;52	-20;17	2;28	-0;03
758	π^6 Ori	1601		46;20	-21;30	43;55	-21;08	2;25	-0;22
759	δ Ori	1852		55;20	-24;10	52;45	-23;50	2;35	-0;20
760	ϵ Ori	1903		57;20	-24;50	53;52	-24;48	3;28	-0;02
761	ζ Ori	1948		58;10	-25;40	55;05	-25;35	3;05	-0;05
762	η Ori	1788		53;50	-25;50	50;33	-25;49	3;17	-0;01
763	Ori	1892	IBl	56;30	-28;20	53;26	-28;25	3;04	0;05
764	θ^2 Ori	1897	I	56;40	-29;10	53;24	-29;00	3;16	-0;10
765	ι Ori	1899	I	57;00	-29;50	53;23	-29;29	3;37	-0;21
766	Ori	1937		57;40	-30;40	54;19	-30;48	3;21	0;08
767	υ Ori	1855	II	56;30	-30;50	52;18	-30;50	4;12	-0;00
768	β Ori	1713		49;50	-31;30	47;12	-31;24	2;38	-0;06
769	τ Ori	1735		51;00	-30;15	48;14	-30;07	2;46	-0;08
770	Ori	1784		53;20	-31;10	49;58	-31;11	3;22	0;01
771	κ Ori	2004		60;10	-33;30	56;48	-33;21	3;22	-0;09
772	λ Eri	1679	b	48;20	-31;50	45;35	-31;49	2;45	-0;01
773	β Eri	1666		48;30	-28;15	45;43	-28;06	2;47	-0;09
774	ψ Eri	1617		48;00	-29;50	43;34	-30;03	4;26	0;13
775	ω Eri	1560		44;40	-28;15	41;25	-28;05	3;15	-0;10
776	μ Eri	1520	lb	43;10	-25;50	39;42	-25;38	3;28	-0;12
777	ν Eri	1463	I	40;10	-25;20	37;10	-25;23	3;00	0;03
778	ξ Eri	1383		36;20	-24;00	33;42	-25;13	2;38	-0;47
779	o^2 Eri	1325		35;30	-25;00	32;23	-26;59	3;07	-0;01
780	o^1 Eri	1298		32;50	-27;50	29;45	-27;45	3;05	-0;05
781	γ Eri	1231		27;00	-32;50	24;09	-33;22	2;51	0;32
782	π Eri	1162		24;20	-29;00	21;13	-31;22	3;07	0;22
783	δ Eri	1136		24;10	-28;50	21;06	-29;20	3;04	0;30
784	ϵ Eri	1084		22;00	-26;00	19;07	-28;06	2;53	0;06
785	ζ Eri	984		17;10	-25;30	14;08	-26;09	3;02	0;39
786	ρ^3 Eri	925	I	14;50	-23;50	11;29	-24;06	3;21	0;16
787	η Eri	874	Ib	12;10	-23;30	9;03	-24;35	3;07	1;05
788		859	I	10;30	-23;15	7;53	-24;58	2;37	1;43
789	τ^1 Eri	818		5;10	-32;10	2;08	-32;51	3;02	0;41
790	τ^2 Eri	850		5;50	-34;50	2;55	-35;40	2;55	0;50
791	τ^3 Eri	919		8;50	-38;30	4;52	-39;04	3;58	0;34
792	τ^4 Eri	1003		13;50	-38;10	10;18	-38;42	3;32	0;32
793	τ^5 Eri	1088		17;30	-37;00	14;25	-39;37	3;05	0;37
794	τ^6 Eri	1173		21;20	-41;20	17;50	-41;50	3;30	0;30
795	τ^7 Eri	1181		21;30	-42;30	17;32	-42;46	3;58	0;16
796	τ^8 Eri	1213		22;10	-43;15	19;05	-43;51	3;05	0;36
797	τ^9 Eri	1240	l	24;40	-43;20	21;12	-43;42	3;28	0;22
798	υ^1 Eri	1453	B	34;10	-50;20	29;52	-51;03	4;18	0;43
799	υ^2 Eri	1464		35;00	-51;45	30;09	-52;04	4;51	0;19
800	Eri	1393		28;10	-53;50	24;35	-54;47	3;35	0;57

No	Name	HR	α_{Ptol}	δ_{Ptol}	α_{-128}	δ_{-128}	m_{Pt}	m_{HR}	Δm
751	Ori	1638	49;05	10;00	46;38	9;45	4	4.68	-0.68
752	o^2 Ori	1580	48;22	7;38	45;03	7;33	4	4.07	-0.07
753	π^1 Ori	1570	47;30	4;42	45;13	4;13	4	4.65	-0.65
754	π^2 Ori	1544	46;49	3;02	44;26	2;44	4	4.36	-0.36
755	π^3 Ori	1543	46;58	1;25	44;19	0;43	3	3.19	-0.19
756	π^4 Ori	1552	47;21	0;08	45;10	-0;30	3	3.69	-0.69
757	π^5 Ori	1567	48;42	-2;46	46;28	-3;29	3	3.72	-0.72
758	π^6 Ori	1601	49;56	-3;37	47;40	-4;00	3	4.47	-1.47
759	δ Ori	1852	58;39	-4;04	56;16	-4;25	2	2.23	-0.23
760	ϵ Ori	1903	60;35	-4;19	57;29	-5;06	2	1.70	0.30
761	ζ Ori	1948	61;30	-4;58	58;45	-5;37	2	1.30	0.70
762	η Ori	1788	57;43	-6;01	54;49	-6;51	3	3.36	-0.36
763	Ori	1892	60;38	-7;53	57;59	-8;43	4	4.59	-0.59
764	θ^2 Ori	1897	60;58	-8;40	58;06	-9;17	< 3	6.39	-3.09
765	ι Ori	1899	61;24	-9;15	58;13	-9;46	3	2.77	0.23
766	Ori	1937	62;09	-9;56	59;20	-10;50	4	4.80	-0.80
767	υ Ori	1855	61;12	-10;20	57;37	-11;18	4	4.62	-0.62
768	β Ori	1713	55;44	-12;23	53;28	-13;01	1	0.12	0.88
769	τ Ori	1735	56;23	-10;55	53;59	-11;32	> 4	3.60	0.10
770	Ori	1784	58;36	-11;18	55;45	-12;09	4	4.14	-0.14
771	κ Ori	2004	64;53	-12;15	62;01	-12;51	> 3	2.06	0.64
772	λ Eri	1679	54;34	-13;03	52;14	-13;48	> 4	4.27	-0.57
773	β Eri	1666	53;42	-9;34	51;16	-10;12	4	2.79	1.21
774	ψ Eri	1617	53;43	-11;12	50;01	-12;37	4	4.81	-0.81
775	ω Eri	1560	50;25	-10;31	47;34	-11;19	4	4.39	-0.39
776	μ Eri	1520	48;24	-8;36	45;18	-9;28	4	4.02	-0.02
777	ν Eri	1463	45;38	-8;57	43;02	-9;57	4	3.93	0.07
778	ζ Eri	1383	42;32	-10;42	39;58	-10;50	5	5.17	-0.17
779	o^2 Eri	1325	42;10	-11;53	39;28	-12;54	4	4.43	-0.43
780	o^1 Eri	1298	40;10	-13;29	37;29	-14;27	4	4.04	-0.04
781	γ Eri	1231	37;12	-19;59	35;01	-21;28	3	2.95	0.05
782	π Eri	1162	34;14	-19;10	31;45	-20;37	4	4.42	-0.42
783	δ Eri	1136	33;12	-17;13	30;47	-18;48	3	3.54	-0.54
784	ϵ Eri	1084	31;01	-17;12	28;34	-18;22	3	3.73	-0.73
785	ζ Eri	984	25;51	-16;37	23;28	-18;23	3	4.80	-1.80
786	ρ^3 Eri	925	22;50	-16;05	20;17	-17;30	4	5.26	-1.26
787	η Eri	874	20;40	-16;39	18;23	-18;52	3	3.89	-0.89
788		859	19;06	-17;03	17;32	-19;40	4	6.32	-2.32
789	τ^1 Eri	818	18;41	-27;08	16;23	-28;57	4	4.47	-0.47
790	τ^2 Eri	850	20;37	-29;15	18;33	-31;09	4	4.75	-0.75
791	τ^3 Eri	919	25;04	-31;23	22;05	-33;24	4	4.09	-0.09
792	τ^4 Eri	1003	28;55	-29;17	26;17	-31;05	4	3.69	0.31
793	τ^5 Eri	1088	32;17	-28;45	30;05	-30;26	4	4.27	-0.27
794	τ^6 Eri	1173	36;28	-29;35	33;56	-31;15	4	4.23	-0.23
795	τ^7 Eri	1181	37;11	-30;34	34;12	-32;11	5	5.24	-0.24
796	τ^8 Eri	1213	38;04	-31;02	35;58	-32;39	4	4.65	-0.65
797	τ^9 Eri	1240	40;01	-30;19	37;30	-31;50	4	4.66	-0.66
798	υ^1 Eri	1453	50;26	-34;00	47;43	-35;52	4	4.51	-0.51
799	υ^2 Eri	1464	51;42	-35;05	48;28	-36;43	4	3.82	0.18
800	Eri	1393	48;10	-38;44	46;17	-40;39	4	3.96	0.04

No	Name	HR	C	λ_{Ptol}	β_{Ptol}	λ_{-128}	β_{-128}	$\Delta\lambda$	$\Delta\beta$
801	υ^4 Eri	1347		25;50	-53;10	22;36	-54;11	3;14	1;01
802		1214	I	17;50	-51;00	14;06	-53;26	3;44	0;26
803		1195	I	14;50	-53;30	11;56	-54;29	2;54	0;59
804		1143	I	11;50	-52;30	9;07	-55;00	2;43	2;30
805	θ^1 Eri	897		0;10	-53;30	353;21	-53;53	6;49	0;23
806	ι Lep	1696		49;40	-33;00	46;06	-34;59	3;34	-0;01
807	κ Lep	1705		49;50	-36;30	46;17	-36;05	3;33	-0;25
808	ν Lep	1757		51;20	-35;40	48;22	-35;38	2;58	-0;02
809	λ Lep	1756		51;20	-36;40	48;09	-36;28	3;11	-0;12
810	μ Lep	1702		49;10	-39;15	45;43	-39;19	3;27	0;04
811	ϵ Lep	1654		46;10	-45;15	42;22	-45;12	3;48	-0;03
812	α Lep	1865	l	55;50	-41;30	51;45	-41;21	4;05	-0;09
813	β Lep	1829	L	54;50	-44;20	50;02	-44;09	4;48	-0;11
814	δ Lep	2035	b	61;00	-42;00	57;23	-44;12	3;37	0;12
815	γ Lep	1983		59;00	-45;50	55;29	-45;54	3;31	0;04
816	ζ Lep	1998		60;00	-38;20	56;24	-38;30	3;36	0;10
817	η Lep	2085		62;40	-38;10	59;20	-37;59	3;20	-0;11
818	α CMa	2491		77;40	-39;10	74;52	-39;09	2;48	-0;01
819	θ CMa	2574		79;40	-33;00	76;45	-34;59	2;55	-0;01
820	μ CMa	2593		81;20	-36;30	77;31	-36;56	3;49	0;26
821	γ CMa	2657		83;20	-37;45	80;06	-38;16	3;14	0;31
822	ι CMa	2596	L	85;20	-38;00	78;01	-39;56	7;19	-0;04
823	π CMa	2590		80;30	-42;40	78;21	-43;01	2;09	0;21
824	ν^3 CMa	2443		76;10	-41;15	72;29	-41;34	3;41	0;19
825	ν^2 CMa	2429		76;00	-42;30	72;09	-42;34	3;51	0;04
826	β CMa	2294		71;00	-41;20	67;38	-41;32	3;22	0;12
827	ξ^1 CMa	2387		74;40	-46;30	71;08	-46;51	3;32	0;21
828	ξ^2 CMa	2414		76;10	-45;50	72;08	-46;21	4;02	0;31
829	o^2 CMa	2653	b	84;40	-46;10	81;31	-46;24	3;09	0;14
830	o^1 CMa	2580		81;40	-45;00	78;41	-47;03	2;59	0;03
831	δ CMa	2693		86;40	-48;45	83;56	-48;43	2;44	-0;02
832	ϵ CMa	2618		83;40	-51;30	81;18	-51;38	2;22	0;08
833	κ CMa	2538	L	83;00	-55;10	79;07	-55;26	3;53	0;16
834	ζ CMa	2282		69;40	-53;45	67;50	-53;40	1;50	-0;05
835	η CMa	2827		92;10	-50;40	90;08	-50;52	2;02	0;12
836	Mon	2648	Ib	79;30	-25;15	78;05	-27;00	1;25	1;45
837	θ Col	2177	L	70;00	-61;30	63;30	-60;58	6;30	-0;32
838	κ Col	2256		71;20	-58;45	66;58	-58;50	4;22	0;05
839	δ Col	2296		73;00	-55;00	68;55	-56;58	4;05	-0;02
840	λ CMa	2361		74;10	-54;00	71;05	-56;04	3;05	0;04
841	μ Col	1996		58;00	-55;30	55;07	-55;57	2;53	0;27
842	λ Col	2056		60;20	-57;40	57;45	-57;32	2;35	-0;08
843	γ Col	2106		62;20	-59;50	59;26	-59;01	2;54	-0;49
844	β Col	2040		59;00	-59;40	56;43	-59;42	2;17	0;02
845	α Col	1956		56;00	-57;40	52;31	-57;39	3;29	-0;01
846	ϵ Col	1862		52;10	-59;30	48;59	-58;53	3;11	-0;37
847	β CMi	2845		85;00	-12;00	82;39	-13;44	2;21	-0;16
848	α CMi	2943	L	89;10	-16;10	86;33	-15;37	2;37	-0;33
849	Pup	3102		100;20	-42;30	98;14	-42;49	2;06	0;19
850	ρ Pup	3185		104;20	-43;20	102;04	-43;30	2;16	0;10

No	Name	HR	α_{Ptol}	δ_{Ptol}	α_{-128}	δ_{-128}	m_{Pt}	m_{HR}	Δm
801	υ^4 Eri	1347	46;12	-38;47	44;34	-40;41	4	3.56	0.44
802		1214	40;38	-40;59	38;16	-42;36	4	5.11	-1.11
803		1195	38;55	-42;21	37;35	-44;11	4	4.17	-0.17
804		1143	36;08	-42;28	36;05	-45;31	4	4.59	-0.59
805	θ^1 Eri	897	28;46	-47;16	24;08	-50;05	1	3.42	-2.42
806	ι Lep	1696	56;34	-15;47	53;37	-16;43	5	4.45	0.55
807	κ Lep	1705	57;08	-17;12	54;06	-17;43	5	4.36	0.64
808	ν Lep	1757	58;07	-16;04	55;40	-16;48	5	5.30	-0.30
809	λ Lep	1756	58;23	-17;02	55;44	-17;39	5	4.29	0.71
810	μ Lep	1702	57;24	-19;59	54;40	-20;56	> 4	3.31	0.39
811	ϵ Lep	1654	57;02	-26;22	54;08	-27;18	> 4	3.19	0.51
812	α Lep	1865	63;15	-20;49	60;01	-21;34	3	2.58	0.42
813	β Lep	1829	63;15	-23;45	59;33	-24;36	3	2.84	0.16
814	δ Lep	2035	67;53	-22;33	65;07	-23;17	> 4	3.81	-0.11
815	γ Lep	1983	66;47	-24;30	64;09	-25;15	> 4	3.60	0.10
816	ζ Lep	1998	65;47	-17;01	62;55	-17;55	> 4	3.55	0.15
817	η Lep	2085	67;53	-16;26	65;10	-16;53	> 4	3.71	-0.01
818	α CMa	2491	80;06	-15;45	77;50	-16;04	1	1.46	-0.46
819	θ CMa	2574	81;23	-11;28	78;56	-11;46	4	4.07	-0.07
820	μ CMa	2593	82;52	-12;52	79;46	-13;40	5	5.00	0.00
821	γ CMa	2657	84;34	-14;02	81;58	-14;49	4	4.12	-0.12
822	ι CMa	2596	86;17	-16;13	80;26	-16;36	4	4.37	-0.37
823	π CMa	2590	82;37	-19;04	80;59	-19;40	5	4.68	0.32
824	ν^3 CMa	2443	79;07	-17;56	76;15	-18;41	5	4.43	0.57
825	ν^2 CMa	2429	79;07	-19;11	76;08	-19;43	5	3.95	1.05
826	β CMa	2294	75;04	-18;29	72;27	-19;11	3	1.98	1.02
827	ξ^1 CMa	2387	78;34	-23;16	75;59	-24;04	5	4.33	0.67
828	ξ^2 CMa	2414	79;37	-22;29	76;39	-23;27	5	4.54	0.46
829	o^2 CMa	2653	86;00	-22;24	83;40	-22;52	4	3.02	0.98
830	o^1 CMa	2580	83;49	-23;20	81;36	-23;40	5	3.87	1.13
831	δ CMa	2693	87;35	-24;56	85;35	-25;06	< 3	1.84	1.46
832	ϵ CMa	2618	85;33	-27;45	83;53	-28;06	3	1.50	1.50
833	κ CMa	2538	85;19	-31;26	82;45	-31;59	4	3.96	0.04
834	ζ CMa	2282	76;09	-30;53	74;52	-31;07	3	3.02	-0.02
835	η CMa	2827	91;32	-26;50	90;05	-27;09	< 3	2.45	0.85
836	Mon	2648	80;31	-1;45	79;22	-3;43	4	4.99	-0.99
837	θ Col	2177	77;58	-38;30	73;53	-38;44	4	5.02	-1.02
838	κ Col	2256	78;12	-35;40	75;27	-36;17	4	4.37	-0.37
839	δ Col	2296	78;57	-33;49	76;17	-34;15	4	3.85	0.15
840	λ CMa	2361	79;33	-32;44	77;31	-33;10	4	4.48	-0.48
841	μ Col	1996	68;46	-34;01	66;59	-35;01	4	5.17	-1.17
842	λ Col	2056	70;57	-35;48	69;13	-36;09	4	4.87	-0.87
843	γ Col	2106	72;52	-37;39	70;46	-37;21	4	4.36	-0.36
844	β Col	2040	70;45	-37;54	69;19	-38;23	2	3.12	-1.12
845	α Col	1956	68;11	-36;24	65;55	-37;03	2	2.64	-0.64
846	ϵ Col	1862	66;28	-38;45	64;12	-38;50	4	3.87	0.13
847	β CMi	2845	85;05	9;46	82;46	9;48	4	2.90	1.10
848	α CMi	2943	89;12	7;41	86;38	8;04	1	0.38	0.62
849	Pup	3102	98;02	-18;57	96;24	-19;16	5	4.20	0.80
850	ρ Pup	3185	101;03	-20;02	99;17	-20;11	3	2.81	0.19

No	Name	HR	C	λ_{Ptol}	β_{Ptol}	λ_{-128}	β_{-128}	$\Delta\lambda$	$\Delta\beta$
851	ζ Pup	3045		98;50	-43;00	96;38	-45;10	2;12	0;10
852	o Pup	3034	B	98;40	-44;00	96;40	-46;17	2;00	0;17
853		2944		95;20	-45;30	93;22	-46;17	1;58	0;47
854		2948		96;20	-47;15	94;04	-47;40	2;16	0;25
855		2922	B	95;20	-49;30	93;35	-49;21	1;45	-0;09
856	Pup	2996	B	99;20	-49;30	96;30	-49;27	2;50	-0;03
857	Pup	2993		98;30	-49;15	96;12	-48;57	2;18	-0;18
858		3113		104;00	-49;50	101;34	-49;54	2;26	0;04
859		2834		94;00	-51;00	91;19	-53;16	2;41	0;16
860	π Pup	2773		94;00	-58;40	90;58	-58;47	3;02	0;07
861		2937		100;10	-55;30	97;09	-55;36	3;01	0;06
862		2961	I	102;10	-58;40	99;47	-58;39	2;23	-0;01
863		3017		103;40	-57;15	101;37	-57;57	2;03	0;42
864		3084	b	106;30	-57;45	104;46	-58;17	1;44	0;32
865	ζ Pup	3165		111;10	-58;40	109;21	-58;33	1;49	-0;07
866		3080		108;10	-58;00	105;51	-59;54	2;19	-0;06
867		3162		111;00	-59;20	110;04	-59;48	0;56	0;28
868		3225	L	113;10	-56;40	111;37	-57;35	1;33	0;55
869		3243		114;20	-57;40	112;59	-58;01	1;21	0;21
870		3535	b	125;40	-51;30	123;42	-53;17	1;58	1;47
871		3477		126;10	-55;40	124;32	-57;30	1;38	1;50
872		3426		124;00	-57;10	122;48	-58;24	1;12	1;14
873		3487	I	129;00	-58;00	128;23	-60;15	0;47	0;15
874		3445	I	129;00	-61;15	127;26	-61;16	1;34	0;01
875	β Pyx	3438	B	120;10	-51;50	117;30	-51;19	2;40	-0;31
876	α Pyx	3468		119;20	-47;00	117;12	-49;05	2;08	0;05
877	γ Pyx	3518	b	118;00	-43;20	116;14	-43;29	1;46	0;09
878	δ Pyx	3556		119;00	-43;30	117;24	-42;59	1;36	-0;31
879	λ Vel	3634		134;10	-54;30	132;02	-55;58	2;08	1;28
880	ψ Vel	3786		137;30	-51;15	135;41	-51;13	1;49	-0;02
881	σ Pup	2878	I	101;10	-61;00	99;38	-64;06	1;32	1;06
882		3055	L	109;00	-64;30	109;58	-65;25	-0;58	0;55
883	$\gamma^{2,1}$ Vel	3207		120;00	-63;50	118;17	-64;37	1;43	0;47
884	χ Car	3117		128;30	-69;40	121;55	-70;28	6;35	0;48
885	o Vel	3447	I	135;10	-65;40	135;50	-66;21	-0;40	0;41
886	δ Vel	3485		141;20	-65;50	139;56	-67;14	1;24	1;24
887		3498		146;00	-67;20	144;27	-68;31	1;33	1;11
888	κ Vel	3734		151;00	-62;50	149;55	-63;44	1;05	0;54
889		3803		158;00	-62;15	155;15	-64;12	2;45	1;57
890	η Col	2120		64;00	-65;50	59;59	-66;32	4;01	0;42
891	ν Pup	2451		80;10	-65;40	77;45	-66;21	2;25	0;41
892	α Car	2326		77;10	-73;00	75;36	-76;07	1;34	1;07
893	τ Pup	2553	b	89;00	-71;45	88;28	-73;05	0;32	1;20
894	σ Hya	3418		104;00	-13;00	101;41	-14;49	2;19	-0;11
895	δ Hya	3410		103;20	-13;10	100;48	-12;36	2;32	-0;34
896	ε Hya	3482		105;20	-11;30	102;54	-11;16	2;26	-0;14
897	η Hya	3454	B	105;30	-14;15	102;47	-14;28	2;43	0;13
898	ζ Hya	3547	B	107;30	-12;15	105;05	-11;10	2;25	-1;05
899	ω Hya	3613	lb	110;20	-11;50	107;51	-11;13	2;29	-0;37
900	θ Hya	3665	b	113;20	-13;40	110;37	-13;05	2;43	-0;35

No	Name	HR	α_{Ptol}	δ_{Ptol}	α_{-128}	δ_{-128}	m_{Pt}	m_{HR}	Δm
851	ζ Pup	3045	96;42	-21;22	95;01	-21;34	4	3.34	0.66
852	o Pup	3034	96;30	-22;21	94;59	-22;41	4	4.50	-0.50
853		2944	94;01	-21;44	92;31	-22;36	4	4.70	-0.70
854		2948	94;41	-23;30	92;59	-23;59	4	3.75	0.25
855		2922	93;51	-25;43	92;35	-25;39	4	4.64	-0.64
856	Pup	2996	96;43	-25;52	94;41	-25;50	4	3.96	0.04
857	Pup	2993	96;08	-25;35	94;30	-25;20	4	4.59	-0.59
858		3113	100;02	-26;29	98;18	-26;31	4	4.79	-0.79
859		2834	92;45	-29;11	90;55	-29;33	4	5.35	-1.35
860	π Pup	2773	92;32	-34;51	90;37	-35;04	3	2.70	0.30
861		2937	96;46	-31;54	94;46	-32;00	5	4.53	0.47
862		2961	97;42	-35;09	96;12	-35;08	5	4.84	0.16
863		3017	98;51	-33;50	97;27	-34;32	4	3.61	0.39
864		3084	100;35	-34;36	99;25	-35;03	4	4.49	-0.49
865	ζ Pup	3165	103;23	-35;49	102;17	-35;39	2	2.25	-0.25
866		3080	101;14	-36;52	99;51	-36;44	5	3.73	1.27
867		3162	103;08	-36;27	102;28	-36;58	5	5.52	-0.52
868		3225	105;07	-34;03	103;56	-34;55	5	4.45	0.55
869		3243	105;38	-35;09	104;43	-35;29	5	4.44	0.56
870		3535	114;59	-30;45	113;06	-32;16	> 4	5.82	-2.12
871		3477	113;55	-34;50	112;15	-36;27	> 4	4.07	-0.37
872		3426	111;59	-35;56	110;50	-37;03	> 4	4.14	-0.44
873		3487	114;08	-39;27	113;36	-39;40	> 4	3.91	-0.21
874		3445	113;30	-40;36	112;36	-40;29	> 4	3.84	-0.14
875	β Pyx	3438	111;03	-30;12	109;21	-29;27	3	3.97	-0.97
876	α Pyx	3468	111;12	-27;19	109;41	-27;14	3	3.68	-0.68
877	γ Pyx	3518	111;33	-21;35	110;11	-21;36	4	4.01	-0.01
878	δ Pyx	3556	112;16	-21;54	111;11	-21;16	4	4.89	-0.89
879	λ Vel	3634	119;40	-35;11	117;42	-36;16	2	2.21	-0.21
880	ψ Vel	3786	123;19	-32;50	122;06	-32;32	< 2	3.60	-1.30
881	σ Pup	2878	96;32	-39;24	95;31	-40;34	4	3.25	0.75
882		3055	100;46	-41;22	101;06	-42;29	6	4.11	1.89
883	$\gamma^{2,1}$ Vel	3207	107;12	-41;48	105;59	-42;29	2	1.03	0.97
884	χ Car	3117	109;01	-48;23	105;28	-48;29	2	3.47	-1.47
885	o Vel	3447	114;44	-45;43	114;44	-46;34	3	3.62	-0.62
886	δ Vel	3485	117;56	-46;58	116;19	-48;04	3	1.96	1.04
887		3498	119;16	-49;11	117;38	-50;01	2	4.49	-2.49
888	κ Vel	3734	125;23	-46;24	124;11	-47;02	3	2.50	0.50
889		3803	129;51	-47;38	126;46	-48;41	3	3.13	-0.13
890	η Col	2120	75;44	-43;17	73;47	-44;31	> 4	3.96	-0.26
891	ν Pup	2451	84;34	-42;00	83;20	-42;55	> 3	3.17	-0.47
892	α Car	2326	84;43	-51;23	84;21	-52;41	1	0.72	0.28
893	τ Pup	2553	89;32	-47;54	89;19	-49;22	> 3	2.93	-0.23
894	σ Hya	3418	103;39	8;11	101;25	8;27	4	4.44	-0.44
895	δ Hya	3410	103;11	10;04	100;43	10;43	4	4.16	-0.16
896	ϵ Hya	3482	105;20	11;31	102;56	11;53	4	3.38	0.62
897	η Hya	3454	105;12	8;46	102;31	8;42	4	4.30	-0.30
898	ζ Hya	3547	107;23	10;32	105;07	11;45	4	3.11	0.89
899	ω Hya	3613	110;14	10;35	107;52	11;23	5	4.97	0.03
900	θ Hya	3665	112;53	8;19	110;20	9;10	4	3.88	0.12

No	Name	HR	C	λ_{Ptol}	β_{Ptol}	λ_{-128}	β_{-128}	$\Delta\lambda$	$\Delta\beta$
901	τ^2 Hya	3787		118;50	-15;20	116;13	-15;08	2;37	-0;12
902	ι Hya	3845		120;40	-14;50	118;05	-14;24	2;35	-0;26
903	τ^1 Hya	3759		118;30	-17;10	115;59	-16;53	2;31	-0;17
904		3750	Ib	119;10	-19;45	117;01	-20;00	2;09	0;15
905	α Hya	3748	B	120;00	-20;30	117;49	-22;33	2;11	2;03
906	κ Hya	3849		126;00	-26;30	123;14	-26;43	2;46	0;13
907	υ^1 Hya	3903	b	128;40	-24;00	126;13	-26;11	2;27	0;11
908	υ^2 Hya	3970	B	131;10	-23;15	128;53	-23;17	2;17	0;02
909	μ Hya	4094		138;00	-24;40	135;38	-24;41	2;22	0;01
910	φ^3 Hya	4171		140;00	-23;15	138;39	-23;33	1;21	0;18
911	ν Hya	4232	lb	143;00	-22;10	140;54	-21;59	2;06	-0;11
912	β Crt	4343	b	151;30	-25;45	149;05	-25;36	2;25	-0;09
913	χ^1 Hya	4314		152;20	-30;10	150;01	-30;14	2;19	0;04
914	ξ Hya	4450		162;10	-31;10	158;41	-31;29	3;29	0;19
915	o Hya	4494		164;30	-33;10	161;48	-33;22	2;42	0;12
916	β Hya	4552		166;10	-31;20	164;04	-31;24	2;06	0;04
917	γ Hya	5020		180;00	-13;40	177;27	-13;36	2;33	-0;04
918	π Hya	5287	B	193;30	-17;40	189;02	-12;48	4;28	-4;52
919		3314		102;30	-23;15	100;23	-22;39	2;07	-0;36
920	ϵ Sex	4042	IB	131;00	-16;20	129;56	-17;28	1;04	1;08
921	α Crt	4287		146;20	-21;00	144;32	-22;42	1;48	-0;18
922	γ Crt	4405		152;30	-19;30	149;48	-19;39	2;42	0;09
923	δ Crt	4382		150;00	-16;00	147;19	-17;40	2;41	-0;20
924	ζ Crt	4514		157;00	-18;30	154;33	-18;16	2;27	-0;14
925	ϵ Crt	4402		149;20	-13;40	146;45	-13;30	2;35	-0;10
926	η Crt	4567		159;10	-16;10	156;37	-16;03	2;33	-0;07
927	θ Crt	4468	L	151;40	-11;30	149;06	-11;18	2;34	-0;12
928	α Crv	4623		165;20	-21;40	162;43	-21;41	2;37	0;01
929	ϵ Crv	4630		164;20	-19;40	162;13	-19;36	2;07	-0;04
930	ζ Crv	4696		166;40	-18;10	164;21	-18;11	2;19	0;01
931	γ Crv	4662		163;30	-14;50	161;18	-14;25	2;12	-0;25
932	δ Crv	4757		166;40	-12;30	164;00	-12;00	2;40	-0;30
933	η Crv	4775		167;00	-11;45	164;30	-11;29	2;30	-0;16
934	β Crv	4786		170;30	-18;10	167;51	-17;56	2;39	-0;14
935	Cen	5192		190;30	-21;40	188;32	-21;23	1;58	-0;17
936	Cen	5221		190;00	-18;50	188;18	-18;47	1;42	-0;03
937	Cen	5168		189;10	-20;30	187;25	-20;15	1;45	-0;15
938	Cen	5210		190;00	-18;00	188;26	-19;52	1;34	-0;08
939	ι Cen	5028		186;10	-25;40	183;50	-25;44	2;20	0;04
940	θ Cen	5288		195;40	-22;30	193;00	-21;29	2;40	-1;01
941		5089		189;10	-27;30	187;01	-27;26	2;09	-0;04
942	ψ Cen	5367		198;10	-22;20	196;13	-22;16	1;57	-0;04
943		5378		199;10	-23;45	197;19	-23;37	1;51	-0;08
944		5485		202;00	-18;15	199;52	-18;04	2;08	-0;11
945		5471	b	202;30	-20;50	200;25	-20;45	2;05	-0;05
946	ν Cen	5190		193;20	-28;20	191;42	-28;04	1;38	-0;16
947	μ Cen	5193		194;00	-29;20	192;05	-28;46	1;55	-0;34
948	φ Cen	5248		195;10	-26;00	193;34	-27;47	1;36	-0;13
949	χ Cen	5285		196;20	-26;30	194;40	-26;22	1;40	-0;08
950	η Cen	5440		202;50	-25;15	200;46	-25;16	2;04	0;01

No	Name	HR	α_{Ptol}	δ_{Ptol}	α_{-128}	δ_{-128}	m_{Pt}	m_{HR}	Δm
901	τ^2 Hya	3787	117;52	5;43	115;25	6;17	4	4.57	-0.57
902	ι Hya	3845	119;43	5;51	117;19	6;40	4	3.91	0.09
903	τ^1 Hya	3759	117;12	3;59	114;53	4;36	4	4.60	-0.60
904		3750	117;19	1;20	115;16	1;21	6	5.38	0.62
905	α Hya	3748	117;25	-2;00	115;32	-1;18	2	1.98	0.02
906	κ Hya	3849	121;59	-6;37	119;31	-6;22	4	5.06	-1.06
907	υ^1 Hya	3903	124;26	-6;44	122;15	-6;29	4	4.12	-0.12
908	υ^2 Hya	3970	127;22	-4;40	125;19	-4;16	4	4.60	-0.60
909	μ Hya	4094	132;58	-7;48	130;55	-7;16	3	3.81	-0.81
910	φ^3 Hya	4171	135;10	-7;01	133;54	-7;01	4	4.91	-0.91
911	ν Hya	4232	138;09	-6;53	136;22	-6;10	3	3.11	-0.11
912	β Crt	4343	144;18	-12;55	142;18	-12;04	> 4	4.48	-0.78
913	χ^1 Hya	4314	143;19	-17;18	141;23	-16;42	4	4.94	-0.94
914	ξ Hya	4450	151;08	-21;33	148;08	-20;42	4	3.54	0.46
915	o Hya	4494	152;10	-24;12	149;53	-23;30	4	4.70	-0.70
916	β Hya	4552	154;25	-23;09	152;41	-22;30	3	4.28	-1.28
917	γ Hya	5020	174;23	-12;29	172;08	-11;25	> 4	3.00	0.70
918	π Hya	5287	184;59	-21;34	183;02	-15;20	> 4	3.27	0.43
919		3314	101;28	0;06	99;34	0;43	3	3.90	-0.90
920	ϵ Sex	4042	129;03	2;03	127;46	1;07	3	5.24	-2.24
921	α Crt	4287	140;48	-8;41	139;20	-7;56	4	4.08	-0.08
922	γ Crt	4405	147;29	-7;26	145;03	-6;44	4	4.08	-0.08
923	δ Crt	4382	145;48	-5;11	143;31	-4;04	4	3.56	0.44
924	ζ Crt	4514	151;51	-8;04	149;46	-7;03	> 4	4.73	-1.03
925	ϵ Crt	4402	146;43	-0;54	144;25	0;03	4	4.83	-0.83
926	η Crt	4567	154;40	-6;42	152;27	-5;43	< 4	5.18	-0.88
927	θ Crt	4468	149;36	0;20	147;19	1;20	4	4.70	-0.70
928	α Crv	4623	157;56	-14;02	155;39	-13;08	3	4.02	-1.02
929	ϵ Crv	4630	157;52	-11;50	156;03	-11;02	3	3.00	0.00
930	ζ Crv	4696	160;33	-11;20	158;31	-10;32	5	5.21	-0.21
931	γ Crv	4662	159;04	-7;04	157;16	-5;55	3	2.59	0.41
932	δ Crv	4757	162;50	-6;08	160;38	-4;42	3	2.95	0.05
933	η Crv	4775	163;26	-5;35	161;18	-4;25	4	4.31	-0.31
934	β Crv	4786	163;57	-12;49	161;44	-11;37	3	2.65	0.35
935	Cen	5192	180;21	-23;58	178;45	-22;56	> 5	4.19	0.51
936	Cen	5221	181;13	-21;12	179;43	-20;28	> 5	4.73	-0.03
937	Cen	5168	179;41	-22;22	178;15	-21;26	> 4	4.23	-0.53
938	Cen	5210	180;41	-22;16	179;22	-21;31	> 5	4.56	0.14
939	ι Cen	5028	174;29	-25;48	172;26	-24;57	3	2.75	0.25
940	θ Cen	5288	184;43	-26;48	182;47	-24;48	3	2.06	0.94
941		5089	176;15	-28;39	174;25	-27;44	4	3.88	0.12
942	ψ Cen	5367	187;09	-27;39	185;25	-26;48	4	4.05	-0.05
943		5378	187;23	-29;20	185;46	-28;27	4	4.42	-0.42
944		5485	192;43	-25;29	190;50	-24;27	4	4.05	-0.05
945		5471	191;59	-28;02	190;07	-27;06	4	4.00	0.00
946	ν Cen	5190	179;35	-31;04	178;19	-30;10	> 4	3.41	0.29
947	μ Cen	5193	179;39	-32;14	178;17	-30;58	> 4	3.04	0.66
948	φ Cen	5248	181;27	-31;31	180;10	-30;41	> 4	3.83	-0.13
949	χ Cen	5285	183;19	-30;39	181;55	-29;52	> 4	4.36	-0.66
950	η Cen	5440	190;06	-32;09	188;12	-31;19	3	2.31	0.69

No	Name	HR	C	λ_{Ptol}	β_{Ptol}	λ_{-128}	β_{-128}	$\Delta\lambda$	$\Delta\beta$
951	κ Cen	5576		207;30	-24;15	205;17	-23;47	2;13	-0;28
952	ζ Cen	5231		198;00	-33;30	195;31	-32;42	2;29	-0;48
953	υ^2 Cen	5260		197;40	-29;00	195;49	-30;45	1;51	-0;15
954	υ^1 Cen	5249	B	196;50	-30;20	194;53	-30;14	1;57	-0;06
955		N5139		192;10	-34;50	190;15	-34;54	1;55	0;04
956		4940	I	189;00	-37;40	187;10	-37;31	1;50	-0;09
957	γ Cen	4819		185;50	-38;00	183;04	-39;57	2;46	-0;03
958	τ Cen	4802	B	185;00	-40;20	182;04	-39;55	2;56	-0;25
959	σ Cen	4743		182;40	-39;00	181;23	-42;13	1;17	1;13
960	δ Cen	4621		182;40	-46;10	178;10	-44;22	4;30	-1;48
961	ρ Cen	4638		183;30	-46;45	180;05	-45;25	3;25	-1;20
962		5172		198;20	-40;45	196;05	-37;05	2;15	-3;40
963	ϵ Cen	5132		196;20	-41;00	196;09	-39;22	0;11	-3;38
964		5141		197;40	-43;45	197;08	-40;11	0;32	-3;34
965	γ Cru	4763	b	190;00	-51;10	187;15	-47;33	2;45	-3;37
966	β Cru	4853		195;20	-51;40	192;20	-48;26	3;00	-3;14
967	δ Cru	4656		186;20	-55;10	186;23	-50;14	-0;03	-4;56
968	$\alpha^{1,2}$ Cru	4730		191;10	-55;20	192;35	-52;40	-1;25	-2;40
969	α^1 Cen	5459	B	218;20	-41;10	212;45	-41;54	5;35	0;44
970	β Cen	5267		204;10	-45;20	204;23	-43;53	-0;13	-1;27
971	$\mu^{1,2}$ Cru	4898	L	194;40	-49;10	191;16	-45;52	3;24	-3;18
972	β Lup	5571		208;00	-24;50	205;32	-24;47	2;28	-0;03
973	α Lup	5469	b	205;50	-29;10	204;01	-29;47	1;49	0;37
974	δ Lup	5695		211;00	-21;15	209;08	-21;10	1;52	-0;05
975	γ Lup	5776		214;10	-19;00	211;58	-20;58	2;12	-0;02
976	ϵ Lup	5708		213;00	-25;10	210;37	-24;59	2;23	-0;11
977	λ Lup	5626		210;10	-25;00	208;12	-26;16	1;58	-0;44
978	π Lup	5605		210;30	-27;00	208;08	-28;09	2;22	-0;51
979	μ Lup	5683		214;40	-28;30	210;52	-28;13	3;48	-0;17
980	κ^1 Lup	5646		213;40	-30;10	210;01	-29;22	3;39	-0;48
981	ζ Lup	5649		215;40	-33;10	211;19	-32;31	4;21	-0;39
982	ρ Lup	5453	IL	202;00	-31;20	204;09	-31;55	-2;09	0;35
983	ι Lup	5354		201;50	-30;30	199;19	-29;58	2;31	-0;32
984	τ^2 Lup	5396		203;00	-29;20	200;16	-28;54	2;44	-0;26
985	η Lup	5948		218;50	-15;00	216;14	-17;10	2;36	0;10
986	θ Lup	5987		219;20	-15;20	217;12	-15;20	2;08	0;00
987	χ Lup	5883	I	215;40	-13;20	213;17	-12;54	2;23	-0;26
988	$\xi^{1,2}$ Lup	5925	I	216;40	-11;50	214;35	-12;58	2;05	1;08
989	Lup	5660	lb	207;10	-11;50	205;09	-12;46	2;01	0;56
990	Lup	5686		206;30	-8;00	205;27	-11;16	1;03	1;16
991	σ Ara	6537		237;40	-22;40	235;53	-22;51	1;47	0;11
992	θ Ara	6743	l	243;00	-25;45	241;37	-26;21	1;23	0;36
993	α Ara	6510		236;10	-26;30	235;22	-26;14	0;48	-0;16
994	ϵ^1 Ara	6295		230;40	-30;20	230;01	-29;59	0;39	-0;21
995	γ Ara	6462		235;10	-34;10	234;43	-32;49	0;26	-1;21
996	β Ara	6461		235;00	-33;20	234;38	-31;58	0;22	-1;22
997	ζ Ara	6285	b	230;50	-34;15	230;17	-32;47	0;33	-1;28
998	α Tel	6897		249;10	-21;30	245;29	-22;20	3;41	0;50
999	η^1 CrA	7062		251;40	-19;00	249;47	-20;20	1;53	-0;40
1000		7122	b	253;10	-20;20	251;22	-19;30	1;48	-0;50

No	Name	HR	α_{Ptol}	δ_{Ptol}	α_{-128}	δ_{-128}	m_{Pt}	m_{HR}	Δm
951	κ Cen	5576	195;09	-33;05	193;18	-31;46	4	3.13	0.87
952	ζ Cen	5231	180;54	-37;31	179;13	-35;49	> 3	2.55	0.15
953	v^2 Cen	5260	182;05	-35;11	180;36	-34;13	5	4.34	0.66
954	v^1 Cen	5249	181;41	-34;16	180;03	-33;23	5	3.87	1.13
955		N5139	174;49	-36;19	173;11	-35;37	5	3.00	2.00
956		4940	170;17	-37;31	168;54	-36;40	5	4.71	0.29
957	γ Cen	4819	166;06	-38;16	163;54	-37;11	3	2.17	0.83
958	τ Cen	4802	165;11	-38;14	163;07	-36;46	4	3.86	0.14
959	σ Cen	4743	162;49	-37;54	161;04	-38;30	5	3.91	1.09
960	δ Cen	4621	159;15	-42;17	157;04	-39;07	3	2.60	0.40
961	ρ Cen	4638	159;28	-43;05	157;51	-40;44	4	3.96	0.04
962		5172	176;20	-43;54	177;00	-39;52	4	4.65	-0.65
963	ϵ Cen	5132	172;53	-44;59	175;31	-41;51	2	2.30	-0.30
964		5141	173;27	-46;09	175;49	-42;56	3	5.01	-2.01
965	γ Cru	4763	160;46	-49;09	161;54	-45;13	2	1.63	0.37
966	β Cru	4853	164;23	-51;36	165;12	-47;54	2	1.25	0.75
967	δ Cru	4656	154;13	-50;55	159;00	-47;05	4	2.80	1.20
968	$\alpha^{1,2}$ Cru	4730	157;26	-52;49	161;25	-51;22	2	0.83	1.17
969	α^1 Cen	5459	195;04	-52;12	189;05	-50;40	1	0.01	0.99
970	β Cen	5267	177;50	-50;04	179;26	-48;58	2	0.61	1.39
971	$\mu^{1,2}$ Cru	4898	166;16	-49;22	166;29	-45;23	4	3.28	0.72
972	β Lup	5571	195;21	-33;48	193;02	-32;46	3	2.68	0.32
973	α Lup	5469	190;52	-36;50	188;52	-36;39	3	2.30	0.70
974	δ Lup	5695	200;06	-31;42	198;22	-30;52	4	3.22	0.78
975	γ Lup	5776	203;25	-32;40	201;17	-31;46	4	2.78	1.22
976	ϵ Lup	5708	200;10	-36;02	197;57	-34;55	4	3.37	0.63
977	λ Lup	5626	196;21	-36;36	194;53	-35;09	5	4.05	0.95
978	π Lup	5605	195;34	-38;32	193;48	-36;49	5	3.97	1.03
979	μ Lup	5683	200;05	-39;41	196;29	-37;57	5	4.27	0.73
980	κ^1 Lup	5646	198;06	-40;48	194;59	-38;38	5	3.87	1.13
981	ζ Lup	5649	198;19	-44;15	194;24	-41;57	5	3.41	1.59
982	ρ Lup	5453	185;56	-37;14	187;45	-38;36	5	4.05	0.95
983	ι Lup	5354	186;16	-36;25	184;18	-34;56	4	3.55	0.45
984	τ^2 Lup	5396	188;03	-35;52	185;47	-34;22	> 4	4.35	-0.65
985	η Lup	5948	210;00	-30;39	207;20	-29;49	4	3.41	0.59
986	θ Lup	5987	211;13	-29;17	209;05	-28;28	> 4	4.23	-0.53
987	χ Lup	5883	208;18	-26;07	206;08	-24;49	4	3.95	0.05
988	$\xi^{1,2}$ Lup	5925	209;55	-25;05	207;25	-25;20	4	4.62	-0.62
989	Lup	5660	200;31	-21;36	198;13	-21;40	> 4	4.91	-1.21
990	Lup	5686	200;37	-19;40	199;08	-20;23	> 4	4.34	-0.64
991	σ Ara	6537	228;28	-41;53	226;23	-41;29	5	4.59	0.41
992	θ Ara	6743	233;47	-46;12	231;53	-46;21	4	3.66	0.34
993	α Ara	6510	225;04	-45;08	224;20	-44;33	> 4	2.95	0.75
994	ϵ^1 Ara	6295	216;36	-47;02	216;09	-46;26	5	4.06	0.94
995	γ Ara	6462	219;49	-52;01	220;12	-50;33	> 4	3.34	0.36
996	β Ara	6461	220;07	-51;12	220;35	-49;44	4	2.85	1.15
997	ζ Ara	6285	214;29	-50;42	214;51	-49;06	4	3.13	0.87
998	α Tel	6897	242;55	-43;23	238;09	-43;21	4	3.51	0.49
999	η^1 CrA	7062	246;12	-43;19	244;03	-42;14	5	5.49	-0.49
1000		7122	247;38	-45;32	246;13	-41;41	5	5.36	-0.36

No	Name	HR	C	λ_{Ptol}	β_{Ptol}	λ_{-128}	β_{-128}	$\Delta\lambda$	$\Delta\beta$
1001	ζ CrA	7188		254;50	-18;00	252;43	-19;01	2;07	-0;59
1002	δ CrA	7242		256;10	-18;30	253;57	-17;33	2;13	-0;57
1003	β CrA	7259		257;00	-17;10	254;27	-16;27	2;33	-0;43
1004	α CrA	7254	1	256;50	-14;00	254;29	-14;58	2;21	-1;02
1005	γ CrA	7226		256;30	-15;10	253;59	-13;57	2;31	-1;13
1006	ϵ CrA	7152		255;10	-15;20	252;31	-13;57	2;39	-1;23
1007		7129		254;40	-14;50	251;59	-14;10	2;41	-0;40
1008	λ CrA	7021		251;50	-14;40	249;17	-14;54	2;33	0;14
1009		6942		249;40	-15;50	246;52	-16;08	2;48	0;18
1010	θ CrA	6951		249;10	-18;30	246;56	-18;46	2;14	0;16
1011	α PsA	8728	D	307;00	-20;20	304;01	-20;50	2;59	0;30
1012	β PsA	8576		300;40	-20;20	297;28	-21;11	3;12	0;51
1013	γ PsA	8695		304;10	-22;15	301;39	-23;30	2;31	1;15
1014	δ PsA	8720		305;20	-22;30	302;29	-23;31	2;51	1;01
1015	ϵ PsA	8628		304;20	-16;15	301;39	-17;07	2;41	0;52
1016	μ PsA	8431		295;10	-19;30	292;24	-19;52	2;46	0;22
1017	ζ PsA	8570		301;10	-15;10	299;58	-15;20	1;12	0;10
1018	λ PsA	8478	b	298;50	-14;40	295;44	-15;32	3;06	0;52
1019	η PsA	8386		295;10	-13;00	292;37	-15;05	2;33	0;05
1020	θ PsA	8326		291;50	-16;30	288;58	-16;22	2;52	-0;08
1021	ι PsA	8305		291;00	-18;10	287;36	-18;05	3;24	-0;05
1022	γ Gru	8353	1	290;10	-22;15	287;41	-22;49	2;29	0;34
1023	η Mic	8069	I	278;00	-22;20	276;51	-23;26	1;09	1;06
1024	θ^1 Mic	8151	I	281;10	-22;10	279;49	-23;46	1;21	1;36
1025	ξ Gru	8229	I	284;00	-21;10	281;55	-24;52	2;05	3;42
1026	θ^2 Mic	8180	I	282;00	-20;50	280;29	-24;12	1;31	3;22
1027	α Mic	7965	I	283;50	-15;00	275;59	-15;12	7;51	-1;48
1028	γ Mic	8039	I	283;50	-14;50	278;48	-14;26	5;02	-0;24

No	Name	HR	α_{Ptol}	δ_{Ptol}	α_{-128}	δ_{-128}	m_{Pt}	m_{HR}	Δm
1001	ζ CrA	7188	250;25	-42;48	248;00	-41;24	4	4.75	-0.75
1002	δ CrA	7242	252;23	-41;30	249;50	-40;09	5	4.59	0.41
1003	β CrA	7259	253;38	-40;16	250;39	-39;07	4	4.11	-0.11
1004	α CrA	7254	253;37	-39;05	250;57	-37;39	4	4.11	-0.11
1005	γ CrA	7226	253;20	-38;14	250;31	-36;35	4	4.26	-0.26
1006	ε CrA	7152	251;41	-38;14	248;46	-36;22	6	4.87	1.13
1007		7129	251;10	-37;40	248;06	-36;31	6	5.38	0.62
1008	λ CrA	7021	247;47	-37;06	244;44	-36;48	5	5.13	-0.13
1009		6942	244;56	-37;54	241;34	-37;34	5	5.16	-0.16
1010	θ CrA	6951	243;42	-40;26	240;57	-40;09	5	4.64	0.36
1011	α PsA	8728	316;02	-38;22	312;44	-39;35	1	1.16	-0.16
1012	β PsA	8576	308;41	-40;05	305;05	-41;35	4	4.29	-0.29
1013	γ PsA	8695	313;31	-41;00	310;58	-42;46	4	4.46	-0.46
1014	δ PsA	8720	314;59	-40;54	311;57	-42;34	4	4.21	-0.21
1015	ε PsA	8628	311;30	-35;12	308;41	-36;39	> 4	4.17	-0.47
1016	μ PsA	8431	301;51	-40;34	298;32	-41;22	5	4.50	0.50
1017	ζ PsA	8570	307;34	-34;59	306;11	-35;20	5	6.43	-1.43
1018	λ PsA	8478	304;45	-35;03	301;21	-36;28	4	5.43	-1.43
1019	η PsA	8386	300;35	-36;10	297;34	-36;39	4	5.42	-1.42
1020	θ PsA	8326	297;01	-38;17	293;30	-38;32	4	5.01	-1.01
1021	ι PsA	8305	296;25	-40;05	292;11	-40;27	4	4.34	-0.34
1022	γ Gru	8353	296;27	-44;14	293;22	-45;06	4	3.01	0.99
1023	η Mic	8069	280;39	-45;53	279;14	-46;55	< 3	5.53	-2.23
1024	θ¹ Mic	8151	284;48	-45;26	283;14	-47;02	< 3	4.82	-1.52
1025	ξ Gru	8229	288;19	-44;07	286;13	-47;55	< 3	5.29	-1.99
1026	θ² Mic	8180	285;41	-44;01	284;11	-47;24	5	5.77	-0.77
1027	α Mic	7965	287;22	-40;00	277;25	-38;45	4	4.90	-0.90
1028	γ Mic	8039	287;01	-37;51	280;48	-37;50	4	4.67	-0.67

9. Appendix C

The appendix catalogues the degrees of the phenomena in the second part of Hipparchus's Commentary on Aratus and Eudoxus.[1] It is ordered according to the reference number of the star in the Almagest, Appendix B.

9.1 Column Headings

No.: Number of star according to the star catalogue of the Almagest as in Appendix B.

Name: Modern Name of the star.

p.: Page number of the first instance of a particular phenomenon in Hipparchus (1894).

v: Numerical value of the phenomenon according to the translation of Manitius. Full degrees are understood as ordinal numbers, while the phenomena with half degrees are noted in the usual cardinal number system. Hence, one has to subtract one degree in the case of full degrees in order to derive the value of the phenomenon Φ_i.

a: Occasionally Hipparchus indicates that a star is only close to a simultaneous culmination with a particular degree on the ecliptic. In total Hipparchus uses 8 different positional characterization listed in the following table.

Type No.	positional description	: number of cases
1	1 diameter of the moon west of the meridian :	26
2	nearly :	1
3	1 diameter of the moon east of the meridian :	18
4	2 diameters of the moon east of the meridian :	2
5	a little west of the meridian :	7
6	a little east of the meridian :	8
7	1 1/3 diameter of the moon east :	7
8	1 1/3 diameter of the moon west :	1

---

[1] Hipparchus (1894).

c: Comment number. Cases of philological or astronomical uncertainty are discussed by Manitius in extended footnotes. Their number is listed in column c.
   Usually it indicates an obscure value, e.g. λ Dra.

k: The number represents the kind of the phenomenon tabulated in the following
   with the number of instances.

Type	Symbol	Description
1	Φ_1	The degree of the ecliptic rising simultaneously with a star
2	Φ_2	The degree of the ecliptic culminating when a particular star rises
3	Φ_3	The degree of the ecliptic setting simultaneously with a star
4	Φ_4	The degree of the ecliptic culminating when a particular star sets
5	Φ_5	The degree of the ecliptic culminating together with a particular star

Type of numerical entry	: number
Phenomena in total	: 619
Phenomena of type 1	: 84
Phenomena of type 2	: 82
Phenomena of type 3	: 87
Phenomena of type 4	: 85
Phenomena of type 5	: 281

Φ_H: The Hipparchan value of the phenomenon after adjusting for the different
   numerical system in the Commentary.

Φ_P: The phenomenon is calculated with the basis of

- The ecliptical longitude of the Almagest is decreased by $2°40'$.
- with the rigorous formulae the phenomenon is calculated for the parameters $\epsilon = 23°51'$ and $\phi = 36°$, the geographical latitude of Rhodes.

Φ_C: The theoretical phenomena are derived on the basis of true stellar coordinates
   for the epoch of the year -128.

Δp: The difference of the "Ptolemaic" phenomenon from the "true" one: $\Delta p = \Phi_P - \Phi_C$.

Δh: $\Delta p = \Phi_H - \Phi_C$.

nr	n		p.	v	a	c	k	Π_h	Π_p	Π_c	Δ_p	Δ_h
9	o	UMa	211	80.5			5	80.50	78.47	79.55	-1.08	0.95
19	ϑ	UMa	239	104.5		31	5	104.50	101.51	101.60	-0.09	2.90
20	ι	UMa	267	96.0			5	95.00	93.75	94.47	-0.71	0.53
20	ι	UMa	187	94.5			5	94.50	93.75	94.47	-0.71	0.03
24	α	UMa	241	120.0			5	119.00	119.31	120.63	-1.33	-1.63
25	β	UMa	251	127.5			5	127.50	123.13	123.59	-0.47	3.91
28	λ	UMa	267	119.0	1		5	118.00	116.24	116.30	-0.05	1.70
30	ψ	UMa	239	132.0			5	131.00	130.64	130.86	-0.22	0.14
30	ψ	UMa	241	129.5	3		5	129.50	130.64	130.86	-0.22	-1.36
31	ν	UMa	243	137.0			5	136.00	135.93	135.78	0.14	0.22
35	η	UMa	243	184.0	6		5	183.00	184.26	184.78	-0.52	-1.78
35	η	UMa	245	185.0			5	184.00	184.26	184.78	-0.52	-0.78
45	ν	Dra	235	250.0		i	5	249.00	254.49	254.56	-0.07	-5.56
47	ξ	Dra	261	260.5			5	260.50	260.34	260.49	-0.15	0.01
48	γ	Dra	233	258.5			5	258.50	257.90	258.24	-0.34	0.26
49		Dra	235	275.5		i	5	275.50	268.13	268.41	-0.28	7.09
72	α	Dra	233	198.5	1		5	198.50	196.27	198.65	-2.38	-0.15
73	κ	Dra	209	155.0			5	154.00	150.89	152.33	-1.44	1.67
74	λ	Dra	211	122.0			5	121.00	119.19	120.72	-1.52	0.28
74	λ	Dra	217	152.0		36	5	151.00	119.19	120.72	-1.52	30.28
74	λ	Dra	241	120.0			5	119.00	119.19	120.72	-1.52	-1.72
75	κ	Cep	227	309.0			5	308.00	305.47	306.76	-1.29	1.24
75	κ	Cep	205	308.0			5	307.00	305.47	306.76	-1.29	0.24
76	γ	Cep	229	334.0			5	333.00	334.79	334.74	0.05	-1.74
76	γ	Cep	265	334.5			5	334.50	334.79	334.74	0.05	-0.24
77	β	Cep	257	310.5			5	310.50	309.52	310.54	-1.02	-0.04
79	η	Cep	201	299.0			5	298.00	297.68	297.34	0.34	0.66
82	ι	Cep	203	323.0			5	322.00	322.64	322.36	0.28	-0.36
83	ϵ	Cep	207	7.5			3	7.50	6.39	6.01	0.38	1.49
83	ϵ	Cep	207	99.0			4	98.00	97.07	96.65	0.42	1.35
83	ϵ	Cep	193	245.5			1	245.50	247.10	247.63	-0.54	-2.13
83	ϵ	Cep	193	171.0			2	170.00	171.45	172.16	-0.71	-2.16
85	λ	Cep	209	104.0			4	103.00	102.01	103.57	-1.56	-0.57
85	λ	Cep	209	14.0			3	13.00	10.83	12.22	-1.39	0.78
86	μ	Cep	257	309.0	1		5	308.00	307.60	307.31	0.29	0.69
86	μ	Cep	193	236.5			1	236.50	237.40	237.06	0.34	-0.56
86	μ	Cep	193	158.5			2	158.50	158.62	158.18	0.44	0.32
93	β	Boo	187	151.0			1	150.00	151.38	151.26	0.12	-1.26
93	β	Boo	187	56.5			2	56.50	57.59	57.45	0.15	-0.95
93	β	Boo	253	207.0			5	206.00	206.56	207.27	-0.71	-1.27

nr	n	p.	v	a	c	k	Π_h	Π_p	Π_c	Δ_p	Δ_h
96	ν Boo	201	4.0			4	3.00	3.22	2.03	1.19	0.97
96	ν Boo	201	288.5			3	288.50	288.85	287.94	0.91	0.56
103	ϵ Boo	233	198.5			5	198.50	198.06	199.30	-1.24	-0.80
106	ζ Boo	259	197.0			5	196.00	195.68	196.28	-0.60	-0.28
106	ζ Boo	187	87.0			2	86.00	85.56	86.25	-0.69	-0.25
106	ζ Boo	187	177.0			1	176.00	175.98	176.61	-0.63	-0.61
107	η Boo	245	185.0	1		5	184.00	182.17	183.25	-1.09	0.75
107	η Boo	243	184.0			5	183.00	182.17	183.25	-1.09	-0.25
108	τ Boo	243	181.0	6		5	180.00	180.48	181.27	-0.78	-1.27
109	υ Boo	223	182.0			5	181.00	180.49	181.53	-1.04	-0.53
109	υ Boo	201	216.0			3	215.00	214.73	215.95	-1.22	-0.95
109	υ Boo	201	292.0			4	291.00	290.71	291.51	-0.80	-0.51
110	α Boo	195	191.0			5	190.00	189.73	190.43	-0.70	-0.43
110	α Boo	193	189.5	3		5	189.50	189.73	190.43	-0.70	-0.93
111	α Crb	201	331.0			4	330.00	330.08	330.28	-0.20	-0.28
111	α Crb	201	263.0			3	262.00	261.52	261.70	-0.18	0.30
112	β Crb	187	85.5			2	85.50	85.19	85.88	-0.70	-0.38
112	β Crb	187	177.0			1	176.00	175.65	176.28	-0.63	-0.28
112	β Crb	195	212.5			5	212.50	212.12	212.31	-0.19	0.19
115	γ Crb	195	215.5			5	215.50	215.35	215.86	-0.51	-0.36
116	δ Crb	251	218.5	1		5	218.50	216.96	217.69	-0.73	0.81
117	ϵ Crb	187	184.5			1	184.50	185.42	186.03	-0.61	-1.53
117	ϵ Crb	187	94.5			2	94.50	95.99	96.67	-0.68	-2.17
118	ι Crb	201	343.5			4	343.50	341.45	341.60	-0.15	1.90
118	ι Crb	203	273.5			3	273.50	271.54	271.67	-0.13	1.83
118	ι Crb	217	152.0		36	5	151.00	221.55	221.74	-0.20	-70.74
119	α Her	255	236.0	7		5	235.00	235.92	237.12	-1.20	-2.12
120	β Her	253	228.0			5	227.00	227.20	227.54	-0.33	-0.54
121	γ Her	199	225.5			5	225.50	224.40	224.86	-0.46	0.64
122	κ Her	259	222.0			5	221.00	220.13	220.82	-0.70	0.18
122	κ Her	235	221.5	1		5	221.50	220.13	220.82	-0.70	0.68
125	μ Her	195	247.5			5	247.50	247.47	247.99	-0.52	-0.49
125	μ Her	235	250.0	1		5	249.00	247.47	247.99	-0.52	1.01
126	o Her	189	217.5			1	217.50	217.33	217.78	-0.45	-0.28
126	o Her	189	134.0			2	133.00	132.93	133.48	-0.55	-0.48
131	d Her	235	241.0			5	240.00	237.98	238.39	-0.41	1.61
133	π Her	253	247.0	1		5	246.00	242.09	242.69	-0.60	3.31
133	π Her	195	245.5	5		5	245.50	242.09	242.69	-0.60	2.81
133	π Her	237	245.0			5	244.00	242.09	242.69	-0.60	1.31
137	ι Her	203	316.0			3	315.00	314.90	315.48	-0.59	-0.48

nr	n	p.	v	a	c	k	Π_h	Π_p	Π_c	Δ_p	Δ_h
137	ι Her	203	37.5			4	37.50	37.34	38.08	-0.74	-0.58
141	η Her	255	236.0			5	235.00	234.54	235.27	-0.73	-0.27
143	τ Her	189	161.0			1	160.00	159.87	160.81	-0.94	-0.81
143	τ Her	189	67.5			2	67.50	67.47	68.55	-1.07	-1.05
145	υ Her	253	228.0			5	227.00	226.44	227.12	-0.68	-0.12
146	χ Her	189	67.5		24	2	67.50	72.50	71.88	0.62	-4.38
146	χ Her	261	223.0			5	222.00	222.61	222.40	0.21	-0.40
146	χ Her	189	161.0		24	1	160.00	164.29	163.74	0.55	-3.74
150	ε Lyr	191	218.5			1	218.50	217.98	218.84	-0.86	-0.34
150	ε Lyr	191	132.5		13	2	132.50	133.73	134.79	-1.06	-2.29
150	ε Lyr	205	33.0			4	32.00	31.38	32.13	-0.74	-0.13
150	ε Lyr	205	312.0		13	3	311.00	310.25	310.83	-0.58	0.17
155	β Lyr	205	304.0			3	303.00	302.99	303.62	-0.62	-0.62
155	β Lyr	205	22.5			4	22.50	21.89	22.71	-0.82	-0.21
155	β Lyr	227	263.5			5	263.50	262.95	263.63	-0.67	-0.13
155	β Lyr	225	264.5	1		5	264.50	262.95	263.63	-0.67	0.87
157	γ Lyr	191	228.0			1	227.00	227.10	227.72	-0.62	-0.72
157	γ Lyr	191	146.0			2	145.00	145.21	146.01	-0.80	-1.01
159	β Cyg	207	23.5			4	23.50	23.20	23.41	-0.21	0.09
159	β Cyg	207	304.5			3	304.50	303.99	304.14	-0.16	0.36
161	η Cyg	255	279.0			5	278.00	278.45	278.45	0.01	-0.45
162	γ Cyg	261	286.0			5	285.00	285.35	285.23	0.12	-0.23
163	α Cyg	257	291.5			5	291.50	290.79	290.60	0.19	0.90
163	α Cyg	227	291.5			5	291.50	290.79	290.60	0.19	0.90
163	α Cyg	201	292.0			5	291.00	290.79	290.60	0.19	0.40
164	δ Cyg	255	279.0			5	278.00	278.63	278.83	-0.19	-0.83
165	ϑ Cyg	263	280.0			5	279.00	279.05	278.88	0.17	0.12
166	ι Cyg	263	276.5	3		5	276.50	277.98	278.11	-0.14	-1.61
167	κ Cyg	207	344.0			3	343.00	341.94	342.68	-0.74	0.32
167	κ Cyg	207	72.0			4	71.00	69.84	70.68	-0.84	0.32
167	κ Cyg	235	275.5	6		5	275.50	275.51	276.23	-0.72	-0.73
167	κ Cyg	193	121.0			2	120.00	121.60	122.23	-0.64	-2.23
167	κ Cyg	199	277.5	1		5	277.50	275.51	276.23	-0.72	1.27
167	κ Cyg	193	206.5			1	206.50	207.93	208.47	-0.54	-1.97
168	ε Cyg	263	290.0			5	289.00	288.66	288.57	0.09	0.43
170	ζ Cyg	205	295.0			5	294.00	294.02	293.85	0.17	0.15
170	ζ Cyg	193	189.5			2	189.50	188.43	188.42	0.00	1.08
170	ζ Cyg	193	262.0			1	261.00	260.20	260.20	0.00	0.80
174	Cyg	201	299.0		i	5	298.00	285.61	286.01	-0.40	11.99
178	ζ Cas	209	113.5			4	113.50	111.99	112.18	-0.19	1.32

nr	n	p.	v	a	c	k	Π_h	Π_p	Π_c	Δ_p	Δ_h
178	ζ Cas	193	212.5			2	212.50	212.34	214.18	-1.84	-1.68
178	ζ Cas	193	282.0			1	281.00	280.75	282.51	-1.76	-1.51
178	ζ Cas	205	340.5			5	340.50	340.39	341.45	-1.06	-0.95
178	ζ Cas	209	21.0			3	20.00	19.66	19.83	-0.17	0.17
181	γ Cas	203	343.5			5	343.50	344.79	345.64	-0.86	-2.14
181	γ Cas	231	345.0			5	344.00	344.79	345.64	-0.86	-1.64
182	δ Cas	229	352.0			5	351.00	351.73	351.30	0.43	-0.30
182	δ Cas	205	349.5			5	349.50	351.73	351.30	0.43	-1.80
183	ϵ Cas	209	53.5			3	53.50	52.98	54.33	-1.36	-0.83
183	ϵ Cas	229	358.0			5	357.00	355.49	356.91	-1.41	0.09
183	ϵ Cas	209	155.0			4	154.00	152.82	154.60	-1.77	-0.60
188	κ Cas	205	340.5			5	340.50	339.62	341.01	-1.39	-0.51
188	κ Cas	265	340.5		27	5	340.50	339.62	341.01	-1.39	-0.51
188	κ Cas	227	343.0		27	5	342.00	339.62	341.01	-1.39	0.99
188	κ Cas	193	191.0		27	2	190.00	186.90	188.74	-1.84	1.26
188	κ Cas	193	262.0		27	1	261.00	258.99	260.45	-1.46	0.55
191	χ Per	265	3.0			5	2.00	1.97	2.79	-0.82	-0.79
191	χ Per	199	295.0			1	294.00	294.35	294.30	0.05	-0.30
191	χ Per	201	4.0			5	3.00	1.97	2.79	-0.82	0.21
191	χ Per	219	2.5			5	2.50	1.97	2.79	-0.82	-0.29
193	γ Per	211	14.0			5	13.00	12.89	13.62	-0.73	-0.62
194	ϑ Per	265	9.0	7		5	8.00	8.89	9.60	-0.71	-1.60
197	α Per	215	19.5	5		5	19.50	18.24	18.74	-0.50	0.76
200	δ Per	229	23.5			5	23.50	22.95	23.40	-0.45	0.10
205	π Per	215	32.0			3	31.00	31.01	30.97	0.04	0.03
205	π Per	215	125.5			4	125.50	125.26	125.21	0.05	0.29
206	b Per	215	59.0			3	58.00	56.44	56.75	-0.31	1.25
206	b Per	215	159.5			4	159.50	157.37	157.77	-0.41	1.73
212	v Per	231	25.5			5	25.50	25.01	25.12	-0.11	0.38
213	ϵ Per	215	29.0			5	28.00	28.25	28.57	-0.31	-0.57
215	o Per	199	13.5		24	1	13.50	10.40	10.62	-0.21	2.88
215	o Per	199	277.5		24	2	277.50	275.93	276.05	-0.12	1.45
216	ζ Per	199	13.5		24	1	13.50	14.65	14.38	0.28	-0.88
216	ζ Per	199	277.5		24	2	277.50	278.39	278.23	0.16	-0.73
220	δ Aur	239	52.0			5	51.00	50.59	50.84	-0.26	0.16
220	δ Aur	199	340.5		24	1	340.50	344.33	340.91	3.42	-0.41
220	δ Aur	199	260.0		24	2	259.00	261.02	259.00	2.01	-0.00
220	δ Aur	217	200.0		24	4	199.00	197.32	200.23	-2.91	-1.23
220	δ Aur	217	91.5		24	3	91.50	87.43	89.89	-2.45	1.61
221	ξ Aur	199	260.0		24	2	259.00	254.26	253.67	0.59	5.33

nr	n		p.	v	a	c	k	Π_h	Π_p	Π_c	Δ_p	Δ_h
221	ζ	Aur	221	50.5			5	50.50	49.48	49.29	0.19	1.21
221	ζ	Aur	199	340.5		24	1	340.50	333.08	332.13	0.95	8.37
221	ζ	Aur	217	91.5		24	3	91.50	92.07	92.30	-0.23	-0.80
221	ζ	Aur	217	202.0		24	4	201.00	202.77	203.04	-0.27	-2.04
222	α	Aur	249	44.5			5	44.50	44.96	44.73	0.22	-0.23
222	α	Aur	267	47.0	1		5	46.00	44.96	44.73	0.22	1.27
226	ε	Aur	247	41.5			5	41.50	42.32	42.25	0.06	-0.75
229	ι	Aur	217	152.0			4	151.00	150.86	150.73	0.13	0.27
229	ι	Aur	217	53.0			3	52.00	51.47	51.37	0.10	0.63
234	α	Oph	203	272.0			3	271.00	269.91	270.48	-0.57	0.52
234	α	Oph	203	340.5			4	340.50	339.53	340.19	-0.67	0.31
236	γ	Oph	225	244.0	3		5	243.00	242.22	242.81	-0.59	0.19
240	δ	Oph	251	218.5			5	218.50	218.41	218.93	-0.53	-0.43
240	δ	Oph	205	235.5		24	3	235.50	235.92	236.91	-0.99	-1.41
240	δ	Oph	189	209.0		24	1	208.00	207.47	207.63	-0.17	0.37
240	δ	Oph	189	123.0		24	2	122.00	121.05	121.25	-0.20	0.75
241	ε	Oph	189	123.0		24	2	122.00	122.31	122.82	-0.51	-0.82
241	ε	Oph	205	235.5		24	3	235.50	236.10	236.70	-0.60	-1.20
241	ε	Oph	189	209.0		24	1	208.00	208.54	208.96	-0.43	-0.96
248	ϑ	Oph	189	153.0		25	2	152.00	151.88	152.10	-0.22	-0.10
248	ϑ	Oph	189	233.0		25	1	232.00	232.26	232.43	-0.17	-0.43
248	ϑ	Oph	205	349.5		25	4	349.50	301.25	301.31	-0.06	48.19
248	ϑ	Oph	205	279.0		25	3	278.00	229.75	229.83	-0.08	48.17
252	ζ	Oph	261	223.0			5	222.00	223.37	223.42	-0.05	-1.42
255	ψ	Oph	203	295.0			4	294.00	294.10	294.82	-0.72	-0.82
255	ψ	Oph	203	221.0			3	220.00	219.79	220.84	-1.05	-0.84
263	ι	Ser	191	97.5			2	97.50	96.89	98.15	-1.25	-0.65
263	ι	Ser	191	188.0			1	187.00	186.23	187.36	-1.13	-0.36
265	γ	Ser	203	323.0			4	322.00	321.20	321.82	-0.62	0.18
265	γ	Ser	203	254.0			3	253.00	252.90	253.53	-0.63	-0.53
279	η	Ser	261	251.0			5	250.00	249.45	249.84	-0.40	0.16
280	ϑ	Ser	205	349.5			4	349.50	349.17	349.74	-0.57	-0.24
280	ϑ	Ser	191	164.0			2	163.00	162.99	163.43	-0.44	-0.43
280	ϑ	Ser	191	240.5			1	240.50	240.70	241.03	-0.33	-0.53
280	ϑ	Ser	199	260.0		29	5	259.00	258.28	258.71	-0.43	0.29
280	ϑ	Ser	233	258.5		29	5	258.50	258.28	258.71	-0.43	-0.21
280	ϑ	Ser	205	279.0			3	278.00	277.88	278.34	-0.46	-0.34
280	ϑ	Ser	225	264.5		29	5	264.50	258.28	258.71	-0.43	5.79
281	γ	Sge	211	19.5			4	19.50	19.41	19.51	-0.10	-0.01
281	γ	Sge	235	275.5			5	275.50	275.51	275.54	-0.03	-0.04

nr	n		p.	v	a	c	k	Π_h	Π_p	Π_c	Δ_p	Δ_h
281	γ	Sge	211	301.5			3	301.50	301.11	301.19	-0.07	0.31
281	γ	Sge	197	249.5			1	249.50	248.76	248.73	0.02	0.77
281	γ	Sge	197	175.0			2	174.00	173.64	173.60	0.03	0.40
284	α	Sge	211	14.0			4	13.00	14.24	13.92	0.32	-0.92
284	α	Sge	211	296.5			3	296.50	297.20	296.96	0.24	-0.46
284	α	Sge	197	169.0			2	168.00	169.25	168.86	0.39	-0.86
284	α	Sge	197	245.0			1	244.00	245.43	245.14	0.29	-1.14
285	β	Sge	211	14.0			4	13.00	12.78	13.56	-0.78	-0.56
285	β	Sge	197	169.0			2	168.00	168.38	169.49	-1.10	-1.49
285	β	Sge	211	296.5			3	296.50	296.10	296.69	-0.59	-0.19
285	β	Sge	197	245.0			1	244.00	244.78	245.61	-0.83	-1.61
287	β	Aql	197	253.5			1	253.50	253.12	253.70	-0.58	-0.20
287	β	Aql	197	179.5			2	179.50	179.35	180.11	-0.75	-0.61
290	γ	Aql	211	293.0			3	292.00	291.49	292.15	-0.66	-0.15
290	γ	Aql	211	8.0			4	7.00	6.69	7.56	-0.86	-0.56
292	μ	Aql	197	174.0			2	173.00	173.15	173.04	0.11	-0.04
292	μ	Aql	211	2.1			4	2.10	1.07	1.53	-0.45	0.57
292	μ	Aql	211	286.5			3	286.50	287.21	287.55	-0.35	-1.05
292	μ	Aql	197	249.0			1	248.00	248.39	248.31	0.08	-0.31
293	σ	Aql	211	2.0			4	1.00	0.88	0.72	0.16	0.28
293	σ	Aql	211	286.5			3	286.50	287.06	286.93	0.12	-0.43
301	ε	Del	215	19.5			4	19.50	19.15	18.80	0.34	0.70
301	ε	Del	215	302.0			3	301.00	300.91	300.65	0.26	0.35
301	ε	Del	255	284.0			5	283.00	282.05	281.70	0.35	1.30
303	κ	Del	197	193.0			2	192.00	191.62	191.25	0.37	0.75
303	κ	Del	197	263.5			1	263.50	262.76	262.46	0.30	1.04
304	β	Del	197	259.5		24	1	259.50	259.51	260.06	-0.55	-0.56
305	β	Del	197	187.5		24	2	187.50	187.22	188.13	-0.91	-0.63
305	α	Del	197	187.5		24	2	187.50	187.22	188.13	-0.91	-0.63
305	α	Del	197	259.5		24	1	259.50	259.24	259.96	-0.72	-0.46
306	δ	Del	261	286.0			5	285.00	284.66	284.66	0.00	0.34
307	γ	Del	215	29.0			4	28.00	27.27	26.65	0.62	1.35
307	γ	Del	215	307.5			3	307.50	307.09	306.61	0.47	0.89
315	α	And	209	351.5			3	351.50	351.42	351.02	0.40	0.48
315	α	And	265	334.5			5	334.50	334.03	333.84	0.19	0.66
315	α	And	245	335.0			5	334.00	334.03	333.84	0.19	0.16
315	α	And	209	80.5			4	80.50	80.50	80.05	0.45	0.45
316	γ	Peg	229	334.0			5	333.00	334.43	334.54	-0.11	-1.54
316	γ	Peg	211	343.0		i	3	342.00	342.57	342.72	-0.15	-0.72
316	γ	Peg	211	70.5		i	4	70.50	70.56	70.72	-0.17	-0.22

nr	n		p.	v	a	c	k	Π_h	Π_p	Π_c	Δ_p	Δ_h
316	γ	Peg	195	321.0			1	320.00	320.08	320.12	-0.04	-0.12
316	γ	Peg	195	245.5			2	245.50	245.81	245.84	-0.03	-0.34
319	τ	Peg	203	323.0			5	322.00	322.26	322.01	0.24	-0.01
331	ϵ	Peg	211	33.5			4	33.50	33.82	33.37	0.45	0.13
331	ϵ	Peg	201	299.0			5	298.00	297.77	297.58	0.19	0.42
331	ϵ	Peg	211	312.5			3	312.50	312.15	311.80	0.35	0.70
331	ϵ	Peg	263	297.5			5	297.50	297.77	297.58	0.19	-0.08
336	π	And	205	340.5			5	340.50	341.39	340.96	0.44	-0.46
336	π	And	265	340.5			5	340.50	341.39	340.96	0.44	-0.46
336	π	And	227	343.0			5	342.00	341.39	340.96	0.44	1.04
338	σ	And	259	338.5			5	338.50	336.40	336.68	-0.28	1.82
343	λ	And	195	215.5			2	215.50	211.32	211.28	0.05	4.22
343	λ	And	195	285.0			1	284.00	279.79	279.74	0.04	4.26
344	ζ	And	203	343.5			5	343.50	343.83	343.48	0.35	0.02
344	ζ	And	195	247.5			2	247.50	247.64	247.41	0.23	0.09
344	ζ	And	195	323.5			1	323.50	322.78	322.44	0.34	1.06
346	β	And	231	349.5			5	349.50	348.83	348.77	0.07	0.73
346	β	And	247	350.0			5	349.00	348.83	348.77	0.07	0.23
349	γ	And	211	2.0			5	1.00	0.65	1.43	-0.77	-0.43
349	γ	And	265	3.0	5		5	2.00	0.65	1.43	-0.77	0.57
349	γ	And	245	0.5	6		5	0.50	0.65	1.43	-0.77	-0.93
350	ϕ	Per	211	27.5			3	27.50	26.45	26.88	-0.43	0.62
350	ϕ	Per	211	122.0			4	121.00	119.86	120.36	-0.50	0.64
350	ϕ	Per	229	358.0			5	357.00	354.99	356.10	-1.11	0.90
361	γ	Tri	213	8.0	3		5	7.00	5.51	5.43	0.07	1.57
362	γ	Ari	245	0.5			5	0.50	0.76	0.48	0.28	0.02
363	β	Ari	265	3.0	5		5	2.00	1.31	0.67	0.64	1.33
363	β	Ari	211	2.0			5	1.00	1.31	0.67	0.64	0.33
367	ν	Ari	211	14.0	1		5	13.00	12.43	12.01	0.41	0.99
370	ζ	Ari	205	22.5			5	22.50	21.66	21.30	0.36	1.20
371	τ	Ari	267	119.0			4	118.00	117.92	117.66	0.27	0.34
371	τ	Ari	255	284.0			2	283.00	282.37	282.16	0.22	0.84
371	τ	Ari	231	23.5			5	23.50	23.62	23.34	0.27	0.16
371	τ	Ari	255	21.0			1	20.00	21.40	21.04	0.36	-1.04
371	τ	Ari	267	26.0			3	25.00	24.80	24.57	0.23	0.43
375	α	Ari	219	2.5	3		5	2.50	3.57	3.65	-0.08	-1.15
375	α	Ari	201	4.0			5	3.00	3.57	3.65	-0.08	-0.65
383	o	Tau	269	112.0			4	111.00	111.63	111.49	0.15	-0.49
383	o	Tau	269	20.0			3	19.00	19.35	19.22	0.13	-0.22
383	o	Tau	257	37.0			1	36.00	36.20	36.54	-0.34	-0.54

nr		n	p.	v	a	c	k	Π_h	Π_p	Π_c	Δ_p	Δ_h
383	o	Tau	257	291.5			2	291.50	291.68	291.90	-0.23	-0.40
383	o	Tau	231	25.5			5	25.50	25.40	25.44	-0.05	0.06
384	c	Tau	207	33.0	8	i	5	32.00	30.65	31.11	-0.46	0.89
385	λ	Tau	267	36.5	1		5	36.50	33.96	34.07	-0.11	2.43
385	λ	Tau	211	33.5			5	33.50	33.96	34.07	-0.11	-0.57
390	γ	Tau	267	36.5	3		5	36.50	38.35	38.24	0.12	-1.74
393	α	Tau	247	41.5			5	41.50	41.73	42.04	-0.31	-0.54
398	ζ	Tau	257	309.0			2	308.00	308.03	308.15	-0.11	-0.15
398	ζ	Tau	257	59.0			1	58.00	58.33	58.47	-0.14	-0.47
400	β	Tau	269	157.0			4	156.00	155.76	155.88	-0.11	0.12
400	β	Tau	199	45.5			1	45.50	46.00	45.67	0.33	-0.17
400	β	Tau	269	56.0			3	55.00	55.22	55.31	-0.09	-0.31
400	β	Tau	199	299.0			2	298.00	298.47	298.23	0.24	-0.23
400	β	Tau	239	52.0			5	51.00	51.66	51.59	0.07	-0.59
403	A	Tau	211	33.5		i	5	33.50	34.12	33.47	0.65	0.03
403	A	Tau	207	33.0	7	i	5	32.00	34.12	33.47	0.65	-1.47
414	ι	Tau	267	47.0			5	46.00	47.95	47.62	0.33	-1.62
424	α	Gem	211	80.5			5	80.50	79.94	79.97	-0.03	0.53
426	ϑ	Gem	257	61.5			1	61.50	61.69	57.60	4.09	3.90
426	ϑ	Gem	221	74.0			5	73.00	72.69	69.89	2.80	3.11
426	ϑ	Gem	257	310.5			2	310.50	310.87	307.43	3.44	3.07
429	υ	Gem	221	82.0			5	81.00	81.01	81.44	-0.43	-0.44
433	ε	Gem	211	70.5			5	70.50	70.13	70.10	0.02	0.40
437	η	Gem	269	167.0			4	166.00	165.99	166.30	-0.31	-0.30
437	η	Gem	269	64.0			3	63.00	62.97	63.21	-0.24	-0.21
437	η	Gem	241	64.0	6		5	63.00	64.14	64.14	0.01	-1.14
439	ν	Gem	189	67.5			5	67.50	68.09	67.80	0.30	-0.30
440	γ	Gem	211	70.5			5	70.50	70.46	70.54	-0.08	-0.04
441	ξ	Gem	221	74.0			5	73.00	73.35	73.04	0.31	-0.04
450	η	Cnc	189	94.5	2		5	94.50	95.04	95.89	-0.85	-1.39
450	η	Cnc	267	96.0			5	95.00	95.04	95.89	-0.85	-0.89
452	ϑ	Cnc	189	94.5	1		5	94.50	97.81	98.17	-0.36	-3.67
452	ϑ	Cnc	267	96.0			5	95.00	97.81	98.17	-0.36	-3.17
453	δ	Cnc	209	99.0	6		5	98.00	98.64	99.10	-0.45	-1.10
454	α	Cnc	245	0.5			2	0.50	1.30	1.38	-0.08	-0.88
454	α	Cnc	245	108.0			1	107.00	107.38	107.44	-0.06	-0.44
455	ι	Cnc	245	335.0			2	334.00	333.72	337.14	-3.42	-3.14
455	ι	Cnc	259	222.0			4	221.00	220.42	219.50	0.92	1.50
455	ι	Cnc	245	83.0			1	82.00	84.83	87.85	-3.02	-5.85
455	ι	Cnc	259	109.5			3	109.50	108.76	107.81	0.95	1.69

nr	n	p.	v	a	c	k	Π_h	Π_p	Π_c	Δ_p	Δ_h
456	μ Cnc	257	90.0			1	89.00	89.27	89.02	0.25	-0.02
456	μ Cnc	269	201.5			4	201.50	201.21	201.20	0.01	0.30
456	μ Cnc	257	338.5			2	338.50	338.78	338.49	0.29	0.01
456	μ Cnc	269	91.5			3	91.50	90.72	90.71	0.01	0.79
457	β Cnc	267	96.0			5	95.00	94.25	94.36	-0.11	0.64
457	β Cnc	233	93.5			5	93.50	94.25	94.36	-0.11	-0.86
457	β Cnc	187	94.5			5	94.50	94.25	94.36	-0.11	0.14
457	β Cnc	259	86.5			3	86.50	88.21	86.20	2.01	0.30
457	β Cnc	259	197.0			4	196.00	198.25	195.84	2.42	0.16
462	κ Leo	247	97.5			1	97.50	98.34	98.14	0.20	-0.64
462	κ Leo	247	350.0			2	349.00	349.74	349.49	0.25	-0.49
463	λ Leo	219	108.0			5	107.00	109.59	109.37	0.22	-2.37
463	λ Leo	243	107.0	3		5	106.00	109.59	109.37	0.22	-3.37
464	μ Leo	209	113.5		32	5	113.50	113.80	114.11	-0.31	-0.61
467	γ Leo	211	122.0			5	121.00	121.43	121.79	-0.35	-0.79
467	γ Leo	189	123.0			5	122.00	121.43	121.79	-0.35	0.21
468	η Leo	267	119.0	7		5	118.00	118.93	119.28	-0.35	-1.28
469	α Leo	241	120.0			5	119.00	119.84	120.43	-0.59	-1.43
469	α Leo	193	121.0			5	120.00	119.84	120.43	-0.59	-0.43
474	o Leo	259	110.5			3	110.50	109.87	110.26	-0.39	0.24
474	o Leo	209	113.5			5	113.50	113.90	114.03	-0.13	-0.53
474	o Leo	259	223.0			4	222.00	221.49	221.86	-0.37	0.14
480	b Leo	239	132.0			5	131.00	132.32	133.10	-0.78	-2.10
480	b Leo	191	132.5			5	132.50	132.32	133.10	-0.78	-0.60
480	b Leo	189	134.0			5	133.00	132.32	133.10	-0.78	-0.10
481	δ Leo	243	137.0			5	136.00	135.86	136.14	-0.28	-0.14
484	ι Leo	241	141.0			5	140.00	139.59	139.84	-0.25	0.16
485	σ Leo	241	141.0			5	140.00	139.38	139.68	-0.30	0.32
485	σ Leo	247	41.5			2	41.50	42.00	42.02	-0.02	-0.52
485	σ Leo	247	138.5			1	138.50	138.59	138.60	-0.01	-0.10
488	β Leo	259	164.0			3	163.00	162.18	163.75	-1.56	-0.75
488	β Leo	259	260.5			4	260.50	259.76	260.68	-0.92	-0.18
490	Leo	241	129.5			5	129.50	129.96	130.68	-0.73	-1.18
490	Leo	249	131.0			5	130.00	129.96	130.68	-0.73	-0.68
498	ξ Vir	247	44.5			2	44.50	46.86	45.85	1.01	-1.35
498	ξ Vir	247	142.0			1	141.00	142.50	141.68	0.82	-0.68
501	β Vir	261	147.0			3	146.00	146.46	148.00	-1.54	-2.00
501	β Vir	261	251.0			4	250.00	250.07	251.06	-0.99	-1.06
501	β Vir	191	146.0	3		5	145.00	146.36	147.32	-0.96	-2.32
502	η Vir	269	157.0	5		5	156.00	156.02	155.81	0.22	0.19

nr	n		p.	v	a	c	k	Π_h	Π_p	Π_c	Δ_p	Δ_h
503	γ	Vir	243	161.0	7		5	160.00	161.66	162.07	-0.41	-2.07
503	γ	Vir	239	163.0			5	162.00	161.66	162.07	-0.41	-0.07
505	ϑ	Vir	193	171.0	1		5	170.00	169.04	169.44	-0.40	0.56
505	ϑ	Vir	197	169.0	6		5	168.00	169.04	169.44	-0.40	-1.44
506	δ	Vir	271	167.0	1		5	166.00	165.29	165.84	-0.55	0.16
506	δ	Vir	191	164.0	4		5	163.00	165.29	165.84	-0.55	-2.84
509	ϵ	Vir	191	164.0	4		5	163.00	166.12	167.63	-1.51	-4.63
509	ϵ	Vir	223	166.5	1		5	166.50	166.12	167.63	-1.51	-1.13
509	ϵ	Vir	269	167.0	1		5	166.00	166.12	167.63	-1.51	-1.63
510	α	Vir	197	175.0	5		5	174.00	173.12	173.44	-0.32	0.56
510	α	Vir	251	172.5			5	172.50	173.12	173.44	-0.32	-0.94
510	α	Vir	197	174.0			5	173.00	173.12	173.44	-0.32	-0.44
520	φ	Vir	199	193.0			5	192.00	190.79	191.14	-0.35	0.86
520	φ	Vir	225	191.0	3		5	190.00	190.79	191.14	-0.35	-1.14
521	λ	Vir	197	187.5			5	187.50	187.56	187.66	-0.11	-0.16
522	μ	Vir	261	286.0			4	285.00	285.37	285.77	-0.40	-0.77
522	μ	Vir	247	188.0			1	187.00	187.78	188.13	-0.35	-1.13
522	μ	Vir	261	206.0			3	205.00	206.33	206.97	-0.64	-1.97
522	μ	Vir	247	99.0			2	98.00	98.62	99.00	-0.38	-1.00
529	α	Lib	249	196.0			1	195.00	195.16	195.39	-0.23	-0.39
529	α	Lib	261	196.5			3	196.50	196.51	196.59	-0.08	-0.09
529	α	Lib	259	197.0			5	196.00	195.63	195.81	-0.18	0.19
529	α	Lib	249	108.0			2	107.00	106.88	107.14	-0.26	-0.14
529	α	Lib	261	280.0			4	279.00	279.48	279.52	-0.04	-0.52
531	β	Lib	271	201.5	7		5	201.50	203.12	203.38	-0.26	-1.88
532	δ	Lib	233	198.5			5	198.50	198.59	199.26	-0.67	-0.76
533	ι	Lib	217	202.0			5	201.00	200.67	200.80	-0.14	0.20
533	ι	Lib	195	200.5			5	200.50	200.67	200.80	-0.14	-0.30
539	ξ	Sco	261	297.5			4	297.50	297.96	298.25	-0.29	-0.75
539	ξ	Sco	261	225.0			3	224.00	225.29	225.69	-0.40	-1.69
546	β	Sco	263	292.0			4	291.00	291.44	291.33	0.11	-0.33
546	β	Sco	249	127.5			2	127.50	127.93	127.88	0.05	-0.38
546	β	Sco	249	216.0			3	215.00	215.84	215.68	0.16	-0.68
546	β	Sco	263	216.0			3	215.00	215.84	215.68	0.16	-0.68
546	β	Sco	249	213.0			1	212.00	213.23	213.19	0.04	-1.19
547	δ	Sco	195	212.5			5	212.50	212.41	212.38	0.03	0.12
548	ψ	Sco	249	216.0			1	215.00	214.53	214.97	-0.45	0.03
548	π	Sco	249	131.0			2	130.00	129.50	130.04	-0.54	-0.04
553	α	Sco	251	218.5			5	218.50	218.65	218.74	-0.09	-0.24
557	ϵ	Sco	235	221.5		46	5	221.50	222.23	222.43	-0.20	-0.93

nr	n		p.	v	a	c	k	Π_h	Π_p	Π_c	Δ_p	Δ_h
557	ϵ	Sco	261	223.0			5	222.00	222.23	222.43	-0.20	-0.43
558	μ	Sco	235	221.5		46	5	221.50	221.10	221.52	-0.41	-0.02
563	ι^1	Sco	249	249.0			1	248.00	247.64	247.58	0.06	0.42
563	ι^1	Sco	249	172.5			2	172.50	172.16	172.09	0.07	0.41
570	γ	Sgr	251	245.5			1	245.50	245.35	245.27	0.08	0.23
570	γ	Sgr	251	169.5			2	169.50	169.14	169.03	0.11	0.47
570	γ	Sgr	235	241.0			5	240.00	240.44	240.29	0.15	-0.29
571	δ	Sgr	237	245.0			5	244.00	243.75	243.79	-0.04	0.21
571	δ	Sgr	225	244.0			5	243.00	243.75	243.79	-0.04	-0.79
573	λ	Sgr	253	247.0			5	246.00	246.09	246.47	-0.39	-0.47
576	φ	Sgr	261	251.0			5	250.00	249.82	250.03	-0.21	-0.03
581		Sgr	233	258.5			5	258.50	258.92	259.08	-0.16	-0.58
581	d	Sgr	199	260.0			5	259.00	258.92	259.08	-0.16	-0.08
583	υ	Sgr	263	334.5			4	334.50	335.16	335.02	0.15	-0.52
583	υ	Sgr	263	266.5			3	266.50	266.12	265.99	0.13	0.51
587	χ	Sgr	199	260.0			5	259.00	259.87	259.56	0.31	-0.56
587	χ	Sgr	261	260.5	1		5	260.50	259.87	259.56	0.31	0.94
592	β	Sgr	263	213.0			3	212.00	214.07	219.47	-5.41	-7.47
592	β	Sgr	263	290.0			4	289.00	290.28	293.88	-3.60	-4.88
601	α	Cap	251	268.5			1	268.50	268.49	268.14	0.35	0.36
601	α	Cap	251	198.5			2	198.50	198.58	198.17	0.41	0.33
603	β	Cap	237	274.5			5	274.50	274.49	274.29	0.20	0.21
607	ρ	Cap	237	275.5			5	275.50	276.09	275.52	0.56	-0.02
612	ω	Cap	263	276.5			5	276.50	278.58	278.01	0.57	-1.51
612	ω	Cap	265	272.0			3	271.00	273.05	272.25	0.81	-1.25
612	ω	Cap	265	340.5			4	340.50	343.26	342.30	0.97	-1.80
612	ψ	Cap	199	277.5			5	277.50	278.58	278.01	0.57	-0.51
612	ω	Cap	255	279.0			5	278.00	278.58	278.01	0.57	-0.01
618	η	Cap	255	284.0			5	283.00	284.29	283.44	0.85	-0.44
620	ι	Cap	265	290.0	1		5	289.00	288.43	288.21	0.21	0.79
621	ϵ	Cap	201	292.0			5	291.00	291.41	291.33	0.07	-0.33
623	γ	Cap	263	292.1	7		5	292.10	292.51	292.46	0.05	-0.36
624	δ	Cap	265	293.5			3	293.50	292.60	292.67	-0.07	0.83
624	δ	Cap	205	295.0			5	294.00	294.00	294.23	-0.23	-0.23
624	δ	Cap	251	228.0			2	227.00	227.01	227.42	-0.41	-0.42
624	δ	Cap	265	9.0			4	8.00	8.15	8.24	-0.09	-0.24
624	δ	Cap	251	297.0			1	296.00	295.89	296.36	-0.46	-0.36
626	μ	Cap	227	297.0			5	296.00	295.97	296.13	-0.16	-0.13
635	μ	Aqr	255	284.0			5	283.00	282.70	282.64	0.06	0.36
636	ϵ	Aqr	265	3.0			4	2.00	1.53	1.44	0.09	0.56

nr	n		p.	v	a	c	k	Π_h	Π_p	Π_c	Δ_p	Δ_h
636	ϵ	Aqr	253	276.0			1	275.00	274.05	274.48	-0.43	0.52
636	ϵ	Aqr	265	287.5			3	287.50	287.55	287.48	0.07	0.02
636	ϵ	Aqr	253	207.0			2	206.00	205.02	205.51	-0.49	0.49
639	ζ	Aqr	205	308.0			5	307.00	306.90	306.82	0.08	0.18
639	ζ	Aqr	227	309.0			5	308.00	306.90	306.82	0.08	1.18
640	η	Aqr	265	315.0			3	314.00	313.98	314.05	-0.07	-0.05
640	η	Aqr	265	36.5			4	36.50	36.17	36.26	-0.09	0.24
646	δ	Aqr	253	247.0			2	246.00	245.88	246.76	-0.88	-0.76
646	δ	Aqr	253	320.5			1	320.50	320.17	321.47	-1.30	-0.97
646	δ	Aqr	257	310.5			5	310.50	311.15	311.62	-0.47	-1.12
674	β	Psc	267	47.0			4	46.00	46.37	46.36	0.01	-0.36
674	β	Psc	267	323.0			3	322.00	322.10	322.09	0.01	-0.09
674	β	Psc	255	236.0			2	235.00	235.74	235.90	-0.17	-0.90
674	β	Psc	255	307.0			1	306.00	306.34	306.55	-0.21	-0.55
681	ω	Psc	203	331.0			5	330.00	330.86	330.47	0.39	-0.47
692	α	Psc	219	2.5	3		5	2.50	3.59	3.81	-0.21	-1.31
692	α	Psc	255	16.0			1	15.00	14.45	15.54	-1.09	-0.54
692	α	Psc	255	279.0			2	278.00	278.27	278.91	-0.64	-0.91
692	α	Psc	201	4.0			5	3.00	3.59	3.81	-0.21	-0.81
695	η	Psc	267	89.0			4	88.00	89.07	88.22	0.86	-0.22
695	η	Psc	255	263.5			2	263.50	263.64	263.28	0.35	0.22
695	η	Psc	255	348.5			1	348.50	348.85	348.23	0.62	0.27
695	η	Psc	267	359.0			3	358.00	359.16	358.39	0.78	-0.39
697	g	Psc	267	5.0			3	4.00	4.34	4.35	-0.01	-0.35
697	g	Psc	267	96.0			4	95.00	94.80	94.81	-0.01	0.19
698	τ	Psc	267	5.0			3	4.00	3.98	3.51	0.47	0.49
698	τ	Psc	267	96.0			4	95.00	94.40	93.88	0.52	1.12
707	χ	Psc	205	349.5			5	349.50	352.13	349.18	2.95	0.32
707	χ	Psc	229	352.0			5	351.00	352.13	349.18	2.95	1.82
710		Psc	203	331.0			5	330.00	330.07	330.57	-0.50	-0.57
712	λ	Cet	239	14.0			3	13.00	13.19	13.60	-0.42	-0.60
712	λ	Cet	239	104.5			4	104.50	104.65	105.12	-0.47	-0.62
713	α	Cet	211	19.5	3		5	19.50	20.19	20.08	0.11	-0.58
715	δ	Cet	211	14.0			5	13.00	13.95	14.31	-0.37	-1.31
718	υ^1	Cet	213	8.0			5	7.00	6.83	6.39	0.43	0.61
719	ρ	Cet	227	297.0			2	296.00	295.85	296.75	-0.91	-0.75
719	ρ	Cet	227	43.0			1	42.00	42.32	43.60	-1.29	-1.60
723	τ	Cet	225	291.5			2	291.50	291.47	291.82	-0.35	-0.32
723	τ	Cet	225	37.0			1	36.00	35.88	36.41	-0.53	-0.41
723	τ	Cet	245	0.5			5	0.50	1.33	1.14	0.19	-0.64

nr	n		p.	v	a	c	k	Π_h	Π_p	Π_c	Δ_p	Δ_h
723	τ	Cet	213	2.0			5	1.00	1.33	1.14	0.19	-0.14
726	ϑ	Cet	229	355.0			5	354.00	354.03	353.73	0.30	0.27
727	η	Cet	231	349.5			5	349.50	349.27	349.13	0.14	0.37
727	η	Cet	247	350.0			5	349.00	349.27	349.13	0.14	-0.13
728	φ^2	Cet	231	345.0			5	344.00	344.21	344.13	0.09	-0.13
732	ι	Cet	225	350.0			1	349.00	348.37	348.63	-0.26	0.37
732	ι	Cet	225	263.5			2	263.50	263.36	263.51	-0.15	-0.01
733	β	Cet	239	52.0			4	51.00	51.79	51.24	0.55	-0.24
733	β	Cet	205	340.5		6	5	340.50	341.88	341.83	0.05	-1.33
733	β	Cet	239	326.5			3	326.50	326.54	326.09	0.45	0.41
733	β	Cet	239	326.5			3	326.50	326.54	326.09	0.45	0.41
736	γ	Ori	187	56.5	1		5	56.50	55.83	55.75	0.08	0.75
768	β	Ori	187	56.5	1		5	56.50	55.89	55.89	-0.00	0.61
768	β	Ori	239	132.0			4	131.00	131.18	131.27	-0.09	-0.27
768	β	Ori	239	37.0			3	36.00	35.90	35.98	-0.07	0.02
771	κ	Ori	227	343.0			2	342.00	340.63	339.95	0.68	2.05
771	κ	Ori	227	93.0			1	92.00	90.85	90.27	0.58	1.73
772	λ	Eri	239	34.5			3	34.50	34.53	34.47	0.06	0.03
772	λ	Eri	239	129.5			4	129.50	129.50	129.43	0.07	0.07
783	δ	Eri	207	33.0	3	i	5	32.00	33.21	33.08	0.13	-1.08
805	ϑ	Eri	227	352.0			2	351.00	350.53	353.87	-3.34	-2.87
805	ϑ	Eri	239	337.0			3	336.00	337.06	330.29	6.77	5.71
805	ϑ	Eri	239	64.0			4	63.00	64.24	56.30	7.94	6.70
805	ϑ	Eri	227	100.0			1	99.00	98.97	101.62	-2.64	-2.62
806	ι	Lep	229	334.0			2	333.00	334.88	334.26	0.62	-1.26
806	ι	Lep	229	87.0			1	86.00	85.87	85.32	0.55	0.68
811	ϵ	Lep	241	26.5			3	26.50	27.33	26.46	0.87	0.04
811	ϵ	Lep	241	120.0			4	119.00	120.89	119.87	1.02	-0.87
812	α	Lep	241	64.0			5	63.00	63.27	62.18	1.09	0.82
813	β	Lep	241	64.0			5	63.00	63.31	61.73	1.58	1.27
815	γ	Lep	229	101.5			1	101.50	101.81	101.44	0.37	0.06
815	γ	Lep	229	355.0			2	354.00	354.11	353.65	0.46	0.35
817	η	Lep	241	44.0			3	43.00	43.34	42.94	0.40	0.06
817	η	Lep	241	141.0			4	140.00	140.43	139.92	0.51	0.08
819	ϑ	CMa	211	80.5			5	80.50	80.06	79.87	0.19	0.63
821	γ	CMa	241	59.0			3	58.00	58.56	57.88	0.68	0.12
821	γ	CMa	241	161.0			4	160.00	160.16	159.26	0.90	0.74
826	β	CMa	229	358.0			2	357.00	357.12	356.81	0.31	0.19
826	β	CMa	229	105.0			1	104.00	104.15	103.91	0.24	0.09
830	o^1	CMa	221	82.0	6		5	81.00	82.52	82.31	0.22	-1.31

nr		n	p.	v	a	c	k	Π_h	Π_p	Π_c	Δ_p	Δ_h
831	δ	CMa	187	87.0			5	86.00	86.00	85.96	0.04	0.04
831	δ	CMa	187	85.5			5	85.50	86.00	85.96	0.04	-0.46
834	ζ	CMa	241	137.0			4	136.00	135.55	136.31	-0.76	-0.31
834	ζ	CMa	241	41.0			3	40.00	39.45	40.06	-0.61	-0.06
835	η	CMa	229	124.5			1	124.50	123.92	124.43	-0.51	0.07
835	η	CMa	269	89.0	3		5	88.00	89.67	90.08	-0.41	-2.08
835	η	CMa	229	23.5			2	23.50	23.11	23.78	-0.67	-0.28
845	α	Col	221	67.5			5	67.50	68.26	67.77	0.50	-0.27
847	β	CMi	231	345.0			2	344.00	344.16	344.25	-0.09	-0.25
847	β	CMi	243	181.0			4	180.00	179.40	179.98	-0.58	0.02
847	β	CMi	243	75.0			3	74.00	73.16	73.60	-0.45	0.40
847	β	CMi	231	93.5			1	93.50	93.80	93.88	-0.07	-0.38
848	α	CMi	243	184.0			4	183.00	182.30	182.81	-0.51	0.19
848	α	CMi	231	349.5			2	349.50	350.29	349.88	0.41	-0.38
848	α	CMi	243	78.0			3	77.00	75.39	75.79	-0.39	1.21
848	α	CMi	231	99.0			1	98.00	98.78	98.45	0.33	-0.45
854		Pup	191	97.5	3		5	97.50	92.48	92.73	-0.25	4.77
865	ζ	Pup	221	102.0			5	101.00	100.74	101.26	-0.52	-0.26
877	γ	Pyx	243	185.0			4	184.00	183.31	183.79	-0.48	0.21
877	γ	Pyx	243	78.0			3	77.00	76.18	76.55	-0.37	0.45
877	γ	Pyx	219	108.0			5	107.00	107.96	108.57	-0.61	-1.57
892	α	Car	243	16.0			3	15.00	14.07	9.95	4.12	5.05
892	α	Car	243	107.0			4	106.00	105.64	101.03	4.61	4.97
894	σ	Hya	233	198.5			4	198.50	198.62	199.14	-0.52	-0.64
894	σ	Hya	233	89.0			3	88.00	88.52	88.96	-0.44	-0.96
895	δ	Hya	249	99.0	3		5	98.00	99.66	99.82	-0.16	-1.82
895	δ	Hya	219	108.5			1	108.50	108.95	108.73	0.22	-0.23
895	δ	Hya	219	2.5			2	2.50	3.35	3.07	0.28	-0.57
897	η	Hya	223	102.0			5	101.00	101.54	101.47	0.07	-0.47
900	ϑ	Hya	243	107.0			5	106.00	108.68	108.72	-0.04	-2.72
903	τ^1	Hya	269	112.0	3		5	111.00	112.79	112.98	-0.19	-1.98
905	α	Hya	209	113.5			5	113.50	113.11	113.59	-0.49	-0.09
908	v^2	Hya	211	122.0			5	121.00	122.62	122.94	-0.32	-1.94
911	v	Hya	233	111.0			3	110.00	112.39	113.02	-0.63	-3.02
911	v	Hya	219	50.5			2	50.50	52.81	53.34	-0.53	-2.84
911	v	Hya	233	221.5			4	221.50	223.85	224.43	-0.58	-2.93
911	v	Hya	219	146.5			1	146.50	147.38	147.82	-0.44	-1.32
912	β	Crt	241	141.0			5	140.00	139.52	139.79	-0.27	0.21
918	π	Hya	233	161.0			3	160.00	159.07	166.46	-7.38	-6.46
918	π	Hya	233	258.5			4	258.50	257.91	262.26	-4.35	-3.76

nr	n	p.	v	a	c	k	Π_h	Π_p	Π_c	Δ_p	Δ_h
918	π Hya	245	185.0	5		5	184.00	182.80	183.32	-0.52	0.68
918	π Hya	243	184.0			5	183.00	182.80	183.32	-0.52	-0.32
918	π Hya	219	108.0			2	107.00	106.83	103.33	3.50	3.67
918	π Hya	219	195.5			1	195.50	195.12	192.01	3.11	3.49
922	ϑ Crt	233	241.0			4	240.00	232.79	232.63	0.16	7.37
922	ϑ Crt	233	132.5			3	132.50	122.67	122.48	0.19	10.02
926	η Crt	219	160.5			1	160.50	160.61	160.70	-0.08	-0.20
926	η Crt	219	67.5			2	67.50	68.32	68.42	-0.10	-0.92
926	η Crt	233	132.5			3	132.50	132.35	132.62	-0.28	-0.12
926	η Crt	233	241.0			4	240.00	240.32	240.52	-0.20	-0.52
928	α Crv	209	155.0			5	154.00	153.64	153.67	-0.03	0.33
929	ε Crv	189	153.0			5	152.00	153.55	154.10	-0.55	-2.10
929	ε Crv	209	155.0			5	154.00	153.55	154.10	-0.55	-0.10
930	ζ Crv	269	157.0			5	156.00	156.38	156.71	-0.33	-0.71
931	γ Crv	221	166.0			1	165.00	164.43	164.79	-0.36	0.21
931	γ Crv	221	74.0			2	73.00	72.66	73.07	-0.40	-0.07
932	δ Crv	193	158.5	3		5	158.50	158.76	158.98	-0.22	-0.48
932	δ Crv	215	159.5			5	159.50	158.76	158.98	-0.22	0.52
934	β Crv	221	82.0			2	81.00	81.17	81.13	0.04	-0.13
934	β Crv	243	161.0			5	160.00	160.02	160.15	-0.13	-0.15
934	β Crv	221	173.0			1	172.00	172.02	171.99	0.04	0.01
934	β Crv	215	159.5			5	159.50	160.02	160.15	-0.13	-0.65
936	h Cen	197	179.5			5	179.50	178.71	179.69	-0.98	-0.19
936	h Cen	243	181.0	1		5	180.00	178.71	179.69	-0.98	0.31
939	ι Cen	221	190.0			1	189.00	189.70	190.07	-0.37	-1.07
939	ι Cen	251	172.5	1		5	172.50	171.44	171.73	-0.30	0.77
939	ι Cen	221	102.0			2	101.00	100.75	101.16	-0.41	-0.16
940	ϑ Cen	245	185.0	5		5	184.00	182.52	183.05	-0.52	0.95
940	ϑ Cen	243	184.0			5	183.00	182.52	183.05	-0.52	-0.05
944	c Cen	199	193.0			5	192.00	191.17	191.81	-0.64	0.19
950	η Cen	195	191.0	1		5	190.00	188.34	188.95	-0.61	1.05
950	η Cen	193	189.5			5	189.50	188.34	188.95	-0.61	0.55
969	α Cen	221	166.5			2	166.50	165.57	159.84	5.74	6.66
969	α Cen	221	244.0			1	243.00	242.66	238.32	4.34	4.68
970	β Cen	223	182.0			5	181.00	175.20	179.37	-4.17	1.63
977	λ Lup	235	275.5		41	4	275.50	253.95	255.42	-1.46	20.08
977	λ Lup	235	191.0		41	1	190.00	216.97	217.45	-0.49	-27.45
981	ζ Lup	235	148.5		41	3	148.50	138.54	139.78	-1.24	8.72
981	ζ Lup	235	250.0		41	4	249.00	244.75	245.61	-0.86	3.39
991	σ Ara	237	274.5			4	274.50	275.62	275.85	-0.23	-1.35

nr	n	p.	v	a	c	k	Π_h	Π_p	Π_c	Δ_p	Δ_h
991	σ Ara	237	189.0			3	188.00	189.87	190.27	-0.40	-2.27
992	ϑ Ara	223	263.0			1	262.00	260.53	263.50	-2.97	-1.50
992	ϑ Ara	223	191.0			2	190.00	188.84	192.54	-3.70	-2.54
993	α Ara	199	225.5			5	225.50	224.54	226.89	-2.35	-1.39
994	ε Ara	223	182.0		42	2	181.00	171.93	175.58	-3.65	5.42
994	ε Ara	223	255.0		42	1	254.00	247.46	250.24	-2.78	3.76
995	γ Ara	237	245.0		24	4	244.00	244.52	249.90	-5.39	-5.90
995	γ Ara	237	138.5		24	3	138.50	138.20	146.21	-8.01	-7.71
996	β Ara	237	245.0		24	4	244.00	247.86	253.00	-5.15	-9.00
996	β Ara	237	138.5		24	3	138.50	143.10	151.06	-7.96	-12.56
1011	α PsA	225	264.5		i	2	264.50	259.89	260.52	-0.63	3.98
1011	α PsA	257	310.5		i	5	310.50	310.35	310.19	0.16	0.31
1011	α PsA	237	2.0		i	4	1.00	7.46	6.34	1.11	-5.34
1011	α PsA	225	350.5		i	1	350.50	342.41	343.49	-1.08	7.01
1011	α PsA	237	287.5		i	3	287.50	292.07	291.23	0.84	-3.73
1021	ι PsA	225	316.5			1	316.50	316.64	315.48	1.16	1.02
1021	ι PsA	225	244.0			2	243.00	243.42	242.60	0.83	0.40
1022	γ Gru	237	333.0			4	332.00	335.41	334.64	0.78	-2.64
1022	γ Gru	237	264.0			3	263.00	266.34	265.65	0.69	-2.65

10. Literature

Aaboe, A. (1960). *On the Tables of Planetary Visibility in the Almagest and the Handy Tables.* Danske Vidensk. Selskab, Hist.–filos. Medd., vol. 37, no. 8.

– (1974). Scientific Instruments in Antiquity. *Philos. Trans. Royal Soc. of London,* Ser. A 276, pp. 21–42.

Aristotle (1984), *The Complete Works of Aristotle.* Ed. J. Barnes. 2 vols., Princeton.

Bailly, J.-S. (1781). *Histoire de l'astronomie ancienne depuis son origine jusqu'à l'établissement de l'école d'Alexandrie.* 2nd edition (1st edition 1775). Paris.

Björnbo, A. A. (1901). Hat Menelaos aus Alexandria einen Fixsternkatalog verfaßt? *Bibliotheca Mathematica,* 3. Folge, Bd. 2, pp. 196–212.

– (1902). *Studien über Menelaos' Sphärik. Beiträge zur Geschichte der Sphärik und Trigonometrie der Griechen.* Abhandlungen zur Geschichte der mathematischen Wissenschaften **14**. Leipzig.

Boll, F. (1894). *Studien über Claudius Ptolemäus.* Leipzig.

– (1901). Die Sternenkataloge des Hipparch und des Ptolemaios. *Bibliotheca Mathematica,* 3. Folge, Bd. 2, pp. 185–195.

– (1916). *Antike Beobachtungen farbiger Sterne.* Abhandlungen der kgl. Bayerischen Akad. d. Wissenschaften, Philos.–philol. u. hist. Kl., Bd. 30, Abh. 1. München.

Brahe, T. (1913–29). *Tychonis Brahe Dani Opera omnia.* Ed. J. L. E. Dreyer. 15 vols., Copenhagen.

Britton, J. P. (1967). *On the Quality of Solar and Lunar Observations and Parameters in Ptolemy's Almagest.* Ph. D. diss., Yale.

– (1969). Ptolemy's Determination of the Obliquity of the Ecliptic. *Centaurus* **14**, pp. 29–41.

Bourgey, L. (1955). *Observation et expérience chez Aristote.* Paris.

Burnet, J. (1950). *Greek Philosophy from Thales to Plato.* London.

Cassirer, E. (1932). Die Antike und die Entdeckung der exakten Wissenschaften. *Die Antike* **8**, pp. 276–300.

Cherniss, H. (1944). *Aristotle's Criticism of Plato and the Academy*. Baltimore.

Claghorn, G. S. (1954). *Aristotle's Criticism of Plato's "Timaeus"*. Den Haag.

Cohen, I. B. (1971). Delambre. In: *Dictionary of Scientific Biography*. Ed. C. C. Gillispie. New York. Vol. IV, pp. 14–18.

Delambre, J. B. J. (1817). *Histoire de l'astronomie ancienne*. 2 vols., Paris.

Dobrzycki, J. (1963). Katalog gwiazd w de Revolutionibus. *Studia i Materialy z Dziejow Nauki Polskiej*. Seria C, Z. 7, pp. 109–153.

– (1983). Astronomical Aspects of the Calendar Reform. In: *The Gregorian Reform of the Calendar*. Roma, pp. 117–127.

Dreyer, J. L. E. (1917). On the Origin of Ptolemy's Catalogue of Stars. *Monthly Notices of the Royal Astronomical Society* **77**, pp. 528–539.

– (1918). On the Origin of Ptolemy's Catalogue of Stars. Second Paper. *Monthly Notices of the Royal Astronomical Society* **78**, pp. 343–349.

– (1953). *A History of Astronomy from Thales to Kepler*. 2nd edition. New York.

Duhem, P. (1913). *Le système du monde*. Paris. Vol. I–II.

Explanatory Supplement to the Astronomical Ephemeris and the American Ephemeris and Nautical Almanac. Her Majesty's Stationery Office (1961). London.

Evans, J. (1987). On the Origin of the Ptolemaic Star Catalogue. *Journal for the History of Astronomy* **18**, pp. 155–172, 233–278.

Fotheringham, J. K. (1918). The Secular Acceleration of the Sun as Determined from Hipparchus' Equinox Observations, with a Note on Ptolemy's False Equinox. *Monthly Notices of the Royal Astronomical Society* **78**, pp. 406–423.

Frey, G. (1958). *Gesetz und Entwicklung in der Natur*. Hamburg.

Geminus (1898). *Elementa Astronomiae*. Ed. C. Manitius. Leipzig.

Gillispie, C. C. (ed.) (1970–80). *Dictionary of Scientific Biography*. 16 vols., New York.

Gingerich, O. (1980). Was Ptolemy a Fraud? *Quarterly Journal of the Royal Astronomical Society* **21**, pp. 253–266.

– (1981). Ptolemy Revisited: A Reply to R. R. Newton. *Quarterly Journal of the Royal Astronomical Society* **22**, pp. 40–44.

Gingerich, O., Welther, B. L. (1984). Some Puzzles of Ptolemy's Star Catalogue. *Sky & Telescope* **67**, pp. 421–423.

Goldstein, B. R. (1972). Theory and Observation in Medieval Astronomy. *Isis* **63**, pp. 39–47.

– (1983). The Obliquity of the Ecliptic in Ancient Greek Astronomy. *Archives Internationales d'Histoires des Sciences* **33**, pp. 3–14.

Gundel, W. (1936). *Neue astrologische Texte des Hermes Trismegistos.* Abhandlungen der Bayerischen Akad. d. Wissenschaften, Philos.–hist. Abteilung, Neue Folge, Heft 12. München.

Hankins, T. L. (1973). Lalande. In: *Dictionary of Scientific Biography.* Ed. C. C. Gillispie. New York. Vol. XII, pp. 579–582.

Hartner, W. (1968). *Oriens–Occidens.* Ausgewählte Schriften. Hildesheim.

– (1977). The Rôle of Observations in Ancient and Medieval Astronomy. *Journal for the History of Astronomy* **8**, pp. 1–11.

– (1980). Ptolemy and Ibn Yūnus on Solar Parallax. *Archives Internationales d'Histoire des Sciences* **30**, pp. 5–25.

Hawkins, G. S., Rosenthal, S. K. (1967). 5000 and 10000–Year Star Catalogs, *Smithsonian Contributions to Astrophysics* **10**, no. 2. Washington, pp. 141–179.

Heath, T. L. (1913). *Aristarchus of Samos, the Ancient Copernicus,* repr. 1983. New York.

– (1921). *A History of Greek Mathematics,* repr. 1981. New York.

– (1932). *Greek Astronomy.* London.

Hertzog, K. P. (1987). Ancient Stellar Anomalies. *Quarterly Journal of the Royal Astronomical Society* **28**, pp. 27–29.

Heiberg, J. L. (1895). Ptolemäus de Analemmate. *Abhandlungen zur Geschichte der Mathematik* **7**, pp. 1–30.

Hill, G. W. (1900). Ptolemy's Problem. *Astronomical Journal* **21**, pp. 33–35.

Hipparchus (1894). *Hipparchi in Arati et Eudoxi Phenomena Commentarium.* Ed. and German trans. C. Manitius. Leipzig.

Hoffleit, D. (1982). *The Bright Star Catalogue.* 4th revised edition. New Haven.

Hultsch, F. (1900). Hipparchos über die Grösse und Entfernung der Sonne. *Berichte über die Verhandlungen der kgl. Sächsischen Ges. d. Wissenschaften,* Philol.–hist. Kl., Bd. 52, pp. 169–200.

Huxley, G. (1964). Aristarchus of Samos and Greco–Babylonian Astronomy. *Greek, Roman and Byzantine Studies* **5**, pp. 123–131.

Ideler, L. (1806). *Historische Untersuchungen über die astronomischen Beobachtungen der Alten.* Berlin.

Kanitschneider, B. (1974). *Philosophisch–historische Grundlagen der physikalischen Kosmologie.* Stuttgart.

Kennedy, E. S. (1956). *A Survey of Islamic Astronomical Tables.* Trans. of the American Philosophical Society, N. S. vol 46.2. Philadelphia.

Kepler, J. (1937–). *Gesammelte Werke.* München.

Knobel, E. B. (1877). The Chronology of Star Catalogues. *Memoirs of the Royal Astronomical Society* **43**, pp. 1–74.

Koyré, A. (1957). *From the Closed World to the Infinite Universe.* Baltimore.

– (1980). *Von der geschlossenen Welt zum unendlichen Universum.* German trans. of Koyré, A. (1957). Frankfurt.

Krause, M. (1936). *Die Sphärik des Menelaos aus Alexandrien in der Verbesserung von Abu Nasr Mansur b. 'Ali b. 'Iraq.* Abhandlungen d. Ges. d. Wissenschaften zu Göttingen, Philol.–hist. Kl., 3. Folge, Nr. 17.

Kunitzsch, P. (1959). *Arabische Sternnamen in Europa.* Wiesbaden.

– (1961). *Untersuchungen zur Sternnomenklatur der Araber.* Wiesbaden.

– (1966). *Typen von Sternverzeichnissen in astronomischen Handschriften des zehnten bis vierzehnten Jahrhunderts.* Wiesbaden.

– (1974). *Der Almagest. Die Syntaxis Mathematica des Claudius Ptolemäus in arabisch–lateinischer Überlieferung.* Wiesbaden.

– (1975). *Ibn as–Salah. Zur Kritik der Koordinatenüberlieferung im Sternenkatalog des Almagest.* Göttingen.

– (1981). Observations on the Arabic Reception of the Astrolabe. *Archives Internationales d'Histoires des Sciences* **31**, pp. 243–252.

Lalande, J. (1764/71/92). *Astronomie.* 3rd edition (1st edition 1764, 2nd edition 1771). 3 vols., Paris.

– (1757). Mémoire sur les équations séculaires, et sur les moyens mouvemens du Soleil, de la Lune, de Saturne, de Jupiter et Mars, avec les observations de Tycho Brahé, faites sur Mars en 1593, tirées des manuscrits de cet Auteur. *Mémoires de mathématique et de physique, tirées des registres de l'Académie Royale des Sciences, de l'Année 1757.* Paris, pp. 411–470.

Laplace, P. S. (1796). *Exposition du Système du Monde.* Paris.

– (1797). *Darstellung des Weltsystems.* German trans. of Laplace, P. S. (1796). 2 vols., Frankfurt.

Lienert, G. A. (1973). *Verteilungsfreie Methoden in der Biostatistik.* 2 vols. and tables, 2nd edition, Meisenheim.

Lieske, J. H., Lederle, T., Fricke, W., Morando, B. (1977). Expressions for the Precession Quantities Based upon the IAU (1976) System of Astronomical Constants. *Astronomy and Astrophysics* **58**, pp. 1–16.

Luckey, P. (1927). Das Analemma von Ptolemäus. *Astronomische Nachrichten* **230**, no. 5498, cols. 17–46.

Lundmark, K. (1932). Luminosities, Colours, Diameters, Densities, Masses of Stars. In: *Handbuch der Astrophysik*. Ed. G. Eberhard, A. Kohlschütter, H. Ludendorff. New York. Vol. V.1, pp. 210–574.

Maass, E. (1883)., *Analecta Eratosthenica*. Philologische Untersuchungen **6**.

– (1892). *Aratea*. Philologische Untersuchungen **12**.

– (1898). *Commentariorum in Aratum reliquiae*. Berlin.

Maeyama, Y. (1984). Ancient Stellar Observations Timocharis, Aristyllus, Hipparchus, Ptolemy – the Dates and Accuracies. *Centaurus* **27**, pp. 280–310.

Mercier, R. (1976/77). Studies in the Medieval Conception of Precession. *Archives Internationales d'Histoire des Sciences* **26**, pp. 197–220, and **27**, pp. 33–71.

Mittelstrass, J. (1962). *Die Rettung der Phänomene*. Berlin.

Nadal, R., Brunet, J. P. (1983/84). Le "Commentaire" d'Hipparque I. La sphère mobile. *Archive for History of Exact Sciences* **29**, pp. 201–236.

Nallino, C. A. (1899/1903/07). *Al-Battani sive Albatenii, Opus astronomicum*. Ed. C. A. Nallino. 3 vols., Milano.

Neugebauer, O. (1949). The Early History of the Astrolabe. *Isis* **40**, pp. 240–256.

– (1957). *The Exact Sciences in Antiquity*. 2nd edition (repr. New York 1969). Providence.

– (1972). On Some Aspects of Early Greek Astronomy. *Proc. Amer. Philos. Soc.* **116**, pp. 243–251.

– (1975). *A History of Ancient Mathematical Astronomy*. 3 vols., Berlin.

– (1982). A Greek Arithmetical Method for Finding Oblique Ascensions. *Journal for the History of Astronomy* **13**, pp. 19–22.

Neugebauer, P. V. (1912/14/22). *Tafeln zur astronomischen Chronologie*. 3 vols., Leipzig.

Newton, R. R. (1974). The Authenticity of Ptolemy's Eclipse and Star Data. *Quarterly Journal of the Royal Astronomical Society* **15**, pp. 7–27.

– (1977). *The Crime of Claudius Ptolemy*. Baltimore.

– (1979). On the Fractions of Degrees in an Ancient Star Catalogue. *Quarterly Journal of the Royal Astronomical Society* **20**, pp. 383–394.

– (1980). Comments on "Was Ptolemy a Fraud?" by Owen Gingerich. *Quarterly Journal of the Royal Astronomical Society* **21**, pp. 388–399.

Pannekoek, A. (1955). Ptolemy's Precession. *Vistas in Astronomy* **1**, pp. 60–66.

– (1961). *A History of Astronomy*. New York:

Pappus (1876–1878). *Pappi Alexandrini Collectionis quae supersunt....* Ed. F. Hultsch. 3 vols., Berlin.

Pedersen, O. (1974). *A Survey of the Almagest*. Odense.

Peters, C. H. F. (1887), Mitteilungen, *Vierteljahresschrift der Astronomischen Gesellschaft* **22**, Bericht über die Versammlung der Astronomischen Gesellschaft zu Kiel 1887, August 29–31, pp. 264–283.

Peters, C. H. F., Knobel, E. B. (1915). *Ptolemy's Catalogue of Stars. A Revision of The Almagest*. The Carnegie Institution of Washington, Publ. No. 86. Washington.

Petersen, V. M. (1966). A Comment on a Comment by Manitius. *Centaurus* **11**, pp. 306–309.

Petersen, V. M., Schmidt, O. (1967). The Determination of the Longitude of the Apogee of the Orbit of the Sun According to Hipparchus and Ptolemy. *Centaurus* **12**, pp. 73–96.

Pfeiffer, E. (1916). *Studien zum antiken Sternglauben*. Leipzig.

Pingree, D. (1973). The Greek Influence on Early Islamic Mathematical Astronomy. *Journal of the American Oriental Society* **93**, pp. 32–43.

Proclus (1909). *Procli Diadochi Hypotyposis astronomicarum positionum*. Ed. and German trans. C. Manitius. Leipzig.

Ptolemy, C. (1898–1903). *Claudii Ptolemaei opera quae exstant omnia*. Ed. J. L. Heiberg et. al. 3 vols., Leipzig.

– (1940). *Tetrabiblos*. Ed. and trans. F. E. Robbins. Harvard.

– (1963). *Handbuch der Astronomie*, German trans. and annot. by K. Manitius, introduction and corr. by O. Neugebauer. 2 vols., Leipzig,

– (1967). *The Arabic Version of Ptolemy's Planetary Hypotheses*. Ed. B. R. Goldstein. Transactions of the American Philosophical Society, N. S. vol. 57.4. Philadelphia.

– (1984) *Ptolemy's Almagest*. Trans. and annot. by G. J. Toomer. London.

– (1986), *Der Sternkatalog des Almagest, die arabisch-mittelalterliche Tradition*. Ed. P. Kunitzsch, Wiesbaden.

Rawlins, D. (1982). An Investigation of the Ancient Star Catalog. *Publication of the Astronomical Society of the Pacific* **94**, pp. 359–373.

Rehm, A. (1899). Zu Hipparch und Eratosthenes. *Hermes* **34**, pp. 251–279.

Rome, A. (1936/43/31). *Commentaires de Pappus et de Théon d'Alexandrie sur l'Almageste*. 3 vols., Biblioteca Apostolica Vaticana, Studi e Testi 72, 106, 54. Roma.

Sachs, A. (1974). Babylonian Observational Astronomy. *Philosophical Transactions of the Royal Society*. London. Ser. A. 276, pp. 43–50.

Saltzer, W. (1970). Zum Problem der inneren Planeten in der vorptolemäischen Theorie. *Sudhoffs Archiv* **54**, pp. 141–172.

Sarton, G. (1927/31/47). *Introduction to the History of Science*. The Carnegie Institution of Washington, Publ. No. 376. 3 vols., Washington.

Schjellerup, H. C. F. C. (1874), *Description des étoiles fixes*, St. Petersburg.

– (1881). Recherches sur l'astronomie des anciens. Part I: *Urania* 1, pp. 25–39. II: *Urania* 1, pp. 42–47. III: *Copernicus* 1, pp. 223–236.

Schnabel, P. (1927). Kidenas, Hipparch und die Entdeckung der Präcession. *Zeitschrift für Assyriologie und verwandte Gebiete* **37**, pp. 1–60.

Sezgin, F. (1978). *Geschichte des arabischen Schrifttums*, vol. VI. Leiden.

Spengler, O. (1923). *Der Untergang des Abendlandes*. Repr. of the 33th edition. München 1970.

Steinschneider, M. (1960). *Die arabischen Übersetzungen aus dem Griechischen*. Graz.

Suter, H. (1900). *Die Mathematiker und Astronomen der Araber und ihre Werke*. Abhandlungen zur Geschichte der mathematischen Wissenschaften **10**. Leipzig.

Swerdlow, N. M. (1969). Hipparchus on the Distance of the Sun. *Centaurus* **14**, pp. 287–305.

– (1979). Ptolemy on Trial. *American Scholar* **48**, pp. 523–531.

– (1980). Hipparchus' Determination of the Length of the Tropical Year and the Rate of Precession. *Archive for History of Exact Sciences* **21**, pp. 291–309.

Swerdlow, N. M., Neugebauer, O. (1984). *Mathematical Astronomy in Copernicus' De Revolutionibus*. 2 vols., Berlin.

Szabó, Á. (1981). Astronomische Messungen bei den Griechen im 5. Jahrhundert v. Chr. und ihr Instrument. *Historia Scientiarum* **21**, pp. 1–26.

Tannery, P. (1893). *Recherches sur l'histoire de l'astronomie ancienne*. Paris.

Thiele, G. (1898). *Antike Himmelsbilder*. Berlin.

Toomer, G. J. (1973). The Chord Table of Hipparchus and the Early History of Greek Trigonometry. *Centaurus* **18**, pp. 6–28.

- (1974). Hipparchus on the Distances of the Sun and Moon. *Archive for History of Exact Sciences* **14**, pp. 126–142.

- (1975). Ptolemy. In: *Dictionary of Scientific Biography*. Ed. C. C. Gillispie. New York. Vol. XI, pp. 186–208.

- (1977). Review of O. Pedersen, A Survey of the Almagest. *Archives Internationales d'Histoire des Sciences* **27**, pp. 137–150.

- (1978). Hipparchus. In: *Dictionary of Scientific Biography*. Ed. C. C. Gillispie. New York. Vol. XV, pp. 207–224.

- (1981). Hipparchus' Empirical Basis for His Lunar Mean Motions. *Centaurus* **24**, pp. 97–109.

- (1984). See Ptolemy, C. (1984).

Tuckerman, B. (1962). *Planetary, Lunar and Solar Positions at Five–day and Ten–day Intervals 601 B.C. to A.D. 1*. Memoirs of the American Philosophical Society **56**. Philadelphia.

- (1964). *Planetary, Lunar, and Solar Positions A.D. 2 to A.D. 1649*. Memoirs of the American Philosophical Society **59**. Philadelphia.

Ulmer, K. (1949). Die Wandlungen des naturwissenschaftlichen Denkens zu Beginn der Neuzeit bei Galilei. *Symposion* **2**, pp. 289–359.

Vogt, H. (1920). Der Kalender des Claudius Ptolemäus. *Sitzungsberichte der Heidelberger Akad. d. Wissenschaften*. Philos.–hist. Kl., Abh. 15, pp. 5–61.

- (1925). Versuch einer Wiederherstellung von Hipparchs Fixsternverzeichnis. *Astronomische Nachrichten* **224**, no. 5354–5355, cols. 2–48.

Waerden, B. L. van der (1954). Die Sichtbarkeit der Sterne in der Nähe des Horizontes. *Vierteljahrsschrift der Naturforschenden Gesellschaft in Zürich* **99**, pp. 20–39.

- (1959). Ptolemaios. In: *Pauly's Realencyclopädie der classischen Altertumswissenschaft*. Ed. G. Wissowa et. al., cols. 1793–1831, 1839–1853, 1858–1859.

- (1982). The Motion of Venus, Mercury and the Sun in Early Greek Astronomy. *Archive for History of Exact Sciences* **26**, pp. 99–113.

Werner, H., Schmeidler, F. (1986). *Synopsis der Nomenklatur der Fixsterne*. Stuttgart.

Wiedemann, E. (1970). *Aufsätze zur Arabischen Wissenschaftsgeschichte*. 2 vols., Hildesheim.

Wieland, W. (1962). *Die aristotelische Physik*. Göttingen.

- (1982). *Platon und die Formen des Wissens*. Göttingen.

Wilson, C. (1984). The Sources of Ptolemy's Parameters. *Journal for the History of Astronomy* **15**, pp. 37–47.

Wooland E., Clemence G. (1966). *Spherical Astronomy*. New York.

Index of Quotations

Index

Sources in the History of
Mathematics and Physical Sciences